Introduction to the Theory of Sound Transmission

AGER • Principles of Paleoecology
BELOUSSOV (Trans. Broneer) • Basic Problems in Geotectonics
CAGNIARD (Trans. Flinn and Dix) • Reflection and Refraction of Progressive Seismic Waves
DE SITTER • Structural Geology
EWING, JARDETZKY, and PRESS • Elastic Waves in Layered Media
GRANT and WEST • Interpretation Theory in Applied Geophysics
GRIFFITHS • Scientific Method in Analysis of Sediments
GRIM • Applied Clay Mineralogy
GRIM • Clay Mineralogy
HEINRICH • Mineralogy and Geology of Radioactive Raw Materials
HEISKANEN and VENING MEINESZ • The Earth and Its Gravity Field
HOWELL • Introduction to Geophysics
JACOBS, RUSSELL, and WILSON • Physics and Geology
KRAUSKOPF • Introduction to Geochemistry
KRUMBEIN and GRAYBILL • An Introduction to Statistical Models in Geology
KUZNETSOV, IVANOV, and LYALIKOVA (Trans. Broneer) • Introduction to Geological Microbiology
LEGGET • Geology and Engineering
MENARD • Marine Geology of the Pacific
MILLER • Photogeology
OFFICER • Introduction to the Theory of Sound Transmission
RAMSAY • Folding and Fracturing of Rocks
SHAW • Time in Stratigraphy
SHROCK and TWENHOFEL • Principles of Invertebrate Paleontology
TURNER and VERHOOGEN • Igneous and Metamorphic Petrology
TURNER and WEISS • Structural Analysis of Metamorphic Tectonites
WHITE • Seismic Waves: Radiation, Transmission, and Attenuation

Introduction to the Theory of

SOUND TRANSMISSION

With Application to the Ocean

C. B. OFFICER

Associate Professor
Department of Geology, The Rice Institute
and
Associate in Geophysics
Woods Hole Oceanographic Institution

McGRAW-HILL BOOK COMPANY
NEW YORK TORONTO LONDON
1958

INTRODUCTION TO THE THEORY OF SOUND TRANSMISSION

Library of Congress Catalog Card Number 58-6693

5 6 7 8 9 10 11 - MP - 1

ISBN 07-047612-8

Preface

It has been my intention in this book to give an introduction to the theory of sound transmission. One might consider any acoustic problem as consisting of some sort of source, transmission, and reception. The transmission itself is, of course, an important and integral part of the whole; this is the portion which is covered here. The level of the book is commensurate with that of a senior undergraduate–first-year graduate course. I have attempted to give the theory in such a manner and to such a degree that following this a reader might feel at ease with the published literature in the field. My specific applications of the theory have been to a description of sound transmission in the ocean. Wherever possible I have given the physical explanations of the theoretical results; this I feel is an important part of a thorough understanding of the theory.

I hope that the book may be of interest to those engaged in various aspects of acoustics and geophysics and particularly to those who are entering into or desire to become acquainted with this subject.

The mathematics involved in the various derivations have been carried through in some detail. For those problems involving the evaluation of integrals, the integrals have been reduced to a familiar form or to a convenient tabulated form. On the other hand, no theory or proof of the various mathematical concepts that are used, such as Fourier integrals, is given. On this account the reader may wish to refer to or study an appropriate mathematics text from time to time.

References to the pertinent books and journal articles covered in a particular chapter are given at the end of the chapter. Specific reference within each chapter has been avoided. Of the more recent theoretical developments, however, special mention should be given to C. L. Pekeris for the normal-mode type of solution. Only general mention is given to experimental work. This has been intentional.

I should like to express my thanks to the Rice Institute and the Woods Hole Oceanographic Institution for their generous policy of allowing an individual to follow the pursuits of his interest and to the Office of Naval Research and the Bureau of Ships, U.S. Navy, who have encouraged and supported in a number of ways the preparation of this book. I should also like to acknowledge the discussions that I have had at various times with Drs. J. B. Hersey and A. D. Voorhis and Mr. A. C. Vine, Woods Hole Oceanographic Institution; Profs. M. Ewing, J. Nafe, and J. L. Worzel and Messrs. J. I. Ewing and G. H. Sutton, Lamont Geological Observatory, Columbia University; Prof. F. Press, Seismological Laboratory, California Institute of Technology; Drs. R. Frosch and I. Tolstoy, Hudson Laboratories, Columbia University; and Dr. E. T. Miller, Houston Research Center, Humble Oil Co., on matters pertinent to this book. Messrs. D. G. Harkrider, T. W. Lawhorn, and D. E. Miller, graduate students at the Rice Institute, have critically examined portions of the material covered in this book.

Charles B. Officer

Contents

1. *Fundamental Relations*

In this chapter the fundamental relations underlying the theory of sound transmission are outlined. The chapter begins with a discussion of the stress and strain relations for an elastic material, the reduction of these to an isotropic material and finally to a liquid. It was considered worthwhile to begin with the more general case of an elastic material as background to the simpler case of a liquid and for its application to sound transmission over a solid bottom. The equations of motion are derived in terms of the particle displacements, and their interpretation in terms of irrotational and equivoluminal waves is discussed. The solution of the wave equation is given for the simple cases of plane, cylindrical, and spherical waves in an unbounded medium.

In all the above it has been implicitly assumed that the changes have taken place under isothermal conditions; for gases and liquids it is more nearly correct to assume adiabatic conditions. The velocity of propagation of a disturbance for these two conditions and the dependence of sound velocity in the ocean on confining pressure, temperature, and salinity are given. Digressing to the measurement of sound, the relations between energy, energy density, energy flow, and intensity and pressure, density, and velocity are discussed. Sound-intensity level is usually measured in decibels. This unit is based on a logarithmic scale, and a brief discussion of its use is given. Concluding the chapter are two short sections on the effects to be expected for deviations from the linear, infinitesimal strain theory.

All the material in this chapter is background for the later chapters. It is necessarily brief and has included only those parts of special interest to sound transmission. The section on the decibel is, in part, out of context with the rest of the chapter, but because of its importance in the analysis of sound-transmission data it has been included here. For a fuller discussion of the various topics of this chapter reference is made to the Bibliography.

Components of stress

Stress is defined as the force acting on a given area δS divided by the area. It is the force per unit area. In general the stress σ acting on any given area δS will be at an angle to the normal to δS. It is convenient to resolve the stress into components in the direction of a set of mutually orthogonal axes and to consider the stresses acting on the three planes perpendicular to the axes. The stress on a plane at any orientation to the axes can then be determined in terms of the nine stress components.

In Fig. 1-1 the first subscript denotes the direction of the stress and the second the plane on which the stress is acting. Thus σ_{zx} is a stress in the

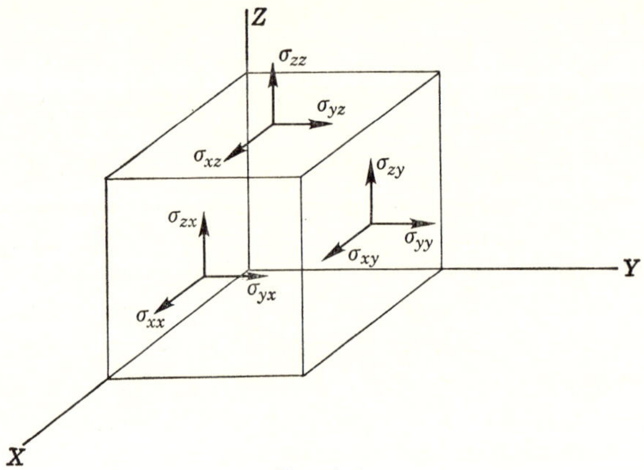

FIG. 1-1.

z direction acting on the plane normal to the x axis. Not all the stress components are independent. Consider the rotational equilibrium of the parallelepiped about an axis through the center of mass parallel to the z axis. The torques about this axis must be equal to the angular momentum, giving

$$(\sigma_{xy}\ \delta x\ \delta z)\delta y - (\sigma_{yx}\ \delta y\ \delta z)\delta x = \rho\ \delta x\ \delta y\ \delta z\ \frac{\overline{\delta x}^2 + \overline{\delta y}^2}{12}\frac{d\omega_z}{dt}$$

$$\sigma_{xy} - \sigma_{yx} = \frac{\overline{\delta x}^2 + \overline{\delta y}^2}{12}\frac{d\omega_z}{dt}$$

where ρ is the density and ω_z the rotational velocity about the z axis. In the limits of reduction to a point the right-hand side approaches zero, and

$$\sigma_{xy} = \sigma_{yx}$$

Similarly, it can be shown that

$$\sigma_{yz} = \sigma_{zy} \qquad \text{and} \qquad \sigma_{zx} = \sigma_{xz} \tag{1-1}$$

The three components σ_{xx}, σ_{yy}, and σ_{zz} normal to the three planes are known as *tensional stresses* or *tensions* if positive and *compressive stresses* or *pressures* if negative. The six components σ_{xy}, σ_{yx}, σ_{yz}, σ_{zy}, σ_{zx}, and σ_{xz} are known as *shearing stresses*. When the shearing stresses are zero and $\sigma_{xx} = \sigma_{yy} = \sigma_{zz}$ and are negative, the stress distribution is known as a *hydrostatic pressure.*

Components of strain

Strain is defined as the change in position of a point with respect to neighboring points divided by the distance between the original point and the reference points. It is the relative change in position per unit distance. Consider a point (x, y, z) which undergoes a displacement (u, v, w); then a neighboring point $(x + \delta x, y + \delta y, z + \delta z)$ will undergo a

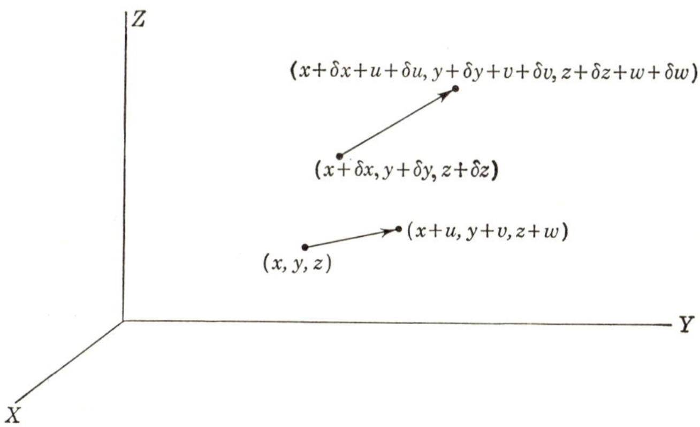

FIG. 1-2.

displacement $(u + \delta u, v + \delta v, w + \delta w)$. To the first order the relative change in displacement will be

$$\begin{aligned}\delta u &= \frac{\partial u}{\partial x}\,\delta x + \frac{\partial u}{\partial y}\,\delta y + \frac{\partial u}{\partial z}\,\delta z \\ \delta v &= \frac{\partial v}{\partial x}\,\delta x + \frac{\partial v}{\partial y}\,\delta y + \frac{\partial v}{\partial z}\,\delta z \\ \delta w &= \frac{\partial w}{\partial x}\,\delta x + \frac{\partial w}{\partial y}\,\delta y + \frac{\partial w}{\partial z}\,\delta z\end{aligned} \tag{1-2}$$

Defining the strain coefficients by

$$\begin{aligned}&\epsilon_{xx} \equiv \frac{\partial u}{\partial x}, \qquad \epsilon_{yy} \equiv \frac{\partial v}{\partial y}, \qquad \epsilon_{zz} \equiv \frac{\partial w}{\partial z} \\ &\epsilon_{yz} \equiv \epsilon_{zy} \equiv \frac{\partial w}{\partial y} + \frac{\partial v}{\partial z}, \quad \epsilon_{zx} \equiv \epsilon_{xz} \equiv \frac{\partial u}{\partial z} + \frac{\partial w}{\partial x}, \quad \epsilon_{xy} \equiv \epsilon_{yx} \equiv \frac{\partial v}{\partial x} + \frac{\partial u}{\partial y} \\ &\theta_x \equiv \frac{\partial w}{\partial y} - \frac{\partial v}{\partial z}, \qquad \theta_y \equiv \frac{\partial u}{\partial z} - \frac{\partial w}{\partial x}, \qquad \theta_z \equiv \frac{\partial v}{\partial x} - \frac{\partial u}{\partial y}\end{aligned} \tag{1-3}$$

the equations become

$$
\begin{aligned}
\delta u &= \epsilon_{xx}\,\delta x + \tfrac{1}{2}(\epsilon_{xy} - \theta_z)\delta y + \tfrac{1}{2}(\epsilon_{xz} + \theta_y)\delta z \\
\delta v &= \tfrac{1}{2}(\epsilon_{yx} + \theta_z)\delta x + \epsilon_{yy}\,\delta y + \tfrac{1}{2}(\epsilon_{yz} - \theta_x)\delta z \\
\delta w &= \tfrac{1}{2}(\epsilon_{zx} - \theta_y)\delta x + \tfrac{1}{2}(\epsilon_{zy} + \theta_x)\delta y + \epsilon_{zz}\,\delta z
\end{aligned}
\tag{1-4}
$$

The coefficients ϵ_{xx}, ϵ_{yy}, and ϵ_{zz} are the fractional changes in length in the direction of measurement. If positive they are termed *extensions* and if negative *contractions*. The coefficients ϵ_{yz}, ϵ_{zy}, ϵ_{zx}, ϵ_{xz}, ϵ_{xy}, and ϵ_{yx} are the *shearing strains*. Their significance may be seen from consideration of a pair of elements $\epsilon_{yz}\,\delta z$, $\epsilon_{zy}\,\delta y$. The first coefficient gives a displacement of point B in the positive y direction of $\epsilon_{yz}\,\delta z$, the second coefficient a displacement of point D in the positive z direction of $\epsilon_{yz}\,\delta y$ as referred to point A. The original square $ABCD$ of Fig. 1-3 is deformed into $AB'C'D'$, the total angle of distortion being simply ϵ_{yz} for small displacements. The coefficients θ_x, θ_y, θ_z produce a rotation without

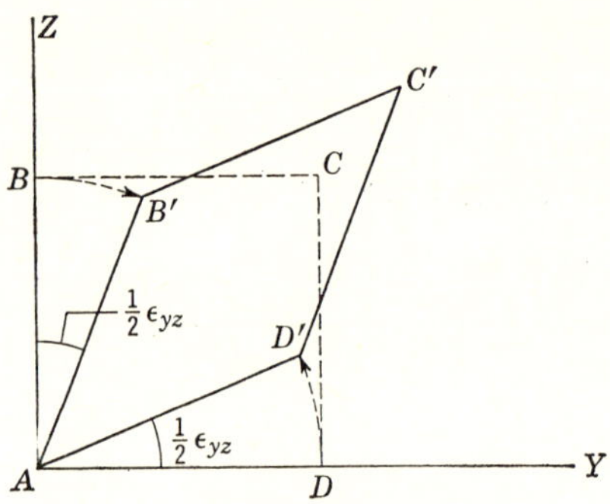

FIG. 1-3.

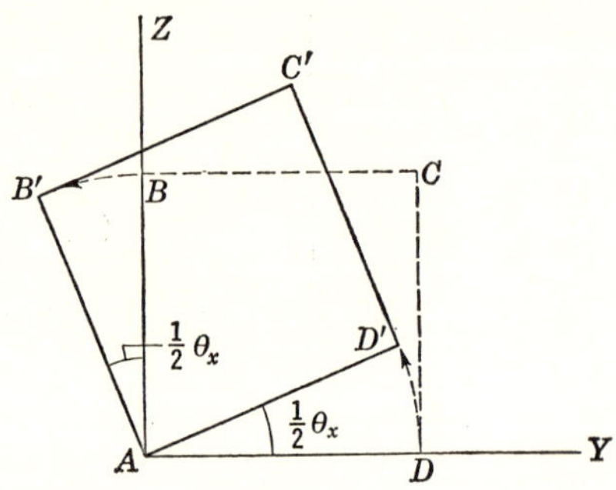

FIG. 1-4.

distortion. The coefficient θ_x gives a displacement of point B in the negative y direction and point D in the positive z direction. The resultant effect is to produce a rotation without distortion of the square $ABCD$ of Fig. 1-4 about the x axis by an amount $\frac{1}{2}\theta_x$ for small displacements. From Eqs. (1-4) the relative displacement of an area can be considered to be made up of three parts, (1) an expansion with or without a change in shape, (2) a rotation, and (3) a distortion. This is illustrated in Fig. 1-5. The ϵ_{xx}, ϵ_{yy}, ϵ_{zz} components represent the first part. The θ_x, θ_y, θ_z components represent the second part. The ϵ_{yz}, ϵ_{zx}, ϵ_{xy} components represent the third part. It is important to remember for what follows later that the ϵ_{xx}, ϵ_{yy}, ϵ_{zz} components represent a change in volume as well as a change in shape. The ϵ_{yz}, ϵ_{zx}, ϵ_{xy} components represent a change in shape only. The θ_x, θ_y, θ_z components represent neither a change in shape nor a change in volume.

A useful strain relation in the later sections on wave propagation is that of *dilatation*. It is defined in the limit as the increase in volume per unit volume. Consider an unstrained cube of sides δx, δy, δz. In a strained state it will have sides $\delta x + \delta u$, $\delta y + \delta v$, $\delta z + \delta w$ or from Eqs. (1-4)

$$\begin{aligned}&(1 + \epsilon_{xx})\delta x + \tfrac{1}{2}(\epsilon_{xy} - \theta_z)\delta y + \tfrac{1}{2}(\epsilon_{xz} + \theta_y)\delta z\\ &\tfrac{1}{2}(\epsilon_{yx} + \theta_z)\delta x + (1 + \epsilon_{yy})\delta y + \tfrac{1}{2}(\epsilon_{yz} - \theta_x)\delta z\\ &\tfrac{1}{2}(\epsilon_{zx} - \theta_y)\delta x + \tfrac{1}{2}(\epsilon_{zy} + \theta_x)\delta y + (1 + \epsilon_{zz})\delta z\end{aligned}$$

The strained volume will be the triple scalar product of the three sides of the parallelepiped. For an original unit cube this will be simply the determinant of the coefficients of the above expression, or

$$\begin{vmatrix} 1 + \epsilon_{xx} & \tfrac{1}{2}(\epsilon_{xy} - \theta_z) & \tfrac{1}{2}(\epsilon_{xz} + \theta_y) \\ \tfrac{1}{2}(\epsilon_{yx} + \theta_z) & 1 + \epsilon_{yy} & \tfrac{1}{2}(\epsilon_{yz} - \theta_x) \\ \tfrac{1}{2}(\epsilon_{zx} - \theta_y) & \tfrac{1}{2}(\epsilon_{zy} + \theta_x) & 1 + \epsilon_{zz} \end{vmatrix} = 1 + \epsilon_{xx} + \epsilon_{yy} + \epsilon_{zz}$$

neglecting higher-order terms in ϵ and θ. Thus, the dilatation Δ is given by

$$\Delta = \epsilon_{xx} + \epsilon_{yy} + \epsilon_{zz} = \frac{\partial u}{\partial x} + \frac{\partial v}{\partial y} + \frac{\partial w}{\partial z} \tag{1-5}$$

A fundamental relation is that of the *equation of conservation*. This expresses the physical fact that mass is conserved. It can be expressed

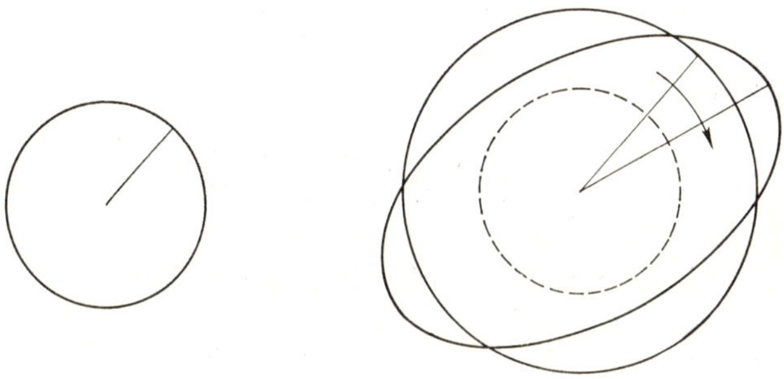

FIG. 1-5.

mathematically by equating the net rate of increase of mass per unit volume to the rate of change in density. Consider the rectangular parallelepiped of Fig. 1-1. The rate of inflow of mass through one face normal to the x axis is $\rho\dot{u}\ \delta y\ \delta z$ and that out the other is $[\rho\dot{u} + (\partial/\partial x)(\rho\dot{u})\ \delta x]\ \delta y\ \delta z$, where $\dot{u} = du/dt$ is the velocity of transport. The net rate of increase of mass from transport across these two faces is $-(\partial/\partial x)(\rho\dot{u})\ \delta x\ \delta y\ \delta z$ and similarly for the faces normal to the y and z axes, giving for the net rate of increase in mass per unit volume $-(\partial/\partial x)(\rho\dot{u}) - (\partial/\partial y)(\rho\dot{v}) - (\partial/\partial z)(\rho\dot{w})$. Equating this to the rate of change of density gives

$$\frac{\partial}{\partial x}(\rho\dot{u}) + \frac{\partial}{\partial y}(\rho\dot{v}) + \frac{\partial}{\partial z}(\rho\dot{w}) + \frac{\partial \rho}{\partial t} = 0 \tag{1-6}$$

If the material is incompressible, the usual case in hydrodynamic flow problems but not in elastic wave propagation, the equation reduces to

$$\frac{\partial \dot{u}}{\partial x} + \frac{\partial \dot{v}}{\partial y} + \frac{\partial \dot{w}}{\partial z} = 0$$

Another form of Eq. (1-6) can be obtained if, instead of considering a description of the displacements and velocities at all points of space at all times, the motion of the particles is followed. Then, a particle originally at position (x, y, z) at time t will be at $(x + \dot{u}\,\delta t, y + \dot{v}\,\delta t, z + \dot{w}\,\delta t)$ at time $t + \delta t$; and a function $F(x, y, z, t)$ associated with this particle will have at time $t + \delta t$ the value

$$F(x + \dot{u}\,\delta t, y + \dot{v}\,\delta t, z + \dot{w}\,\delta t) = F + \frac{\partial F}{\partial x}\dot{u}\,\delta t + \frac{\partial F}{\partial y}\dot{v}\,\delta t + \frac{\partial F}{\partial z}\dot{w}\,\delta t + \frac{\partial F}{\partial t}\delta t$$

Using d/dt to denote differentiation following the motion the new value of F may also be expressed by $F + (dF/dt)\,\delta t$ so that

$$\frac{d}{dt} = \frac{\partial}{\partial t} + \dot{u}\frac{\partial}{\partial x} + \dot{v}\frac{\partial}{\partial y} + \dot{w}\frac{\partial}{\partial z} \tag{1-7}$$

and Eq. (1-6) can be written as

$$\rho\left(\frac{\partial \dot{u}}{\partial x} + \frac{\partial \dot{v}}{\partial y} + \frac{\partial \dot{w}}{\partial z}\right) + \frac{d\rho}{dt} = 0 \tag{1-8}$$

Equation (1-6) is referred to as being in *Eulerian* form and Eq. (1-8) in *Lagrangian* form. In elastic wave propagation the displacements and velocities are usually small so that products of these components may be neglected. Under these conditions the derivatives d/dt and d^2/dt^2 may be replaced by $\partial/\partial t$ and $\partial^2/\partial t^2$. Equation (1-8) becomes

$$\rho\left(\frac{\partial \dot{u}}{\partial x} + \frac{\partial \dot{v}}{\partial y} + \frac{\partial \dot{w}}{\partial z}\right) + \frac{\partial \rho}{\partial t} = 0 \tag{1-9}$$

This is not the case for hydrodynamics.

Stress-strain relations

In practice it is observed that within certain limits the strains produced in an elastic solid are directly proportional to the applied stresses. This is known as *Hooke's law.* Its generalized form states that each of the six components of stress at any point of a body will be a linear function of the six components of strain. Thus,

$$
\begin{aligned}
\sigma_{xx} &= c_{11}\epsilon_{xx} + c_{12}\epsilon_{yy} + c_{13}\epsilon_{zz} + c_{14}\epsilon_{yz} + c_{15}\epsilon_{zx} + c_{16}\epsilon_{xy} \\
\sigma_{yy} &= c_{21}\epsilon_{xx} + c_{22}\epsilon_{yy} + c_{23}\epsilon_{zz} + c_{24}\epsilon_{yz} + c_{25}\epsilon_{zx} + c_{26}\epsilon_{xy} \\
\sigma_{zz} &= c_{31}\epsilon_{xx} + c_{32}\epsilon_{yy} + c_{33}\epsilon_{zz} + c_{34}\epsilon_{yz} + c_{35}\epsilon_{zx} + c_{36}\epsilon_{xy} \\
\sigma_{yz} &= c_{41}\epsilon_{xx} + c_{42}\epsilon_{yy} + c_{43}\epsilon_{zz} + c_{44}\epsilon_{yz} + c_{45}\epsilon_{zx} + c_{46}\epsilon_{xy} \\
\sigma_{zx} &= c_{51}\epsilon_{xx} + c_{52}\epsilon_{yy} + c_{53}\epsilon_{zz} + c_{54}\epsilon_{yz} + c_{55}\epsilon_{zx} + c_{56}\epsilon_{xy} \\
\sigma_{xy} &= c_{61}\epsilon_{xx} + c_{62}\epsilon_{yy} + c_{63}\epsilon_{zz} + c_{64}\epsilon_{yz} + c_{65}\epsilon_{zx} + c_{66}\epsilon_{xy}
\end{aligned}
\tag{1-10}
$$

The coefficients c_{mn} are the *elastic constants* of the material. In order for the strain energy to be a univalued function of the components of strain the coefficients must be related by $c_{mn} = c_{nm}$, reducing the number of independent elastic constants from 36 to 21.

For an *isotropic solid,* i.e., one in which the elastic behavior is independent of direction, the number of independent constants is reduced to two. Briefly, in order for the stress-strain relations to be unaltered by an interchange of axes,

$$
\begin{aligned}
c_{11} &= c_{22} = c_{33} \qquad & c_{44} &= c_{55} = c_{66} \\
c_{12} &= c_{13} = c_{23} \qquad & c_{15} &= c_{16} = c_{26} \\
c_{14} &= c_{25} = c_{36} \qquad & c_{45} &= c_{46} = c_{56}
\end{aligned}
$$

and the relations to be unchanged by a reversal of sense of any of the axes,

$$c_{14} = c_{25} = c_{36} = c_{15} = c_{16} = c_{26} = c_{45} = c_{46} = c_{56} = 0$$

and the relations to be unaltered by a rotation of the axes

$$2c_{44} = c_{11} - c_{12}$$

The two independent elastic constants are usually taken as

$$
\begin{aligned}
c_{11} &= c_{22} = c_{33} = \lambda + 2\mu \\
c_{44} &= c_{55} = c_{66} = \mu \\
c_{12} &= c_{13} = c_{23} = c_{21} = c_{31} = c_{32} = \lambda
\end{aligned}
\tag{1-11}
$$

and are referred to as *Lamé's constants.* The stress-strain relations for an isotropic solid are then

$$
\begin{aligned}
\sigma_{xx} &= (\lambda + 2\mu)\epsilon_{xx} + \lambda\epsilon_{yy} + \lambda\epsilon_{zz} = \lambda\Delta + 2\mu\epsilon_{xx} \\
\sigma_{yy} &= \lambda\epsilon_{xx} + (\lambda + 2\mu)\epsilon_{yy} + \lambda\epsilon_{zz} = \lambda\Delta + 2\mu\epsilon_{yy} \\
\sigma_{zz} &= \lambda\epsilon_{xx} + \lambda\epsilon_{yy} + (\lambda + 2\mu)\epsilon_{zz} = \lambda\Delta + 2\mu\epsilon_{zz} \\
\sigma_{yz} &= \mu\epsilon_{yz} \\
\sigma_{zx} &= \mu\epsilon_{zx} \\
\sigma_{xy} &= \mu\epsilon_{xy}
\end{aligned}
\tag{1-12}
$$

Other elastic constants that are often used for an isotropic solid are Young's modulus, Poisson's ratio, and the bulk modulus. *Young's modulus* E is defined as the ratio of the longitudinal stress on a cylinder to the longitudinal extension produced, all other stresses being zero; and *Poisson's ratio* σ is the ratio of the resultant lateral contraction to the longitudinal extension. Equations (1-12) become

$$\begin{aligned}\sigma_{xx} &= (\lambda + 2\mu)\epsilon_{xx} + \lambda\epsilon_{yy} + \lambda\epsilon_{zz}\\ 0 &= \lambda\epsilon_{xx} + (\lambda + 2\mu)\epsilon_{yy} + \lambda\epsilon_{zz}\\ 0 &= \lambda\epsilon_{xx} + \lambda\epsilon_{yy} + (\lambda + 2\mu)\epsilon_{zz}\end{aligned}$$

Solving the second and third for ϵ_{zz} and ϵ_{yy} in terms of ϵ_{xx} gives

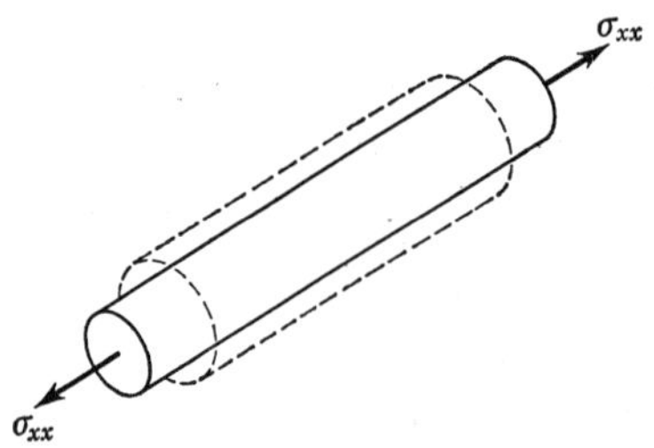

FIG. 1-6.

$$\sigma = -\frac{\epsilon_{yy}}{\epsilon_{xx}} = \frac{\lambda}{2(\lambda + \mu)} \tag{1-13}$$

and substituting in the first gives

$$E = \frac{\sigma_{xx}}{\epsilon_{xx}} = \frac{\mu(3\lambda + 2\mu)}{\lambda + \mu} \tag{1-14}$$

for E and σ in terms of λ and μ. The *bulk modulus*, or *incompressibility* k, is defined as the ratio of the hydrostatic pressure on a body to the resultant fractional change in volume, negative dilatation. In this case Eqs. (1-12) reduce to

$$\begin{aligned}-p &= \lambda\Delta + 2\mu\epsilon_{xx}\\ -p &= \lambda\Delta + 2\mu\epsilon_{yy}\\ -p &= \lambda\Delta + 2\mu\epsilon_{zz}\end{aligned} \tag{1-15}$$

giving

$$k = \frac{p}{-\Delta} = \lambda + \tfrac{2}{3}\mu = \frac{3\lambda + 2\mu}{3} \tag{1-16}$$

where $p = -\sigma_{xx} = -\sigma_{yy} = -\sigma_{zz}$ is a hydrostatic pressure. The *shear modulus* or *rigidity* is defined as the ratio between an applied shear stress and the resultant shear strain and from Eqs. (1-12) is simply the Lamé constant μ.

In many problems in elastic wave propagation it is convenient to assume that $\lambda = \mu$. This relation, known variously as *Poisson's* or *Cauchy's relation*, was originally a result of the hypothesis that matter could be regarded as made up of material points acting on each other at a distance, the force being a function of the distance between the points and acting in a line joining the points, i.e., a central force. This gives one independent elastic constant, the relations between the several above being

$$\lambda = \mu,\ \sigma = \tfrac{1}{4},\ k = \tfrac{5}{3}\mu,\ E = \tfrac{5}{2}\mu \tag{1-17}$$

It is sometimes advantageous to write the stress-strain relations for an isotropic solid (1-12) in terms of the bulk modulus k (1-16) and the rigidity μ

$$\begin{aligned}
\sigma_{xx} &= k\Delta + \tfrac{2}{3}\mu(2\epsilon_{xx} - \epsilon_{yy} - \epsilon_{zz}) \\
\sigma_{yy} &= k\Delta + \tfrac{2}{3}\mu(-\epsilon_{xx} + 2\epsilon_{yy} - \epsilon_{zz}) \\
\sigma_{zz} &= k\Delta + \tfrac{2}{3}\mu(-\epsilon_{xx} - \epsilon_{yy} + 2\epsilon_{zz}) \\
\sigma_{yz} &= \mu\epsilon_{yz} \\
\sigma_{zx} &= \mu\epsilon_{zx} \\
\sigma_{xy} &= \mu\epsilon_{xy}
\end{aligned} \tag{1-18}$$

The bulk modulus k is a coefficient relating to volume changes only. The rigidity μ is a coefficient relating to distortion only. We should like to break down Eqs. (1-18) into two sets of equations for the resultant deformation, one relating to the volume change components only without distortion and the other relating to the distortion components only without volume change. Clearly the last three equations represent only distortion. Adding the first three equations, we obtain

$$\frac{\sigma_{xx} + \sigma_{yy} + \sigma_{zz}}{3} = k\Delta \tag{1-19}$$

This is the equation for the dilatation in terms of the imposed stresses. Let us make the definitions

$$\sigma_H \equiv \sigma'_{xx} \equiv \sigma'_{yy} \equiv \sigma'_{zz} = \frac{\sigma_{xx} + \sigma_{yy} + \sigma_{zz}}{3} \tag{1-20}$$

Then from (1-19) and (1-5) we may write

$$\begin{aligned}
\sigma'_{xx} &= 3k\epsilon'_{xx} \\
\sigma'_{yy} &= 3k\epsilon'_{yy} \\
\sigma'_{zz} &= 3k\epsilon'_{zz}
\end{aligned} \tag{1-21}$$

where

$$\epsilon'_{xx} = \epsilon'_{yy} = \epsilon'_{zz} = \frac{\Delta}{3} \tag{1-22}$$

Equations (1-21) with the subsidiary relation (1-22) are the equations for the components ϵ'_{xx}, ϵ'_{yy}, ϵ'_{zz} representing the volume change without distortion. Subtracting Eqs. (1-21) from (1-18) we obtain

$$\begin{aligned}
\sigma''_{xx} &= \sigma_x - \sigma_H = \tfrac{2}{3}\mu(2\epsilon_{xx} - \epsilon_{yy} - \epsilon_{zz}) = \tfrac{2}{3}\mu(2\epsilon''_{xx} - \epsilon''_{yy} - \epsilon''_{zz}) \\
\sigma''_{yy} &= \sigma_y - \sigma_H = \tfrac{2}{3}\mu(-\epsilon_{xx} + 2\epsilon_{yy} - \epsilon_{zz}) = \tfrac{2}{3}\mu(-\epsilon''_{xx} + 2\epsilon''_{yy} - \epsilon''_{zz}) \\
\sigma''_{zz} &= \sigma_z - \sigma_H = \tfrac{2}{3}\mu(-\epsilon_{xx} - \epsilon_{yy} + 2\epsilon_{zz}) = \tfrac{2}{3}\mu(-\epsilon''_{xx} - \epsilon''_{yy} + 2\epsilon''_{zz}) \\
\sigma_{yz} &= \mu\epsilon_{yz} \\
\sigma_{zx} &= \mu\epsilon_{zx} \\
\sigma_{xy} &= \mu\epsilon_{xy}
\end{aligned} \tag{1-23}$$

where

$$\begin{aligned} \epsilon''_{xx} &= \epsilon_{xx} - \epsilon'_{xx} = \epsilon_{xx} - \frac{\Delta}{3} \\ \epsilon''_{yy} &= \epsilon_{yy} - \epsilon'_{yy} = \epsilon_{yy} - \frac{\Delta}{3} \\ \epsilon''_{zz} &= \epsilon_{zz} - \epsilon'_{zz} = \epsilon_{zz} - \frac{\Delta}{3} \end{aligned} \tag{1-24}$$

and from (1-5) and (1-22)

$$\epsilon''_{xx} + \epsilon''_{yy} + \epsilon''_{zz} = 0 \tag{1-25}$$

Equations (1-23) with the subsidiary relation (1-24) are the equations for the components ϵ''_{xx}, ϵ''_{yy}, ϵ''_{zz}, ϵ_{yz}, ϵ_{zx}, ϵ_{xy} representing the distortion only. As may be seen from the form of Eqs. (1-25), there is no included volume change for the ϵ''_{xx}, ϵ''_{yy}, ϵ''_{zz} components. If in the original equation (1-18) the shear stresses are zero and the tensional stresses are all equal, the resultant deformation will be a volume change with no distortion. This is a hydrostatic condition. If the stress σ_H of Eq. (1-20) is zero, the resultant deformation will be a distortion with no volume change. For any other stress distribution there will be both a volume change and a distortion.

Let us examine what effects occur in the limit of the coefficient μ zero. From Eqs. (1-23) we see that, for an arbitrary initial stress distribution, the strain components approach infinity as μ approaches zero. If equilibrium is to exist, the stress components σ''_{xx}, σ''_{yy}, σ''_{zz}, σ_{yz}, σ_{zx}, σ_{xy} must vanish. In other words a medium for which μ is zero cannot withstand these stresses. This is the condition of a perfect liquid; it is defined as one for which the rigidity μ is zero. The stress-strain equations reduce to (1-21) or more simply

$$\sigma_H = -p = k\Delta \tag{1-26}$$

The stress is hydrostatic.

Conversely, in the limit of k zero, the stress-strain equations reduce to (1-23). These equations are of interest for the stress–rate-of-strain relations in a viscous fluid; this problem is discussed in Chap. 6.

Equations of motion

The equations of motion are obtained from Newton's second law of motion, equating the products of density and acceleration in a given direction to the net forces acting per unit volume in that direction. Referring to Fig. 1-7 it is observed that there are stresses acting in the x direction on each of the six faces of the rectangular parallelepiped, the stress on the back being increased by the appropriate amount $(\partial/\partial r)\ \delta r$ to the front face. The total force in the x direction is then

$$\left(\sigma_{xx} + \frac{\partial \sigma_{xx}}{\partial x}\,\delta x\right)\delta y\,\delta z - \sigma_{xx}\,\delta y\,\delta z + \left(\sigma_{xy} + \frac{\partial \sigma_{xy}}{\partial y}\,\delta y\right)\delta x\,\delta z - \sigma_{xy}\,\delta x\,\delta z$$
$$+ \left(\sigma_{xz} + \frac{\partial \sigma_{xz}}{\partial z}\,\delta z\right)\delta x\,\delta y - \sigma_{xz}\,\delta x\,\delta y = \left(\frac{\partial \sigma_{xx}}{\partial x} + \frac{\partial \sigma_{xy}}{\partial y} + \frac{\partial \sigma_{xz}}{\partial z}\right)\delta x\,\delta y\,\delta z$$

Newton's second law then gives

$$\rho\,\frac{d^2u}{dt^2} = \frac{\partial \sigma_{xx}}{\partial x} + \frac{\partial \sigma_{xy}}{\partial y} + \frac{\partial \sigma_{xz}}{\partial z} + \rho X$$

and similarly

$$\begin{aligned} \rho\,\frac{d^2v}{dt^2} &= \frac{\partial \sigma_{yx}}{\partial x} + \frac{\partial \sigma_{yy}}{\partial y} + \frac{\partial \sigma_{yz}}{\partial z} + \rho Y \\ \rho\,\frac{d^2w}{dt^2} &= \frac{\partial \sigma_{zx}}{\partial x} + \frac{\partial \sigma_{zy}}{\partial y} + \frac{\partial \sigma_{zz}}{\partial z} + \rho Z \end{aligned} \tag{1-27}$$

where X, Y, and Z are the body forces per unit mass. In elastic wave propagation the body forces, such as gravity, can usually be neglected;

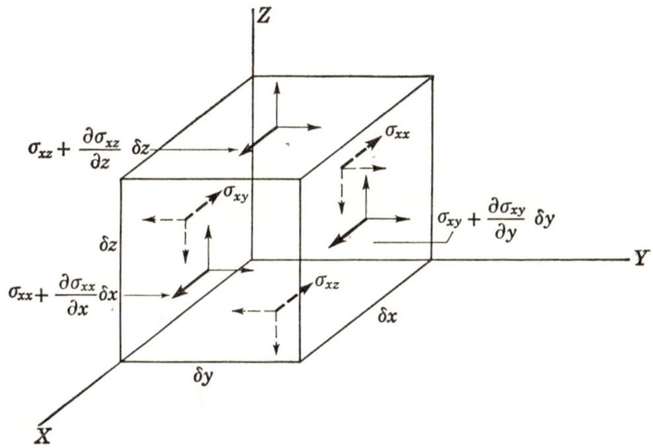

FIG. 1-7.

and for small displacements and velocities d^2/dt^2 replaced by $\partial^2/\partial t^2$, giving

$$\begin{aligned} \rho\,\frac{\partial^2 u}{\partial t^2} &= \frac{\partial \sigma_{xx}}{\partial x} + \frac{\partial \sigma_{xy}}{\partial y} + \frac{\partial \sigma_{xz}}{\partial z} \\ \rho\,\frac{\partial^2 v}{\partial t^2} &= \frac{\partial \sigma_{yx}}{\partial x} + \frac{\partial \sigma_{yy}}{\partial y} + \frac{\partial \sigma_{yz}}{\partial z} \\ \rho\,\frac{\partial^2 w}{\partial t^2} &= \frac{\partial \sigma_{zx}}{\partial x} + \frac{\partial \sigma_{zy}}{\partial y} + \frac{\partial \sigma_{zz}}{\partial z} \end{aligned} \tag{1-28}$$

Substituting for the stresses in terms of the displacements from Eqs. (1-12) and (1-3), we obtain

$$\rho \frac{\partial^2 u}{\partial t^2} = \frac{\partial}{\partial x}\left(\lambda\Delta + 2\mu \frac{\partial u}{\partial x}\right) + \frac{\partial}{\partial y}\left[\mu\left(\frac{\partial v}{\partial x} + \frac{\partial u}{\partial y}\right)\right] + \frac{\partial}{\partial z}\left[\mu\left(\frac{\partial w}{\partial x} + \frac{\partial u}{\partial z}\right)\right] \tag{1-29}$$

which reduces to

$$\rho \frac{\partial^2 u}{\partial t^2} = (\lambda + \mu)\frac{\partial \Delta}{\partial x} + \mu\nabla^2 u$$

and similarly

$$\rho \frac{\partial^2 v}{\partial t^2} = (\lambda + \mu)\frac{\partial \Delta}{\partial y} + \mu\nabla^2 v \tag{1-30}$$

$$\rho \frac{\partial^2 w}{\partial t^2} = (\lambda + \mu)\frac{\partial \Delta}{\partial z} + \mu\nabla^2 w$$

where $\nabla^2 \equiv \partial^2/\partial x^2 + \partial^2/\partial y^2 + \partial^2/\partial z^2$, the *Laplacian operator.*

Equations (1-30) are the equations of motion for an isotropic elastic solid under the assumptions of small displacements (infinitesimal strain theory), perfect elasticity, and no body forces. They define the propagation of two types of waves through the medium. The first can be obtained by differentiating the three equations with respect to $\partial/\partial x$, $\partial/\partial y$, and $\partial/\partial z$, respectively, and adding. This gives

$$\rho \frac{\partial^2 \Delta}{\partial t^2} = (\lambda + 2\mu)\nabla^2 \Delta \tag{1-31}$$

which is the wave equation, whose solution states that the dilatation Δ is propagated through the medium with the velocity $[(\lambda + 2\mu)/\rho]^{1/2}$. The second can be obtained by differentiating the second equation with respect to $\partial/\partial z$ and the third with respect to $\partial/\partial y$ and subtracting to eliminate Δ.

$$\rho \frac{\partial^2}{\partial t^2}\left(\frac{\partial w}{\partial y} - \frac{\partial v}{\partial z}\right) = \mu\nabla^2\left(\frac{\partial w}{\partial y} - \frac{\partial v}{\partial z}\right)$$

or

$$\rho \frac{\partial^2 \theta_x}{\partial t^2} = \mu\nabla^2\theta_x$$

and similarly for the other two combinations.

$$\rho \frac{\partial^2 \theta_y}{\partial t^2} = \mu\nabla^2\theta_y \tag{1-32}$$

$$\rho \frac{\partial^2 \theta_z}{\partial t^2} = \mu\nabla^2\theta_z$$

This again is the wave equation and states that the rotation θ_x, θ_y, θ_z is propagated with the velocity $(\mu/\rho)^{1/2}$.

This reduction gives a complete description of wave propagation in an isotropic medium; for it may be shown that any vector displacement may be reduced along the appropriate set of rectangular axes to a dilatational and a rotational component, the propagation characteristics of which are given separately by Eqs. (1-31) and (1-32). It is important to note, however, that this reduction is not valid for a statistically isotropic medium of two or more components or for an anisotropic medium. The first type of wave, propagated with the higher velocity $(\lambda + 2\mu/\rho)^{1/2}$, is variously known as the *irrotational, dilatational,* or P wave; and the second type of wave, traveling with the slower velocity $(\mu/\rho)^{1/2}$, is known as the *equivoluminal, rotational,* or S wave.

The two different types of propagation can be illustrated in a simple manner by the consideration of plane wave propagation. Assume the wave is propagated in the x direction. Then, the displacement components u, v, w will be functions of x and t only. The equations of motion (1-30) reduce to

$$\begin{aligned} \rho \frac{\partial^2 u}{\partial t^2} &= (\lambda + \mu) \frac{\partial^2 u}{\partial x^2} + \mu \frac{\partial^2 u}{\partial x^2} = (\lambda + 2\mu) \frac{\partial^2 u}{\partial x^2} \\ \rho \frac{\partial^2 v}{\partial t^2} &= \mu \frac{\partial^2 v}{\partial x^2} \\ \rho \frac{\partial^2 w}{\partial t^2} &= \mu \frac{\partial^2 w}{\partial x^2} \end{aligned} \tag{1-33}$$

The displacement in the direction of propagation, *longitudinal* vibration, is propagated with the P velocity $[(\lambda + 2\mu)/\rho]^{1/2}$; and the two displacements at right angles to the direction of propagation, *transverse* vibration, are propagated with the S velocity $(\mu/\rho)^{1/2}$.

For a liquid or a gas, $\mu = 0$, the equations of motion reduce to

$$\rho \frac{\partial^2 \Delta}{\partial t^2} = k\nabla^2 \Delta \tag{1-34}$$

This again is the wave equation. The dilatation Δ is propagated with the velocity

$$c = \left(\frac{k}{\rho}\right)^{1/2} \tag{1-34a}$$

Fundamental solution of the wave equation

Let us consider briefly the properties of the general solution of the wave equation. For one-dimensional motion the wave equation, such as (1-34), becomes

$$\frac{\partial^2 \psi}{\partial t^2} = c^2 \frac{\partial^2 \psi}{\partial x^2} \tag{1-35}$$

where c is a physical constant and $\psi = \psi(x, t)$ is a measure of the disturbance, such as a component of displacement or velocity. The general solution of Eq. (1-35) is

$$\psi = f(x - ct) + F(x + ct) \tag{1-36}$$

as may be seen by substitution back into (1-35), where f and F are arbitrary functions depending only on the initial and boundary conditions.

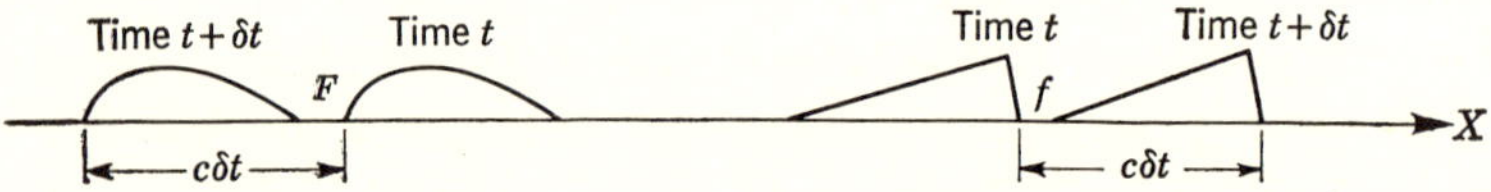

FIG. 1-8.

If the time t is increased by an amount δt, the value of f will be unchanged if x is increased by $c\ \delta t$, that is to say, the function f which was originally measured at position x at time t will be found at position $x + c\ \delta t$ at time $t + \delta t$. f represents a disturbance propagated in the positive x direction with a velocity c. Similarly the function F remains unchanged at time $t + \delta t$ by a decrease in x of $c\ \delta t$; it represents a disturbance propagated in the negative x direction with a velocity c.

Next consider a disturbance which is propagated in three dimensions and is symmetrical about a center O. The Laplacian operator reduces to $\dfrac{\partial^2}{\partial r^2} + \dfrac{2}{r}\dfrac{\partial}{\partial r}$ in spherical coordinates where r is the distance measured from O, and the wave equation becomes

$$\frac{\partial^2\psi}{\partial t^2} = c^2\left(\frac{\partial^2\psi}{\partial r^2} + \frac{2}{r}\frac{\partial\psi}{\partial r}\right) \tag{1-37}$$

or

$$\frac{\partial^2(r\psi)}{\partial t^2} = c^2\frac{\partial^2(r\psi)}{\partial r^2} \tag{1-38}$$

which is the same as Eq. (1-35), giving for ψ

$$\psi = \frac{1}{r}f(r - ct) + \frac{1}{r}F(r + ct) \tag{1-39}$$

$f(r - ct)$ is a spherical wave diverging from O and $F(r + ct)$ is a spherical wave converging on O. The amplitude decreases as r^{-1}.

The amplitude factor would be expected from physical considerations; for as a spherical wave is propagated outward, the area of the wave front (sphere) increases as r^2. Consequently the energy flow per unit area will decrease as r^{-2}. As will be shown later, the energy flow is proportional to ψ^2 so that from Eq. (1-39) its distance dependence is as r^{-2}.

Next consider a disturbance which is propagated in three dimensions and is symmetrical about an axis. The Laplacian operator reduces to

$\dfrac{\partial^2}{\partial r^2} + \dfrac{1}{r}\dfrac{\partial}{\partial r}$ in cylindrical coordinates where r is now the distance measured from the polar axis, and the wave equation becomes

$$\frac{\partial^2 \psi}{\partial t^2} = c^2 \left(\frac{\partial^2 \psi}{\partial r^2} + \frac{1}{r}\frac{\partial \psi}{\partial r} \right) \tag{1-40}$$

The solution of this equation is not so easy to obtain as that for Eq. (1-37). Its solution is given in terms of Bessel's functions of the first and second kind of zero order. An approximate solution can be given as follows; for

$$\frac{\partial^2 (r^{1/2}\psi)}{\partial r^2} = r^{1/2} \left(\frac{\partial^2 \psi}{\partial r^2} + \frac{1}{r}\frac{\partial \psi}{\partial r} - \frac{1}{4r^2}\psi \right)$$

At moderate ranges the third term on the right-hand side may be neglected with respect to the first two so that Eq. (1-40) may be written approximately as

$$\frac{\partial^2 (r^{1/2}\psi)}{\partial t^2} = c^2 \frac{\partial^2 (r^{1/2}\psi)}{\partial r^2} \tag{1-41}$$

whose solution is simply

$$\psi = \frac{1}{r^{1/2}} f(r - ct) + \frac{1}{r^{1/2}} F(r + ct) \tag{1-42}$$

$f(r - ct)$ is a cylindrical wave diverging from the polar axis and $F(r + ct)$ a cylindrical wave converging on the polar axis. The amplitude decreases as $r^{-1/2}$.

Again the amplitude factor is expected from physical considerations; for as a cylindrical wave is propagated outward, the area of the wave front increases as r so that the energy flow per unit area decreases as r^{-1}. From Eq. (1-42) the energy flow is predicted to decrease as r^{-1}.

Adiabatic and isothermal conditions and the velocity of sound in sea water

In all the above it has been implicitly assumed that the disturbances have taken place under conditions of constant temperature. In general this is not the case. We must include the appropriate thermodynamic relations. In particular we must introduce the temperature variations and resultant heat flow associated with sound propagation. In Chap. 6 it will be shown that there is little loss due to heat conduction associated with sound propagation. The disturbance is propagated more nearly under *adiabatic*, no heat transfer, rather than *isothermal* conditions. We shall proceed to derive the velocity of propagation under the incorrect assumption of isothermal conditions and then under the more nearly correct assumption of adiabatic conditions.

For any material there will be in general a thermodynamic equation of state relating the three variables pressure, volume, and temperature. We may write this as

$$f(P, V, T) = 0 \tag{1-42a}$$

where P is the total pressure, V the specific volume, and T the absolute temperature. The specific volume is the volume per unit mass; it is the reciprocal of the density ρ. With the above notation and (1-16) and (1-34a) the velocity of sound may be rewritten in another useful form as

$$c = \left(\frac{k}{\rho}\right)^{1/2} = \left(\frac{-p}{\Delta\rho}\right)^{1/2} = \left(-\frac{dP}{dV}\frac{V_0}{\rho}\right)^{1/2} = \left(\frac{dP}{d\rho}\right)^{1/2} \tag{1-42b}$$

since for small displacements $1/V_0$ is nearly equal to ρ. It must be remembered now, however, that we must stipulate the thermodynamic conditions under which k or $dP/d\rho$ is measured. The derivative $dP/d\rho$ can no longer be considered exact.

For a *perfect gas* the thermodynamic relation (1-42a) is the general gas law,

$$PV = RT \tag{1-43}$$

where R is the gas constant for the particular gas. For isothermal conditions Eq. (1-43) reduces to Boyle's law,

$$PV = P_0V_0 = \text{constant} \tag{1-44}$$

where P and V are the total pressure and specific volume in the disturbed state and P_0 and V_0 the total pressure and specific volume in the undisturbed state. The unbalanced stress, or excess pressure, p, used in the derivation of the equations of motion is equal to $P - P_0$. From Eq. (1-44) we may derive the expression for the isothermal bulk modulus. Taking differentials of (1-44) we obtain

$$\frac{dP}{dV} = -\frac{P}{V} = -\frac{P_0}{V_0}\left(\frac{V_0}{V}\right)^2$$

so that from (1-16)

$$k_T = \frac{-p}{\Delta} = \frac{dP}{-(dV/V_0)} = P_0\left(\frac{V_0}{V}\right)^2 = P_0\left(\frac{\rho}{\rho_0}\right)^2 \tag{1-45}$$

For small displacements (infinitesimal strain) ρ/ρ_0 is nearly equal to unity. The bulk modulus k_T and the velocity of sound propagation c_T can be approximated by

$$k_T = P_0 \tag{1-46}$$

$$c_T = \left(\frac{k_T}{\rho}\right)^{1/2} = \left(\frac{P_0}{\rho_0}\right)^{1/2} \tag{1-47}$$

For adiabatic conditions there is no heat transfer. From the equation of state (1-43) the condition of the gas is defined by any two of three quantities P, V, T. Let us express the heat differential dQ in terms of P and V. We shall have then that

$$dQ = \left(\frac{\partial Q}{\partial V}\right)_P dV + \left(\frac{\partial Q}{\partial P}\right)_V dP \tag{1-48}$$

The subscript of the partial derivative indicates the second variable (constant for the particular derivative). We may rewrite the equation of state (1-43) in terms of differentials of P, V, and T as

$$\begin{aligned} R\,dT &= P\,dV + V\,dP \\ \frac{dT}{T} &= \frac{dV}{V} + \frac{dP}{P} \end{aligned} \tag{1-49}$$

If $dP = 0$, we shall have from (1-49) and (1-48) that

$$dQ = \left(\frac{\partial Q}{\partial V}\right)_P \frac{V}{T}\,dT, \qquad P = \text{constant}$$

or

$$s_P = \left(\frac{\partial Q}{\partial T}\right)_P = \left(\frac{\partial Q}{\partial V}\right)_P \frac{V}{T} \tag{1-50}$$

The quantity s_P is the specific heat at constant pressure. It is defined as the heat necessary to raise the temperature of a unit mass of the gas one degree where the gas is kept at constant pressure during the process. If $dV = 0$, we have from (1-49) and (1-48) that

$$dQ = \left(\frac{\partial Q}{\partial P}\right)_V \frac{P}{T}\,dT, \qquad V = \text{constant}$$

or

$$s_V = \left(\frac{\partial Q}{\partial T}\right)_V = \left(\frac{\partial Q}{\partial P}\right)_V \frac{P}{T} \tag{1-51}$$

The quantity s_V is the specific heat at constant volume. It is defined as the heat necessary to raise the temperature of a unit mass of the gas one degree where the gas is kept at constant volume during the process. Substituting (1-50) and (1-51) into (1-48), we obtain

$$\frac{dQ}{T} = s_P \frac{dV}{V} + s_V \frac{dP}{P} \tag{1-52}$$

We may rewrite (1-52) in the following form from (1-49)

$$\begin{aligned} \frac{dQ}{T} &= s_P \frac{dV}{V} + s_V \left(\frac{dT}{T} - \frac{dV}{V}\right) \\ &= (s_P - s_V)\frac{dV}{V} + s_V \frac{dT}{T} \end{aligned} \tag{1-53}$$

If the volume is increased with no change in temperature, work is done by the gas. For a closed system dQ is increased, and we see then that s_P must be greater than s_V. For an adiabatic change dQ is equal to zero so that from (1-52)

$$
\begin{aligned}
s_P \frac{dV}{V} + s_V \frac{dP}{P} &= 0 \\
PV^\gamma = P_0 V_0{}^\gamma &= \text{constant}
\end{aligned}
\tag{1-54}
$$

where γ is the ratio of the two specific heats

$$\gamma \equiv \frac{s_P}{s_V} \tag{1-55}$$

The adiabatic bulk modulus will then be

$$k_Q = \frac{dP}{-(dV/V_0)} = \gamma P \frac{V_0}{V} = \gamma P_0 \left(\frac{V_0}{V}\right)^{\gamma+1} = \gamma P_0 \left(\frac{\rho}{\rho_0}\right)^{\gamma+1} \tag{1-56}$$

For small displacements the bulk modulus and sound velocity reduce to

$$k_Q = \gamma P_0 \tag{1-57}$$

$$c_Q = \left(\frac{k_Q}{\rho}\right)^{1/2} = \left(\frac{\gamma P_0}{\rho_0}\right)^{1/2} \tag{1-58}$$

The velocity of propagation is higher for an adiabatic change than for an isothermal change. Using (1-43) we may rewrite (1-58) in the form

$$c_Q = (\gamma R T_0)^{1/2} \tag{1-59}$$

This equation states that for a given gas the velocity of sound propagation is dependent only on the ratio of the specific heats and the absolute temperature. The sound velocity varies as the square root of the absolute temperature. It further states that for small amplitudes the velocity is independent of the excess pressure, $P - P_0$.

For a *liquid* it is considerably more difficult to write down the equation of state. However, from the assumption that such a relation must exist [Eq. (1-42*a*)], we may derive a relation between the adiabatic sound velocity c_Q and the isothermal sound velocity c_T. Considering the temperature T as the dependent variable in (1-42*a*) we may write

$$dT = \left(\frac{\partial T}{\partial V}\right)_P dV + \left(\frac{\partial T}{\partial P}\right)_V dP \tag{1-59a}$$

If Q is a constant, $dQ = 0$, we shall have from (1-48)

$$\left(\frac{\partial P}{\partial V}\right)_Q = -\frac{(\partial Q/\partial V)_P}{(\partial Q/\partial P)_V} \tag{1-59b}$$

If T is a constant, we shall have from (1-59a)

$$\left(\frac{\partial P}{\partial V}\right)_T = -\frac{(\partial T/\partial V)_P}{(\partial T/\partial P)_V} \tag{1-59c}$$

If P is a constant, we shall have from (1-48) and (1-59a)

$$dQ = \left(\frac{\partial Q}{\partial V}\right)_P dV = \frac{(\partial Q/\partial V)_P}{(\partial T/\partial V)_P}\,dT$$

or

$$s_P \equiv \left(\frac{\partial Q}{\partial T}\right)_P = \frac{(\partial Q/\partial V)_P}{(\partial T/\partial V)_P} \tag{1-59d}$$

If V is a constant, we shall have from (1-48) and (1-59a)

$$dQ = \left(\frac{\partial Q}{\partial P}\right)_V dP = \frac{(\partial Q/\partial P)_V}{(\partial T/\partial P)_V}\,dT$$

or

$$s_V \equiv \left(\frac{\partial Q}{\partial T}\right)_V = \frac{(\partial Q/\partial P)_V}{(\partial T/\partial P)_V} \tag{1-59e}$$

Then, from (1-34a), (1-16), (1-59b), (1-59d), and (1-59e) we shall have for the ratio of the two sound velocities

$$\frac{c_Q}{c_T} = \left(\frac{k_Q}{k_T}\right)^{1/2} = \left[\frac{(\partial Q/\partial V)_P(\partial T/\partial P)_V}{(\partial Q/\partial P)_V(\partial T/\partial V)_P}\right]^{1/2} = \left(\frac{s_P}{s_V}\right)^{1/2} = \gamma^{1/2}$$

or

$$c_Q = \gamma^{1/2}\,c_T \tag{1-59f}$$

the same as that obtained for a perfect gas [Eqs. (1-58) and (1-47)].

For the ocean the velocity of sound propagation is dependent in an empirical manner on the temperature, salinity, and confining pressure (depth). It has been determined by direct measurement of the velocity as a function of these three parameters and by computations from measurements of k_T, ρ, and γ as a function of these three parameters. The primary dependence, similar to Eq. (1-59), is on temperature. With increase in confining pressure both k and ρ increase, with k increasing more rapidly. The same is true for an increase in salinity. The velocity in sea water varies from about 4,800 to 5,000 ft/sec. It increases with temperature at a variable rate of about 5.7 ft/sec per degree Fahrenheit; it increases with salinity at 4.27 ft/sec per one thousandth part increase in salinity; and it increases with depth at 1.82 ft/sec per 100 ft. In general the sound-velocity profile for a particular part of the ocean is determined from the temperature and salinity versus depth measurements through a set of empirical tables or graphs.

For a *solid* there is little difference between isothermal and adiabatic conditions as they affect sound transmission. The ratio γ of the two specific heats is nearly unity.

Relations for energy, energy density, energy flow, and intensity in terms of density, pressure, and particle velocity

In this section we shall consider motion in a liquid under the restrictions of infinitesimal strain theory. To summarize, the governing equations that have been obtained are first the stress-strain relation (1-26),

$$p = -k\Delta \tag{1-26}$$

second the equations of motion (1-30) rewritten in terms of the velocity components and the excess pressure

$$\begin{aligned} \rho_0 \frac{\partial \dot{u}}{\partial t} &= -\frac{\partial p}{\partial x} = k\frac{\partial \Delta}{\partial x} \\ \rho_0 \frac{\partial \dot{v}}{\partial t} &= -\frac{\partial p}{\partial y} = k\frac{\partial \Delta}{\partial y} \\ \rho_0 \frac{\partial \dot{w}}{\partial t} &= -\frac{\partial p}{\partial z} = k\frac{\partial \Delta}{\partial z} \end{aligned} \tag{1-60}$$

where ρ has been replaced by ρ_0 for small amplitudes; third the wave equation (1-34)

$$\rho_0 \frac{\partial^2 \Delta}{\partial t^2} = k\nabla^2 \Delta \tag{1-34}$$

or also for p

$$\rho_0 \frac{\partial^2 p}{\partial t^2} = k\nabla^2 p \tag{1-61}$$

and fourth the equation of continuity (1-8)

$$\frac{\partial \rho}{\partial t} = -\rho\left(\frac{\partial \dot{u}}{\partial x} + \frac{\partial \dot{v}}{\partial y} + \frac{\partial \dot{w}}{\partial z}\right) \tag{1-62}$$

where d/dt can be approximated by $\partial/\partial t$ for small amplitudes. A new quantity is introduced, the *condensation*, s defined as the change in density during the propagation of a disturbance with respect to the equilibrium density, i.e., the fractional change in density.

$$s = \frac{\rho - \rho_0}{\rho_0} = \frac{1/V - 1/V_0}{1/V_0} = \frac{V_0 - V}{V} \approx \frac{V_0 - V}{V_0} = -\Delta \tag{1-63}$$

The equation of continuity can then be written as

$$-\frac{\partial \Delta}{\partial t} = \frac{\partial s}{\partial t} = -\left(\frac{\partial \dot{u}}{\partial x} + \frac{\partial \dot{v}}{\partial y} + \frac{\partial \dot{w}}{\partial z}\right) \tag{1-64}$$

The kinetic energy of motion is given by

$$\mathrm{KE} = \tfrac{1}{2}\rho_0(\dot{u}^2 + \dot{v}^2 + \dot{w}^2) \tag{1-65}$$

per unit volume. The potential energy is defined as the work done during the volume change from the equilibrium to the compressed state. The potential energy per unit volume can then be given by

$$\mathrm{PE} = -\int_0^\Delta p\,d\Delta = k\int_0^\Delta \Delta\,d\Delta = \tfrac{1}{2}k\Delta^2 = \frac{p^2}{2k} \tag{1-66}$$

where Δ, the dilatation, is the volume change per unit volume. The *energy density*, total energy per unit volume, is the sum of the kinetic and potential energies or

$$E = \mathrm{KE} + \mathrm{PE} = \frac{\rho_0}{2}(\dot{u}^2 + \dot{v}^2 + \dot{w}^2) + \frac{p^2}{2k} \tag{1-67}$$

An equation of continuity for energy flow can be derived in the same manner as the equation of conservation for mass transport [Eq. (1-16)]

$$\frac{\partial E}{\partial t} = -\left(\frac{\partial F_x}{\partial x} + \frac{\partial F_y}{\partial y} + \frac{\partial F_z}{\partial z}\right) \tag{1-68}$$

Equation (1-68) states that the net energy flow per unit time, power, into a unit volume is equal to the rate of increase of energy density. F_x, F_y, and F_z are the components of energy flow across unit areas normal to the respective axes per unit time. From Eqs. (1-64) and (1-26) we can write

$$\frac{1}{k}\frac{\partial p}{\partial t} = -\left(\frac{\partial \dot{u}}{\partial x} + \frac{\partial \dot{v}}{\partial y} + \frac{\partial \dot{w}}{\partial z}\right) \tag{1-69}$$

Multiplying Eqs. (1-60) by $\dot{u}$, $\dot{v}$, and $\dot{w}$, respectively, and Eq. (1-69) by p and adding, we obtain

$$\frac{\partial}{\partial t}\left[\frac{\rho_0}{2}(\dot{u}^2 + \dot{v}^2 + \dot{w}^2) + \frac{p^2}{2k}\right] = -\left[\frac{\partial(p\dot{u})}{\partial x} + \frac{\partial(p\dot{v})}{\partial y} + \frac{\partial(p\dot{w})}{\partial z}\right] \tag{1-70}$$

From Eqs. (1-67) and (1-68) we then obtain

$$\frac{\partial F_x}{\partial x} + \frac{\partial F_y}{\partial y} + \frac{\partial F_z}{\partial z} = \frac{\partial(p\dot{u})}{\partial x} + \frac{\partial(p\dot{v})}{\partial y} + \frac{\partial(p\dot{w})}{\partial z}$$

or

$$\begin{aligned} F_x &= p\dot{u} \\ F_y &= p\dot{v} \\ F_z &= p\dot{w} \end{aligned} \tag{1-71}$$

Thus, we have determined that the instantaneous energy flow per unit time is equal to the product of the excess hydrostatic pressure and the particle velocity and is in the direction of the particle velocity. This is a general relation between energy flow and pressure and velocity.

For plane waves, or spherical waves at a distance, a simplified relation can be obtained. Equation (1-69) and the equations of motion (1-60) become

$$\frac{\partial p}{\partial t} = -k\frac{\partial \dot{u}}{\partial x} \tag{1-72}$$

and

$$\frac{\partial p}{\partial x} = -\rho_0\frac{\partial \dot{u}}{\partial t} \tag{1-73}$$

Operating on (1-72) with $\rho_0(\partial/\partial t)$ and on (1-73) with $-k(\partial/\partial x)$ and adding, the one-dimensional wave equation for p is obtained. Its solution (1-36) is, for a wave traveling in the positive x direction,

$$p = f(x - ct) \qquad \text{where} \qquad c = \left(\frac{k}{\rho_0}\right)^{1/2}$$

Taking derivatives of the above with respect to $\partial/\partial x$ and $\partial/\partial t$ gives

$$\frac{\partial p}{\partial x} = f' \qquad \text{and} \qquad \frac{\partial p}{\partial t} = -cf'$$

or

$$\frac{\partial p}{\partial t} = -c\frac{\partial p}{\partial x} \tag{1-74}$$

Substituting (1-74) into (1-72) and (1-73) then produces the desired result that

$$\frac{\partial}{\partial x}(k\dot{u} - cp) = 0 \tag{1-75}$$

and

$$\frac{\partial}{\partial t}\left(\rho_0\dot{u} - \frac{p}{c}\right) = 0$$

or

$$\frac{\partial}{\partial t}(k\dot{u} - cp) = 0 \tag{1-76}$$

Equation (1-75) states that $(k\dot{u} - cp)$ is not a function of x and (1-76) that it is not a function of t. Therefore, $(k\dot{u} - cp)$ is a constant; and since this relation holds at times of zero pressure or velocity, the constant must be zero. Thus,

$$\dot{u} = \frac{c}{k}p = \frac{1}{\rho_0 c}p \tag{1-77}$$

For plane waves we have this simple relation. The particle velocity is directly related to the excess pressure. When the particles are moving in the direction of motion, they are in a region of positive excess pressure;

when they move in the direction opposite to the motion, they are in a region of negative excess pressure. Analogous to the similar mechanical and electrical relations the quantity $\rho_0 c$ of (1-77) is referred to as the *acoustic impedance* (*resistance*).

Intensity, defined as the time average of the energy flow per unit time across a unit area normal to the motion, can also be given simply from Eqs. (1-71) and (1-77) by

$$I_x = \overline{p\dot{u}} = \frac{\overline{p^2}}{\rho_0 c} \tag{1-78}$$

It is to be emphasized that relations (1-77) and the right-hand member of (1-78) are true for a plane wave, or a spherical wave at distance, but not for the general case.

Relations for particle displacement, particle velocity, dilatation, and stress in terms of velocity and displacement potential

In many problems it would be desirable to obtain a solution for a single quantity φ, which could be related directly to the physically measurable quantities of particle displacement or particle velocity and stress. To define and describe such a quantity we must go through the following arguments.

Consider a quantity A. We generally describe A in terms of an orthogonal set of axes x, y, z as $A(x, y, z)$. If we should desire to describe A in terms of a second set of axes x', y', z' with a different orientation to x, y, z, we would have to transform the coordinates x, y, z to x', y', z'. This would introduce the direction cosines between the two axes, and A would be a function not only of x', y', z' but also of the direction cosines between the axes. Now the use of a particular set of axes is merely an artifice for the description of a physical quantity, and A should be in no way dependent on the orientation of one set of reference axes with respect to another. The quantity A cannot be a function of the direction cosines between the x, y, z and x', y', z' axes. When the axes coincide, $x = x'$, $y = y'$, and $z = z'$ so that A must be the same function of x', y', z' as it is of x, y, z. We have then simply that for A to be a proper space function of the coordinates x, y, z

$$A(x, y, z) = A(x', y', z') \tag{1-79}$$

Consider next a vector V whose components V_x, V_y, V_z are defined by

$$V_x = \frac{\partial \varphi}{\partial x}, \; V_y = \frac{\partial \varphi}{\partial y}, \; V_z = \frac{\partial \varphi}{\partial z} \tag{1-80}$$

where φ is a proper function of x, y, z. We want to determine whether V

is a proper vector function. Transforming the axes x, y, z to x', y', z' we shall have from Fig. 1-9

$$\begin{aligned} x' &= l_1 x + l_2 y + l_3 z = \frac{\partial x}{\partial x'} x + \frac{\partial y}{\partial x'} y + \frac{\partial z}{\partial x'} z \\ y' &= m_1 x + m_2 y + m_3 z = \frac{\partial x}{\partial y'} x + \frac{\partial y}{\partial y'} y + \frac{\partial z}{\partial y'} z \\ z' &= n_1 x + n_2 y + n_3 z = \frac{\partial x}{\partial z'} x + \frac{\partial y}{\partial z'} y + \frac{\partial z}{\partial z'} z \end{aligned} \tag{1-81}$$

where l_1, l_2, l_3 are the direction cosines of x' referred to x, y, z, etc. Then from (1-81) we shall have for V'_x, V'_y, V'_z

$$\begin{aligned} V'_x &= l_1 V_x + l_2 V_y + l_3 V_z = \frac{\partial \varphi}{\partial x}\frac{\partial x}{\partial x'} + \frac{\partial \varphi}{\partial y}\frac{\partial y}{\partial x'} + \frac{\partial \varphi}{\partial z}\frac{\partial z}{\partial x'} = \frac{\partial \varphi}{\partial x'} \\ V'_y &= m_1 V_x + m_2 V_y + m_3 V_z = \frac{\partial \varphi}{\partial x}\frac{\partial x}{\partial y'} + \frac{\partial \varphi}{\partial y}\frac{\partial y}{\partial y'} + \frac{\partial \varphi}{\partial z}\frac{\partial z}{\partial y'} = \frac{\partial \varphi}{\partial y'} \\ V'_z &= n_1 V_x + n_2 V_y + n_3 V_z = \frac{\partial \varphi}{\partial x}\frac{\partial x}{\partial z'} + \frac{\partial \varphi}{\partial y}\frac{\partial y}{\partial z'} + \frac{\partial \varphi}{\partial z}\frac{\partial z}{\partial z'} = \frac{\partial \varphi}{\partial z'} \end{aligned} \tag{1-82}$$

We conclude that V is a proper vector function and can be represented by (1-80). A solution of a problem in terms of φ gives through (1-80) the solution for V_x, V_y, V_z.

	x	y	z
x'	l_1	l_2	l_3
y'	m_1	m_2	m_3
z'	n_1	n_2	n_3

FIG. 1-9.

In many of the more complicated problems in wave propagation involving initial and boundary conditions it is convenient to solve the wave equation for the *velocity potential* φ where the velocity potential is defined by the following

$$\dot{u} = -\frac{\partial \varphi}{\partial x}, \quad \dot{v} = -\frac{\partial \varphi}{\partial y}, \quad \dot{w} = -\frac{\partial \varphi}{\partial z} \tag{1-83}$$

The velocity potential is related to the dilatation and excess pressure by the following

$$\begin{aligned} \Delta &= \frac{\partial u}{\partial x} + \frac{\partial v}{\partial y} + \frac{\partial w}{\partial z} \\ \frac{\partial \Delta}{\partial t} &= \frac{\partial^2 u}{\partial x\,\partial t} + \frac{\partial^2 v}{\partial y\,\partial t} + \frac{\partial^2 w}{\partial z\,\partial t} = -\nabla^2\varphi \\ \frac{\partial^2 \Delta}{\partial t^2} &= -\nabla^2 \frac{\partial \varphi}{\partial t} \end{aligned}$$

where from Eq. (1-34) since φ also satisfies the wave equation

$$\frac{k}{\rho}\nabla^2\Delta = -\nabla^2\frac{\partial \varphi}{\partial t}$$

so that

$$\Delta = -\frac{\rho}{k}\frac{\partial\varphi}{\partial t} \tag{1-84}$$

and from (1-16)

$$p = \rho\frac{\partial\varphi}{\partial t} \tag{1-85}$$

Of alternate use is a solution of the wave equation in terms of the *displacement potential* φ where the displacement potential is generally defined by the following

$$u = \frac{\partial\varphi}{\partial x}, v = \frac{\partial\varphi}{\partial y}, w = \frac{\partial\varphi}{\partial z} \tag{1-86}$$

From (1-12) and (1-15) the unbalanced stresses or excess pressure are then given by

$$-p = \sigma_{xx} = \sigma_{yy} = \sigma_{zz} = \lambda\nabla^2\varphi \tag{1-87}$$

A solution of the wave equation in terms of either of the potential functions above will give the correct solution for sound propagation in a fluid medium. In an elastic medium both irrotational and equivoluminal waves can exist, each traveling with a different velocity. We should desire then to be able to express the displacement in terms of two potentials, φ and ψ, one, φ, characteristic of the irrotational motion and satisfying the wave equation with the irrotational velocity,

$$\alpha^2\nabla^2\varphi = \frac{\partial^2\varphi}{\partial t^2} \tag{1-88}$$

where

$$\alpha^2 \equiv \frac{\lambda + 2\mu}{\rho} \tag{1-89}$$

and the other, ψ, characteristic of the equivoluminal motion and satisfying the wave equation with the equivoluminal velocity.

$$\beta^2\nabla^2\psi = \frac{\partial^2\psi}{\partial t^2} \tag{1-90}$$

where

$$\beta^2 \equiv \frac{\mu}{\rho} \tag{1-91}$$

There is a certain amount of latitude in the definition of the potential functions in terms of the displacement. We shall find the following most useful. Let the displacements u, v, w be given by

$$u = \frac{\partial\varphi}{\partial x} + u', v = \frac{\partial\varphi}{\partial y} + v', w = \frac{\partial\varphi}{\partial z} + w' \tag{1-92}$$

where φ and therefore any derivative of φ with respect to x, y, or z satisfies

Eq. (1-88). The displacements (1-92) must satisfy Eqs. (1-30) which give upon substitution

$$\begin{aligned}
\rho \frac{\partial^2}{\partial t^2}\left(\frac{\partial \varphi}{\partial x} + u'\right) &= (\lambda + 2\mu)\nabla^2 \frac{\partial \varphi}{\partial x} + \mu\nabla^2 u' \\
&\quad + (\lambda + \mu)\frac{\partial}{\partial x}\left(\frac{\partial u'}{\partial x} + \frac{\partial v'}{\partial y} + \frac{\partial w'}{\partial z}\right) \\
\rho \frac{\partial^2}{\partial t^2}\left(\frac{\partial \varphi}{\partial y} + v'\right) &= (\lambda + 2\mu)\nabla^2 \frac{\partial \varphi}{\partial y} + \mu\nabla^2 v' \\
&\quad + (\lambda + \mu)\frac{\partial}{\partial y}\left(\frac{\partial u'}{\partial x} + \frac{\partial v'}{\partial y} + \frac{\partial w'}{\partial z}\right) \\
\rho \frac{\partial^2}{\partial t^2}\left(\frac{\partial \varphi}{\partial z} + w'\right) &= (\lambda + 2\mu)\nabla^2 \frac{\partial \varphi}{\partial z} + \mu\nabla^2 w' \\
&\quad + (\lambda + \mu)\frac{\partial}{\partial z}\left(\frac{\partial u'}{\partial x} + \frac{\partial v'}{\partial y} + \frac{\partial w'}{\partial z}\right)
\end{aligned} \tag{1-93}$$

These equations will be satisfied if u', v', w' satisfy the following relations

$$\beta^2\nabla^2 u' = \frac{\partial^2 u'}{\partial t^2}, \quad \beta^2\nabla^2 v' = \frac{\partial^2 v'}{\partial t^2}, \quad \beta^2\nabla^2 w' = \frac{\partial^2 w'}{\partial t^2} \tag{1-94}$$

and

$$\frac{\partial u'}{\partial x} + \frac{\partial v'}{\partial y} + \frac{\partial w'}{\partial z} = 0 \tag{1-95}$$

As we might have expected, Eqs. (1-95) and (1-94) define an equivoluminal displacement which is propagated with the velocity β. We shall choose for a potential, ψ, to define u', v', w' and satisfy the relations (1-94) and (1-95) the following form

$$u' = \frac{\partial^2 \psi}{\partial x\, \partial z}, \quad v' = \frac{\partial^2 \psi}{\partial y\, \partial z}, \quad w' = \frac{\partial^2 \psi}{\partial z^2} - \nabla^2 \psi \tag{1-96}$$

where ψ satisfies Eq. (1-90). The final form for the displacement u, v, w in terms of the potentials φ and ψ will then be

$$\begin{aligned}
u &= \frac{\partial \varphi}{\partial x} + \frac{\partial^2 \psi}{\partial x\, \partial z} \\
v &= \frac{\partial \varphi}{\partial y} + \frac{\partial^2 \psi}{\partial y\, \partial z} \\
w &= \frac{\partial \varphi}{\partial z} + \frac{\partial^2 \psi}{\partial z^2} - \nabla^2 \psi
\end{aligned} \tag{1-97}$$

The stresses are obtained by substitution of (1-97) into Eqs. (1-12).

It will be convenient to solve many of the problems in cylindrical coordinates where there is symmetry about the polar, z, axis. There are then two independent displacement components, a radial and a polar component. From Fig. 1-10 the rectangular displacement components u, v are in terms of the radial displacement component q

$$u = \frac{x}{r} q \tag{1-98}$$

and

$$v = \frac{y}{r} q$$

where

$$r^2 = x^2 + y^2 \tag{1-99}$$

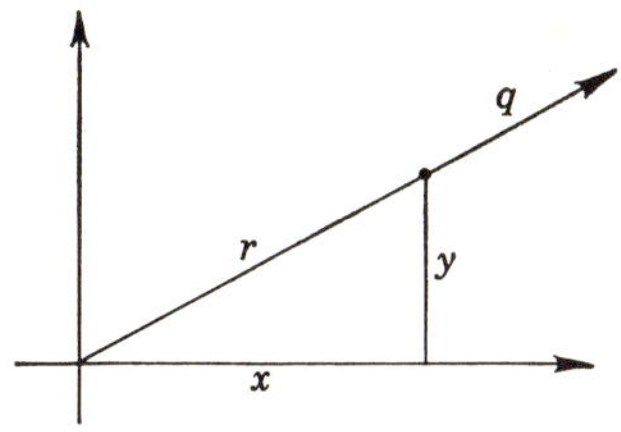

FIG. 1-10.

The partial spatial derivative with respect to x or y of a function $\varphi(r, z)$ in this case will be

$$\frac{\partial \varphi[r(x, y), z]}{\partial x} = \frac{\partial \varphi(r, z)}{\partial r} \frac{\partial r}{\partial x} = \frac{x}{r} \frac{\partial \varphi}{\partial r} \tag{1-100}$$

With the use of (1-98) and (1-99) the displacement (q, w) in symmetrical cylindrical coordinates becomes for the corresponding rectangular formulation (1-97)

$$\begin{aligned} q &= \frac{r}{x} u = \frac{r}{x} \frac{\partial}{\partial x} \left(\varphi + \frac{\partial \psi}{\partial z} \right) = \frac{\partial \varphi}{\partial r} + \frac{\partial^2 \psi}{\partial r\, \partial z} \\ w &= \frac{\partial \varphi}{\partial z} + \frac{\partial^2 \psi}{\partial z^2} - \nabla^2 \psi \end{aligned} \tag{1-101}$$

A similar relation holds for the other formulations.

In some transmission problems there is a dependence on only two rectangular coordinates, x, z. The definition of the potential functions φ and ψ here is taken slightly different from that given above for three-dimensional dependence or symmetrical cylindrical coordinate dependence. We shall start as before and let the displacement be given by

$$u = \frac{\partial \varphi}{\partial x} + u', \; v = v, \; w = \frac{\partial \varphi}{\partial z} + w' \tag{1-102}$$

where all quantities are a function of the spatial coordinates x and z only and the potential φ satisfies the wave equation (1-88) as before. Substitution of Eqs. (1-102) into Eqs. (1-30), which they must satisfy, gives

$$\begin{aligned} \rho \frac{\partial^2}{\partial t^2} \left(\frac{\partial \varphi}{\partial x} + u' \right) &= (\lambda + 2\mu) \nabla^2 \frac{\partial \varphi}{\partial x} + \mu \nabla^2 u' + (\lambda + \mu) \frac{\partial}{\partial x} \left(\frac{\partial u'}{\partial x} + \frac{\partial w'}{\partial z} \right) \\ \rho \frac{\partial^2 v}{\partial t^2} &= \mu \nabla^2 v \\ \rho \frac{\partial^2}{\partial t^2} \left(\frac{\partial \varphi}{\partial z} + w' \right) &= (\lambda + 2\mu) \nabla^2 \frac{\partial \varphi}{\partial z} + \mu \nabla^2 w' + (\lambda + \mu) \frac{\partial}{\partial z} \left(\frac{\partial u'}{\partial x} + \frac{\partial w'}{\partial z} \right) \end{aligned} \tag{1-103}$$

Equations (1-103) will be satisfied if u', w' satisfy the following

$$\beta^2\nabla^2 u' = \frac{\partial^2 u'}{\partial t^2}, \quad \beta^2\nabla^2 w' = \frac{\partial^2 w'}{\partial t^2} \tag{1-104}$$

and

$$\frac{\partial u'}{\partial x} + \frac{\partial w'}{\partial z} = 0 \tag{1-105}$$

It is convenient to define the potential ψ by the following relations to u', w':

$$u' = \frac{\partial \psi}{\partial z}, \quad w' = -\frac{\partial \psi}{\partial x} \tag{1-106}$$

The conditions that ψ satisfy the wave equation (1-90) and the definition (1-106) guarantee that the relations (1-104) and (1-105) will be satisfied. The final form for the displacements in terms of φ and ψ will then be

$$\begin{aligned} u &= \frac{\partial \varphi}{\partial x} + \frac{\partial \psi}{\partial z} \\ v &= v \\ w &= \frac{\partial \varphi}{\partial z} - \frac{\partial \psi}{\partial x} \end{aligned} \tag{1-107}$$

It is seen that the displacement v is independent of the displacement in the other two directions and is propagated with the β velocity whereas the other two displacements are mixed and contain portions propagated with both the α and β velocities.

Unit of measurement of intensity, the decibel

Practically all measurements of intensity from underwater sound-transmission measurements are quoted in decibels. The *decibel* is based on the logarithmic scale to the base 10, and intensity on the decibel scale is equal to ten times the logarithm of the ratio of measured intensity I to a reference intensity I_0.

$$I(\text{db}) = 10 \log \frac{I}{I_0} \tag{1-108}$$

The primary reasons for using the decibel scale are the wide range in values over which intensity is observed and the convenience in determining intensity levels measured through a usual sequence of receiver, amplifiers, and recorder.

Various reference levels are used in stating intensities in decibels. The common practice in underwater sound transmission is to refer intensities to a pressure level of 1 dyne/cm^2 under the assumption of plane waves and the relation (1-78) between intensity and pressure. The reason for using a reference pressure level rather than an intensity level is that in most experiments the receiver (hydrophone) is sensitive to pressure. It is usually quite difficult to measure intensity directly. Mathematically the relation for sound intensity level in terms of pressure is

$$I_1(\text{db}) = 10 \log \frac{p^2/\rho_0 c_0}{p_0{}^2/\rho_0 c_0} = 20 \log \frac{p}{p_0} = 20 \log p_1 \qquad (1\text{-}109)$$

where I_1 is the intensity in decibels referred to a pressure level of 1 dyne/cm^2 and p_1 is the incident excess pressure in units of dyne/cm^2. Other reference levels that are used are a pressure level of 0.0002 dyne/cm^2, the threshold of human hearing of air-borne sounds, and intensity levels of 1 watt/cm^2 and 1 watt/yard^2. The relations between the various systems are shown in the diagram of Fig. 1-11.

In the interpretation of transmission experiments the decibel offers several advantages for convenient analysis of the data graphically. For instance, in a simple experiment where there is a spreading loss such as that given by Eqs. (1-39) or (1-42), the intensity will be

$$I = S \frac{1}{r^n}$$

where S is the source level, or in decibels

$$I(\text{db}) = S(\text{db}) - 10n \log r$$

so that a plot of the intensity in decibels versus log r will be a straight line whose slope is $-10n$. It is easy to see from such a plot whether the measurements do follow the above law, if the value of n is what is predicted theoretically, and by how much it deviates if that is not the case. In another example where the spreading loss is known, say spherical spreading, and there is an additional loss from such causes as viscous absorption or scattering where the energy lost per unit distance is proportional to the incident energy so that

$$dE = -\alpha E \, dr$$

or

$$E = e^{-\alpha r}$$

the intensity will be

$$I = S \frac{1}{r^2} e^{-\alpha r}$$

or in decibels

$$I(\text{db}) + 20 \log r = S(\text{db}) - \beta r$$

Then, a plot of the measured intensity in decibels plus the spreading loss, 20 log r, versus r will be a straight line whose slope is $-\beta$, the excess loss in decibels per unit distance. Again from this plot it is easy to determine

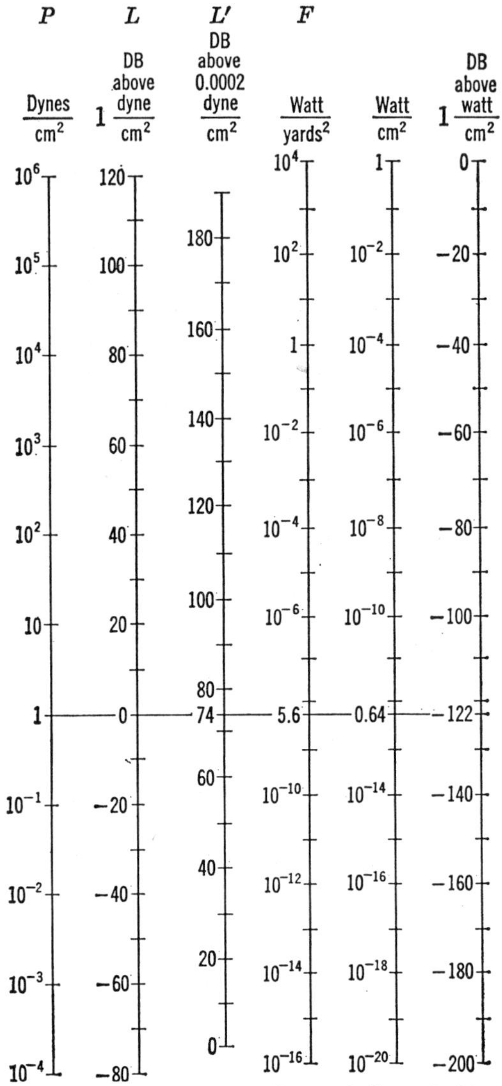

FIG. 1-11. (*From C. Eckert, Natl. Defense Research Council, Div.* 6, *vol.* **7**, 1945.)

if the excess loss does follow such a law and if so what the value of the coefficient β is.

A few simple relations which are useful to remember when thinking of intensities in terms of decibels are that a level of 3 db is an intensity ratio of 2:1, a level of 6 db an intensity ratio of 4:1 (pressure ratio of 2:1), a level of 10 db an intensity ratio of 10:1, and a level of 20 db an intensity ratio of 100:1 (pressure ratio of 10:1).

Waves of finite amplitude

In the previous sections the equations of motion were derived on the assumptions of disturbances of small amplitude and perfect elasticity. These relations are valid for most problems in sound transmission except in the immediate vicinity of the source and will be used in all the succeeding chapters on the development of the theory of sound transmission. In this and the following section, though, the effects to be expected from deviations from these assumptions will be outlined. The Riemann theory for waves of finite amplitude is derived in this section. This theory leads to the consideration of steep-fronted waves of finite amplitude. The Rankine-Hugoniot theory for shock fronts is derived in the following section.

When we deal with waves of finite amplitude, the approximation of replacing d/dt by $\partial/\partial t$ in the equations of motion is no longer valid; and we must consider the exact form. From Eqs. (1-7), (1-27), and (1-60) this will be, for one-dimensional wave propagation in terms of the particle velocity $\dot{u}$ and the excess pressure p,

$$\frac{d\dot{u}}{dt} = \frac{\partial \dot{u}}{\partial t} + \dot{u}\frac{\partial \dot{u}}{\partial x} = -\frac{1}{\rho}\frac{\partial p}{\partial x} \tag{1-110}$$

The equation of continuity (1-6) for one-dimensional propagation is

$$\frac{\partial \rho}{\partial t} + \frac{\partial(\rho \dot{u})}{\partial x} = 0 \tag{1-111}$$

In general, given the appropriate thermodynamic law under which the motion takes place such as that for adiabatic changes, p can be expressed as an explicit function of ρ. We may introduce the variable ψ defined by

$$\psi = \int_{\rho_0}^{\rho} \left(\frac{dp}{d\rho}\right)^{1/2} \frac{d\rho}{\rho} \tag{1-112}$$

From (1-42*b*) the sound velocity a for small displacements will be

$$a^2 \equiv \frac{dp}{d\rho} \tag{1-113}$$

Equations (1-110) and (1-111) can now be written in a useful form by substituting ψ for the variables p and ρ. Equation (1-111) becomes

$$\begin{aligned} \frac{d\rho}{d\psi}\frac{\partial \psi}{\partial t} + \dot{u}\frac{d\rho}{d\psi}\frac{\partial \psi}{\partial x} + \rho\frac{\partial \dot{u}}{\partial x} &= 0 \\ \frac{\partial \psi}{\partial t} + \dot{u}\frac{\partial \psi}{\partial x} &= -\rho\frac{d\psi}{d\rho}\frac{\partial \dot{u}}{\partial x} \\ \frac{\partial \psi}{\partial t} + \dot{u}\frac{\partial \psi}{\partial x} &= -a\frac{\partial \dot{u}}{\partial x} \end{aligned} \tag{1-114}$$

and Eq. (1-110) becomes

$$\begin{aligned} \frac{\partial \dot{u}}{\partial t} + \dot{u}\frac{\partial \dot{u}}{\partial x} &= -\frac{1}{\rho}\frac{dp}{d\rho}\frac{d\rho}{d\psi}\frac{\partial \psi}{\partial x} \\ \frac{\partial \dot{u}}{\partial t} + \dot{u}\frac{\partial \dot{u}}{\partial x} &= -a\frac{\partial \psi}{\partial x} \end{aligned} \tag{1-115}$$

Adding and subtracting Eqs. (1-114) and (1-115) we obtain

$$\left[\frac{\partial}{\partial t} + (\dot{u} + a)\frac{\partial}{\partial x}\right](\psi + \dot{u}) = 0 \tag{1-116}$$

and

$$\left[\frac{\partial}{\partial t} + (\dot{u} - a)\frac{\partial}{\partial x}\right](\psi - \dot{u}) = 0 \tag{1-117}$$

The differential in the brackets of Eqs. (1-116) and (1-117) is in the form of the total time differential of Eq. (1-7) following the motion of the fluid. Thus, Eq. (1-116) states that the quantity $(\psi + \dot{u})$ is constant for a geometrical point moving with the velocity $(\dot{u} + a)$. Similarly Eq. (1-117) states that the quantity $(\psi - \dot{u})$ is constant for a point moving with the velocity $(\dot{u} - a)$.

These equations enable us to understand, in a general way, the nature of the motion. Consider an initial disturbance confined between the planes $x = x_1$ and $x = x_2$ so that for $x < x_1$ and $x > x_2$ both ψ and $\dot{u}$ are zero. The region within which $(\psi + \dot{u})$ is variable will advance in the positive x direction and the quantity $(\psi - \dot{u})$ will recede in the opposite direction. After a time these regions will separate leaving between a region for which both quantities are zero so that $\psi = 0$ and $\dot{u} = 0$, i.e., the fluid is at rest and at its normal density. The original disturbance has been split into two progressive waves traveling in opposite directions. In the advancing wave $\psi - \dot{u} = 0$ so that $\psi = \dot{u}$, and we have from Eq. (1-116) that both the particle velocity and the density and pressure are propagated forward at the rate $(\dot{u} + a) = (\psi + a)$. In the receding wave we have $\psi + \dot{u} = 0$ so that $\psi = -\dot{u}$, and we obtain similar results from Eq. (1-117) for a wave propagating now with the velocity

$$(\dot{u} - a) = -(\psi + a)$$

We may visualize the progress of a wave of finite amplitude from the diagram of Fig. 1-12. Consider a wave as shown at time t_0 advancing in the positive x direction. The point A will have a velocity of propagation $(a_1 + \dot{u}_1)$ different from that of point B, $(a_2 + \dot{u}_2)$. We have as a first approximation from Eq. (1-77) for small-amplitude motion that $\dot{u} = p/\rho_0 c$ so that $\dot{u}_1 > \dot{u}_2$. In general the quantity a_1 will be greater than a_2. Thus the velocity of propagation of point A is greater than that for point

B and point A will advance on point B building up a steep-fronted wave. As A approaches B, the pressure gradient between the two points becomes large; and the effects of heat conduction and viscosity can no longer be neglected. These two effects will prevent A from overtaking B, and what might be called an equilibrium state will be reached at some later time t_2. The wave of finite amplitude has developed a shock front.

An interesting result from this is that the spectrum of a disturbance of finite amplitude, such as that of Fig. 1-12, is gradually shifted toward

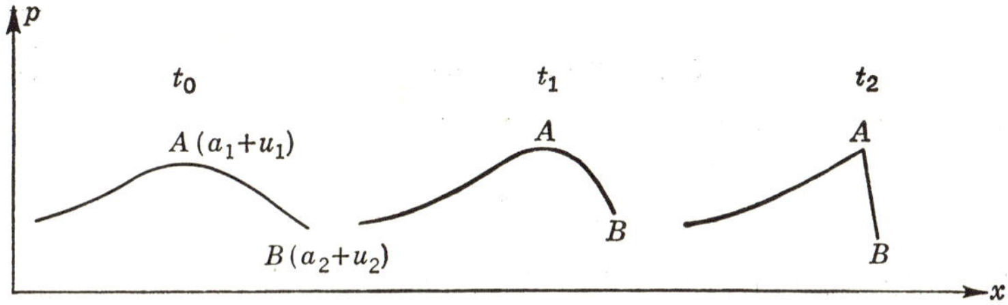

FIG. 1-12.

higher frequencies with distance as the shock front builds up. Similarly, a finite amplitude source of single frequency f_0 will be measured at distance with harmonics $2f_0$, $3f_0$, etc.

Shock fronts

A complete theory of shock fronts must take account of heat conduction and viscosity and the thickness of the front, but useful results can be obtained by the simple expedient of applying the laws of conservation of mass, momentum, and energy across the shock front without regard to its thickness.

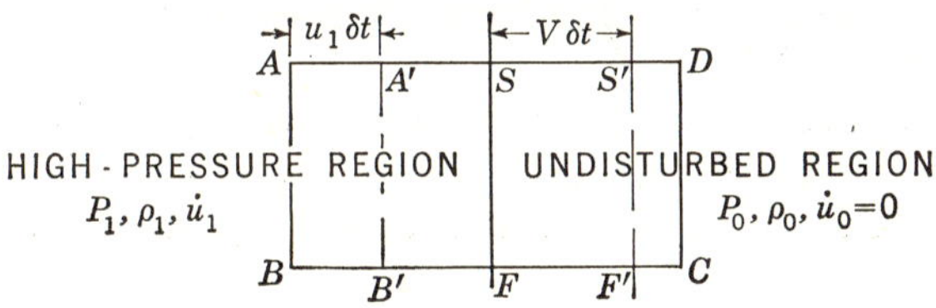

FIG. 1-13.

Consider a prism $ABCD$ across a shock front SF whose length AB and width are of unit dimensions. For a shock wave the change in pressure, density, and velocity will take place in a very narrow region near the plane SF. To the right will be the undisturbed region and to the left the high-pressure region. In Fig. 1-13, $\dot{u}$ is the particle velocity in the high-pressure region and V the velocity of propagation of the shock front.

The material within the prism $ABCD$ will occupy the prism $A'B'CD$ at a later time δt, and the shock front will have moved to $S'F'$. The original mass of the cylinder at time $t = 0$ is

$$\rho_1 \overline{AS} + \rho_0 \overline{SD}$$

and at the later time δt

$$\rho_1 \overline{A'S'} + \rho_0 \overline{S'D}$$

so that from the law of conservation of mass

$$\rho_1(\overline{A'S'} - \overline{AS}) = \rho_0(\overline{SD} - \overline{S'D})$$

or

$$\rho_1(V - \dot{u}_1) = \rho_0 V \tag{1-118}$$

The change in momentum of the prism is

$$\rho_1 \dot{u}_1 \overline{A'S'} - \rho_1 \dot{u}_1 \overline{AS} = \rho_1 \dot{u}_1 (V - \dot{u}_1)\delta t$$

and from the law of conservation of momentum this must be equal to the impulse $(P_1 - P_0)\delta t$ so that

$$\rho_1 \dot{u}_1 (V - \dot{u}_1) = P_1 - P_0 \tag{1-119}$$

Designating by e_0 and e_1 the internal energy per unit mass of the undisturbed and high-pressure regions, the change in energy of the prism will be

$$\rho_1 \left(e_1 + \frac{\dot{u}_1^2}{2}\right)(\overline{A'S'} - \overline{AS}) - \rho_0 e_0 \overline{SS'} = \rho_1 \left(e_1 + \frac{\dot{u}_1^2}{2}\right)(V - \dot{u}_1)\delta t - \rho_0 e_0 V\,\delta t$$

and from the law of conservation of energy this must be equal to the work done by the pressure P_1 in pushing the rear boundary of the prism forward

$$\rho_1 \left(e_1 + \frac{\dot{u}_1^2}{2}\right)(V - \dot{u}_1) - \rho_0 e_0 V = P_1 \dot{u}_1 \tag{1-120}$$

With these three equations and the thermodynamic relation [Eq. (1-54)], the quantities P_1, ρ_1, $\dot{u}_1$, V, and e_1, can be determined in terms of any one and the conditions in the undisturbed region. From Eq. (1-118) we have

$$\dot{u}_1 = \frac{\rho_1 - \rho_0}{\rho_1} V \tag{1-121}$$

Substituting this and Eq. (1-118) into Eq. (1-119) gives

$$\rho_0 \frac{\rho_1 - \rho_0}{\rho_1} V^2 = P_1 - P_0$$

or

$$V = \left[\frac{\rho_1}{\rho_0}\left(\frac{P_1 - P_0}{\rho_1 - \rho_0}\right)\right]^{1/2} \tag{1-122}$$

and from Eq. (1-121)

$$\dot{u}_1 = \left[\frac{(P_1 - P_0)(\rho_1 - \rho_0)}{\rho_0\rho_1}\right]^{1/2} \tag{1-123}$$

For small amplitudes these equations for a shock front reduce to the conventional relations. The shock-front velocity [Eq. (1-122)] reduces to

$$V = \left[\frac{\rho_1}{\rho_0}\left(\frac{P_1 - P_0}{\rho_1 - \rho_0}\right)\right]^{1/2} \approx \left(\frac{1}{\rho_0}\frac{dP}{d\rho/\rho_0}\right)^{1/2} = \left(\frac{k}{\rho}\right)^{1/2} = c$$

and the particle velocity [Eq. (1-119)] reduces to

$$\dot{u}_1 = \frac{P_1 - P_0}{\rho_1(V - \dot{u}_1)} = \frac{p}{\rho_0 V} = \frac{p}{\rho_0 c}$$

BIBLIOGRAPHY

Bateman, H.: "Partial Differential Equations of Mathematical Physics," Cambridge University Press, New York, 1932, and Dover Publications, New York, 1944.

Bullen, K. E.: "Introduction to the Theory of Seismology," Cambridge University Press, New York, 1947.

Del Grosso, V. A.: The Velocity of Sound in Sea Water at Zero Depth, *Naval Research Lab. Rept.* 4002, Washington, 1952.

Eckart, C. (ed.): Principles of Underwater Sound, Div. 6, vol. 7, *Natl. Defense Research Council Summary Tech. Rept.*, Washington, 1945.

Kolsky, H.: "Stress Waves in Solids," Oxford University Press, New York, 1953.

Kuwahara, S.: Velocity of Sound in Sea-water and Calculation of the Velocity for Use in Sonic Sounding, *Hydrographic Rev.*, 1938, pp. 124–140.

Lamb, H.: "Hydrodynamics," Cambridge University Press, New York, 1932, and Dover Publications, New York, 1945.

Love, A. E. H.: "The Mathematical Theory of Elasticity," Cambridge University Press, New York, 1927, and Dover Publications, New York, 1944.

Matthews, D. J.: Tables of the Velocity of Sound in Pure Water and Sea Water, *British Admiralty, Hydrographic Dept. Rept.* 282, London, 1939.

McLachlan, N. W.: "Theory of Vibrations," Dover Publications, New York, 1951.

Page, L.: "Introduction to Theoretical Physics," D. Van Nostrand Company, Inc., Princeton, N.J., 1935.

Spitzer, L., P. G. Bergmann, and A. Yaspan (ed.): Physics of Sound in the Sea, Div. 6, vol. 8, pt. I, Transmission, *Natl. Defense Research Council Summary Tech. Rept.*, Washington, 1945.

Strutt, J. W. (Lord Rayleigh): "The Theory of Sound," The Macmillan Company, London, 1894, and Dover Publications, New York, 1945.

Wilson, W.: "Theoretical Physics," Methuen & Co., Ltd., London, 1931.

2. *General Theory*

Most of the developments in the later chapters are based in one way or another on solutions of the *wave equation.* In general there are a variety of different types of solutions for any given partial differential equation. For a particular problem one or more of these may be formalized in various ways to meet the environment, boundary, and initial conditions and give a unique solution. For some of the more elementary problems an explicit solution satisfying the above conditions can be given; more often this is not the case. For these later cases there are two types of solution to the wave equation which are of particular importance in sound transmission. One is a transformation of the wave equation to the *eikonal equation* and a solution in terms of *wave surfaces* and *rays.* The other is a development through specific boundary conditions into a solution in terms of *normal modes.* In some instances the physical conditions of the problem lead to a simpler solution in terms of rays; in others a solution in terms of normal modes is more satisfactory.

The wave surfaces are the loci of points which are undergoing the same motion in a one-to-one correspondence at a given instant of time. The rays are the normals to the wave surfaces and give the direction of propagation of energy through the medium. A normal mode defines a preferred frequency of vibration for the system, and a solution in terms of normal modes is a summation of the contributions from the various preferred frequencies of vibration of the system.

In general both these solutions are complicated in application to specific sound-transmission problems through the introduction of variation in sound velocity and specific boundary conditions into the wave equation. The general nature of these complications is discussed for the ray and normal-mode solutions, and illustrative examples are cited.

Development of solution in terms of rays

A linear homogeneous partial differential equation of the form

$$f\left(\frac{\partial}{\partial x}, \frac{\partial}{\partial y}, \frac{\partial}{\partial z}, \frac{\partial}{\partial t}\right) \psi = 0 \tag{2-1}$$

will have a primary solution of the form

$$\psi = f(\Theta) \tag{2-2}$$

where

$$\Theta = c_1x + c_2y + c_3z + c_4t$$

and the coefficients satisfy the equation

$$f(c_1, c_2, c_3, c_4) = 0$$

as can be seen by substitution back into Eq. (2-1). Solution (2-2), however, will also be a solution of the equation

$$f\left(\frac{\partial\psi}{\partial x}, \frac{\partial\psi}{\partial y}, \frac{\partial\psi}{\partial z}, \frac{\partial\psi}{\partial t}\right) = 0 \tag{2-3}$$

Equation (2-3) is referred to as the *equation of characteristics* of (2-1).

For the general three-dimensional wave equation

$$c^2\left(\frac{\partial^2\psi}{\partial x^2} + \frac{\partial^2\psi}{\partial y^2} + \frac{\partial^2\psi}{\partial z^2}\right) = \frac{\partial^2\psi}{\partial t^2} \tag{2-4}$$

the primary solution will be

$$\psi = f(\alpha x + \beta y + \gamma z - ct) \tag{2-5}$$

where

$$\alpha = \frac{c_1}{\sqrt{c_1^2 + c_2^2 + c_3^2}}, \quad \beta = \frac{c_2}{\sqrt{c_1^2 + c_2^2 + c_3^2}}, \quad \gamma = \frac{c_3}{\sqrt{c_1^2 + c_2^2 + c_3^2}}$$

are the three direction cosines. The equation of characteristics becomes

$$\left(\frac{\partial\psi}{\partial x}\right)^2 + \left(\frac{\partial\psi}{\partial y}\right)^2 + \left(\frac{\partial\psi}{\partial z}\right)^2 = \frac{1}{c^2}\left(\frac{\partial\psi}{\partial t}\right)^2 \tag{2-6}$$

Without loss of generality we can take as a form of Θ,

$$\Theta(x, y, z, t) = W(x, y, z) - c_0t \tag{2-7}$$

Substitution of the solution (2-7) into (2-6) gives the time-independent conditional equation

$$\left(\frac{\partial W}{\partial x}\right)^2 + \left(\frac{\partial W}{\partial y}\right)^2 + \left(\frac{\partial W}{\partial z}\right)^2 = \frac{c_0^2}{c^2} = n^2 \tag{2-8}$$

where n is the *index of refraction* and is defined as

$$n \equiv \frac{c_0}{c} \tag{2-9}$$

c_0 being a constant, reference velocity. Equation (2-8) is the *eikonal equation.* This equation is of fundamental importance, for it leads directly to the concept of rays. It is useful in the solution of problems where c is not a constant but a function of the space coordinates.

In those cases where c is not a constant, $c = c(x, y, z)$, such as the ocean where the velocity is a function of the temperature, salinity, and depth, a solution of the eikonal equation (2-8) will not in general be a solution of the wave equation (2-4). It should also be mentioned that the wave equation (2-4) is itself an approximation, for under the conditions that the elastic parameters are functions of the space coordinates we should not be able to make the reduction from Eq. (1-29) to (1-30). It is valid only under the assumption that the space rate of change of these parameters is small with respect to the parameters themselves. Let us examine here under what conditions we can consider the eikonal equation (2-8) a good approximation to the approximate wave equation (2-4). In most problems we shall be interested in solutions in terms of simple harmonic motion of a particular frequency f or a Fourier synthesis of this to a given initial pulse. The properties of simple harmonic motion are discussed later in this chapter; here we shall use such a solution written in the form (2-7). The amplitude of vibration cannot be assumed to be constant and is taken as an undetermined function of the space coordinates, giving

$$\psi = A(x, y, z) \exp \left\{2\pi i f \left[t - \frac{W(x, y, z)}{c_0}\right]\right\}$$
$$= A(x, y, z) \exp \left\{i\omega \left[t - \frac{W(x, y, z)}{c_0}\right]\right\} \qquad (2\text{-}10)$$

where the wavelength λ_0 is

$$\lambda_0 = \frac{c_0}{f} = \frac{2\pi c_0}{\omega} \qquad (2\text{-}11)$$

For this to be a solution of the wave equation the conditional relations between A and W will be, substituting (2-10) into (2-4) and equating real and imaginary parts,

$$\frac{\partial^2 A}{\partial x^2} + \frac{\partial^2 A}{\partial y^2} + \frac{\partial^2 A}{\partial z^2} - A\frac{\omega^2}{c_0^2}\left[\left(\frac{\partial W}{\partial x}\right)^2 + \left(\frac{\partial W}{\partial y}\right)^2 + \left(\frac{\partial W}{\partial z}\right)^2\right] = -\frac{\omega^2}{c^2}A$$

and

$$2\left(\frac{\partial W}{\partial x}\frac{\partial A}{\partial x} + \frac{\partial W}{\partial y}\frac{\partial A}{\partial y} + \frac{\partial W}{\partial z}\frac{\partial A}{\partial z}\right) + A\left(\frac{\partial^2 W}{\partial x^2} + \frac{\partial^2 W}{\partial y^2} + \frac{\partial^2 W}{\partial z^2}\right) = 0$$

or

$$\left(\frac{\partial W}{\partial x}\right)^2 + \left(\frac{\partial W}{\partial y}\right)^2 + \left(\frac{\partial W}{\partial z}\right)^2 - n^2 - \frac{\lambda_0^2}{4\pi}\left[\frac{1}{A}\left(\frac{\partial^2 A}{\partial x^2} + \frac{\partial^2 A}{\partial y^2} + \frac{\partial^2 A}{\partial z^2}\right)\right] = 0 \qquad (2\text{-}12)$$

and

$$\frac{\partial^2 W}{\partial x^2} + \frac{\partial^2 W}{\partial y^2} + \frac{\partial^2 W}{\partial z^2} + \frac{2}{A}\left(\frac{\partial W}{\partial x}\frac{\partial A}{\partial x} + \frac{\partial W}{\partial y}\frac{\partial A}{\partial y} + \frac{\partial W}{\partial z}\frac{\partial A}{\partial z}\right) = 0 \quad (2\text{-}13)$$

For W also to be a solution of the eikonal equation (2-8) the last expression in (2-12) must be zero. In general the expression in the parentheses is not zero so that this condition will be met only if λ_0 equals zero, i.e., in the limit of high frequencies. If W is a solution of the eikonal equation, it will be a good approximation to the wave equation if the last expression of (2-13) is small compared with the sum of the first three terms. However, it is not sufficient to say that this will be true for small λ_0; we should like to know how small λ_0 must be with regard to the physical conditions of a particular problem. Such a relation can be obtained from order-of-magnitude considerations. Let W be an exact solution of the eikonal equation (2-8). Then, designating by the prime symbol a general space derivative, ψ in the form (2-10) will be a good approximation to the solution of the wave equation (2-4) if the condition

$$\lambda_0^2 \frac{A''}{A} \ll (W')^2 \quad (2\text{-}14)$$

is met. From the eikonal equation (2-8) we have that

$$(W')^2 \sim n^2$$

or

$$W' \sim n \sim 1 \quad (2\text{-}15)$$

since $n = c_0/c \sim 1$. Taking a second space derivative of (2-8) we obtain

$$W'W'' \sim nn'$$

or from (2-15)

$$W'' \sim n' \quad (2\text{-}16)$$

Further from (2-13) we may write that

$$W'' \sim W' \frac{A'}{A}$$

or from (2-15) and (2-16) that

$$W'' \sim \frac{A'}{A} \sim n' \quad (2\text{-}17)$$

From (2-15) we may rewrite the condition (2-14) as

$$\lambda_0^2 \frac{A''}{A} \ll 1 \quad (2\text{-}18)$$

From (2-14) or (2-18) we see that the unit of measurement is a wavelength λ_0 and that the inequality relates to changes in A' with respect to A over the distance of one wavelength. For convenience we shall rewrite (2-18) in the form

$$\lambda_0 \frac{\delta A'}{A} \ll 1 \tag{2-19}$$

where $\delta A'$ is taken to mean the change in A' over a distance λ_0. From (2-17) this is equivalent to

$$\lambda_0 \, \delta W'' \ll 1 \tag{2-20}$$

or

$$\lambda_0 \, \delta n' \ll 1 \tag{2-21}$$

From (2-9) the condition (2-21) may be written as

$$\lambda_0 \frac{\delta c'}{c} \ll 1 \tag{2-22}$$

without regard to sign. The inequality (2-22) is the desired result. It states that a solution to the eikonal equation will be a good approximation to the wave equation if the fractional change in the velocity gradient $\delta c'$ over a wavelength is small compared with the gradient c/λ_0. Thus, if in a particular problem there is a boundary between two media of different velocities, the condition (2-22) will not be met; or if there is a region in which there is a rapid change in velocity over the dimensions of a wavelength, condition (2-22) will not be met. Such problems must be treated separately. The inequalities (2-20) and (2-19) are equivalent to (2-22) for the measurable quantities of the wave front W and the amplitude A. Condition (2-20) states that the change in the curvature of the wave front must be small and (2-19) that the fractional change in space rate of change of the amplitude must be small.

The eikonal equation is a first-order partial differential equation. Its solutions,

$$W(x, y, z) = \text{constant} \tag{2-23}$$

represent surfaces in three-dimensional space. For a given value of W and a given instant of time t_0 the phase Θ of (2-7) or (2-10) will be a constant. All points along the surface will be in phase although now in the general case not necessarily of the same amplitude. There will be a one-to-one correspondence of the motion along this surface, and such a surface is called a *wave surface* or *wave front*. At a later time t the motion along the surface will be in a different phase of motion. A picture of the motion throughout space at a given time t_0 can be found by taking successive increments of the constant of (2-23). The normals to this surface define the direction of propagation, the direction in which the motion of a particle on one wave surface is passed on to the next. They are called the *rays* or *ray paths.* We shall see later, as might be expected, that the rays define the direction of energy propagation so that the energy contained within a narrow bundle of rays is confined to that bundle throughout all space. The ray solution then is a complete solution to any particular propagation problem within the validity of the approximation of the eikonal equation to the wave equation. The variations in intensity are given by the diminutions or expansions of the ray bundles and the time of arrival by the integral

$$\int \frac{ds}{c}$$

along the ray path.

The equation for the normals to the wave surface is

$$\frac{dx}{\partial W/\partial x} = \frac{dy}{\partial W/\partial y} = \frac{dz}{\partial W/\partial z} \tag{2-24}$$

where $\partial W/\partial x$, $\partial W/\partial y$, $\partial W/\partial z$ are the direction numbers of the normal. The direction cosines dx/ds, dy/ds, dz/ds are proportional to the direction numbers and satisfy the subsidiary relation

$$\left(\frac{dx}{ds}\right)^2 + \left(\frac{dy}{ds}\right)^2 + \left(\frac{dz}{ds}\right)^2 = 1 \tag{2-25}$$

This proportionality may be expressed by the three equations

$$\begin{aligned} \frac{dx}{ds} &= a\,\frac{\partial W}{\partial x} \\ \frac{dy}{ds} &= a\,\frac{\partial W}{\partial y} \\ \frac{dz}{ds} &= a\,\frac{\partial W}{\partial z} \end{aligned} \tag{2-26}$$

where a is a constant. Squaring and adding gives for the value of a, using the relations (2-25) and the eikonal equation (2-8),

$$a = \frac{1}{n}$$

so that (2-26) becomes

$$\begin{aligned} n\,\frac{dx}{ds} &= \frac{\partial W}{\partial x} \\ n\,\frac{dy}{ds} &= \frac{\partial W}{\partial y} \\ n\,\frac{dz}{ds} &= \frac{\partial W}{\partial z} \end{aligned} \tag{2-27}$$

Let us take a derivative d/ds along the ray path of any one of these equations; we obtain

$$\frac{d}{ds}\left(n\frac{dx}{ds}\right) = \frac{d}{ds}\left(\frac{\partial W}{\partial x}\right) = \frac{\partial}{\partial x}\left(\frac{\partial W}{\partial x}\frac{dx}{ds} + \frac{\partial W}{\partial y}\frac{dy}{ds} + \frac{\partial W}{\partial z}\frac{dz}{ds}\right)$$

or substituting from (2-27) with the relation (2-25)

$$\frac{d}{ds}\left(n\frac{dx}{ds}\right) = \frac{\partial}{\partial x}\left\{n\left[\left(\frac{dx}{ds}\right)^2 + \left(\frac{dy}{ds}\right)^2 + \left(\frac{dz}{ds}\right)^2\right]\right\} = \frac{\partial n}{\partial x}$$

so that our three equations (2-27) reduce to

$$\begin{aligned}\frac{d}{ds}\left(n\frac{dx}{ds}\right) &= \frac{\partial n}{\partial x}\\ \frac{d}{ds}\left(n\frac{dy}{ds}\right) &= \frac{\partial n}{\partial y}\\ \frac{d}{ds}\left(n\frac{dz}{ds}\right) &= \frac{\partial n}{\partial z}\end{aligned} \tag{2-28}$$

which are the equations that determine the rays in terms of the index of refraction, $n = n(x, y, z)$. These equations state that the variation along the ray path of the product of the index of refraction and a direction cosine is equal to the space rate of variation of the index of refraction with respect to the appropriate coordinate. They may be considered as a generalized form of Snell's law.

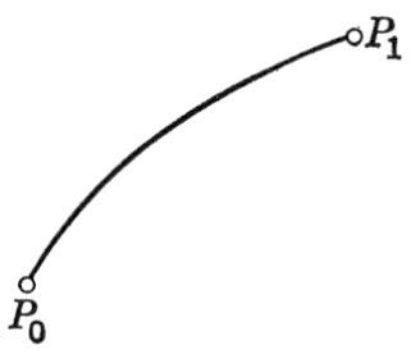

FIG. 2-1.

It is of interest at this point to prove *Fermat's principle*. Fermat's principle states that a ray path between any two points P_0 and P_1 is also a path of stationary time between these points. We want to prove that the integral

$$F = c_0 \int_{P_0}^{P_1} dt = c_0 \int_{P_0}^{P_1} \frac{ds}{c} = \int_{P_0}^{P_1} n\, ds \tag{2-29}$$

has an extremum value, minimum or maximum, along the ray path. It is convenient to express Eq. (2-29) in parametric form. The differential ds of the path is given by

$$ds = \sqrt{x'^2 + y'^2 + z'^2}\, d\sigma \tag{2-30}$$

where the prime symbol indicates derivatives with respect to the parameter σ. Substituting into (2-29) gives

$$F = \int_{P_0}^{P_1} n(x, y, z)\sqrt{x'^2 + y'^2 + z'^2}\, d\sigma = \int_{P_0}^{P_1} G(x, y, z, x', y', z')d\sigma \tag{2-31}$$

This is now in the form of a line integral with respect to σ of the three space variables and their derivatives with respect to σ. From the calculus of variations this line integral will have an extremum value if Euler's equations

$$\begin{aligned}\frac{\partial G}{\partial x} - \frac{d}{d\sigma}\left(\frac{\partial G}{\partial x'}\right) &= 0\\ \frac{\partial G}{\partial y} - \frac{d}{d\sigma}\left(\frac{\partial G}{\partial y'}\right) &= 0\\ \frac{\partial G}{\partial z} - \frac{d}{d\sigma}\left(\frac{\partial G}{\partial z'}\right) &= 0\end{aligned} \tag{2-32}$$

are satisfied, or for this case,

$$\begin{aligned}\sqrt{x'^2 + y'^2 + z'^2}\,\frac{\partial n}{\partial x} - \frac{d}{d\sigma}\left(\frac{nx'}{\sqrt{x'^2 + y'^2 + z'^2}}\right) &= 0\\ \sqrt{x'^2 + y'^2 + z'^2}\,\frac{\partial n}{\partial y} - \frac{d}{d\sigma}\left(\frac{ny'}{\sqrt{x'^2 + y'^2 + z'^2}}\right) &= 0\\ \sqrt{x'^2 + y'^2 + z'^2}\,\frac{\partial n}{\partial z} - \frac{d}{d\sigma}\left(\frac{nz'}{\sqrt{x'^2 + y'^2 + z'^2}}\right) &= 0\end{aligned} \tag{2-33}$$

Substituting for $d\sigma$ in terms of ds from (2-30) gives

$$\begin{aligned}\frac{d}{ds}\left(n\frac{dx}{ds}\right) &= \frac{\partial n}{\partial x}\\ \frac{d}{ds}\left(n\frac{dy}{ds}\right) &= \frac{\partial n}{\partial y}\\ \frac{d}{ds}\left(n\frac{dz}{ds}\right) &= \frac{\partial n}{\partial z}\end{aligned} \tag{2-34}$$

which are simply the original ray equations. Thus we have proved that the time along the ray path has an extremum value. For most physical problems this is a minimum or least time.

To demonstrate that the ray paths also define the direction of energy flow, let us take the real part of the generalized simple harmonic solution (2-10) where we shall consider ψ to be the velocity potential. Then the excess pressure p and the particle velocities $\dot{u}$, $\dot{v}$, $\dot{w}$ will be given by (1-83) and (1-85).

$$\begin{aligned}\psi &= A\cos\left[2\pi f\left(t - \frac{W}{c_0}\right)\right]\\ p &= \rho\frac{\partial \psi}{\partial t} = -2\pi f\rho A\sin\left[2\pi f\left(t - \frac{W}{c_0}\right)\right]\\ \dot{u} &= -\frac{\partial \psi}{\partial x} = -2\pi f\frac{\partial W}{\partial x}\frac{A}{c_0}\sin\left[2\pi f\left(t - \frac{W}{c_0}\right)\right] - \frac{\partial A}{\partial x}\cos\left[2\pi f\left(t - \frac{W}{c_0}\right)\right]\end{aligned}$$

The average energy flow per unit area per unit time, intensity, in the x direction will be from (1-71) and (1-78) the average of $p\dot{u}$ over one cycle of the argument. Thus,

$$I_x = \overline{pu} = 4\pi^2 \frac{\rho A^2 c_0}{\lambda_0^2} \left[\frac{1}{\pi} \left(\int_0^\pi \frac{\partial W}{\partial x} \sin^2 \xi \, d\xi + \frac{\lambda_0}{2\pi} \int_0^\pi \frac{\partial A/\partial x}{A} \sin \xi \cos \xi \, d\xi \right) \right] \tag{2-35}$$

where we have substituted for λ_0 from (2-11) and designated ξ as the argument, $2\pi f[t - (W/c_0)]$. Since

$$\frac{1}{\pi} \int_0^\pi \sin^2 \xi \, d\xi = \tfrac{1}{2}$$

and

$$\frac{1}{\pi} \int_0^\pi \sin \xi \cos \xi \, d\xi = 0$$

the second term of (2-35) will be zero if the variation of $\dfrac{\partial A/\partial x}{A}$ is small over one period. We are assured that this is the case from (2-19) for the ray solution to be a valid approximation. Equation (2-35) then reduces to

$$I_x = 2\pi^2 \frac{\rho A^2 c_0}{\lambda_0^2} \overline{\frac{\partial W}{\partial x}}$$

and similarly for I_y and I_z

$$I_y = 2\pi^2 \frac{\rho A^2 c_0}{\lambda_0^2} \overline{\frac{\partial W}{\partial y}} \qquad I_z = 2\pi^2 \frac{\rho A^2 c_0}{\lambda_0^2} \overline{\frac{\partial W}{\partial z}} \tag{2-36}$$

The total intensity is simply the square root of the sum of the squares of the three quantities (2-36). Its direction is given from (2-36) by the direction numbers $\overline{\partial W}/\partial x$, $\overline{\partial W}/\partial y$, $\overline{\partial W}/\partial z$. These direction numbers are the same as those for the ray paths; therefore the direction of average energy flow is along the ray paths. The total intensity will be

$$I = 2\pi^2 \frac{\rho A^2 c_0}{\lambda_0^2} \sqrt{\left(\overline{\frac{\partial W}{\partial x}}\right)^2 + \left(\overline{\frac{\partial W}{\partial y}}\right)^2 + \left(\overline{\frac{\partial W}{\partial z}}\right)^2} \tag{2-37}$$

From the eikonal equation (2-8), Eq. (2-37) will reduce approximately to

$$I \approx 2\pi^2 \frac{\rho A^2 c_0^2}{\lambda_0^2 c} \tag{2-38}$$

or in terms of the pressure amplitude $a = -2\pi f \rho A$

$$I \approx \frac{a^2}{2\rho c} \tag{2-39}$$

It is instructive to give an alternate derivation of the eikonal equation from somewhat more physical arguments through the use of Hugyens' principle. *Huygens' principle* states that the disturbance at some later time, $t = t_0 + dt$, can be obtained by considering the effect produced by each point on a wave surface at a given time, $t = t_0$, acting as a secondary source. Each point on the wave surface will produce a wavelet propagating outward with the velocity c at that point; the resultant wave front at some later time, $t = t_0 + dt$, will be the envelope of all the wavelets taken at that time. If c is a sufficiently slowly varying function over the original wave surface, the wave surface at $t_0 + dt$ will be defined approximately by the normals $c\,dt$ to the original surface. The wave surface,

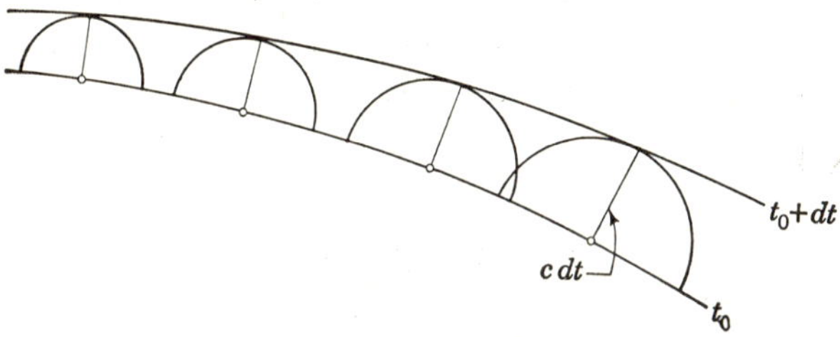

FIG. 2-2.

$W(x, y, z) =$ constant, is the locus of points which are in the same phase of motion at a given time t_0 so that

$$W(x, y, z) - c_0 t_0 = \text{constant}$$

where c_0 is a constant. At a later time, $t_0 + dt$, the wavelets have defined a new wave surface which is in the same phase of motion at $t_0 + dt$ that the original wave surface was at t_0 so that

$$W(x + \alpha c\,dt, y + \beta c\,dt, z + \gamma c\,dt) - c_0(t_0 + dt) = W(x, y, z) - c_0 t_0 \quad (2\text{-}40)$$

where α, β, γ are the direction cosines of the normals to the original wave surface. Since the increments $\alpha c\,dt$, $\beta c\,dt$, $\gamma c\,dt$ are small quantities, the new wave surface can be approximated by

$$W(x + \alpha c\,dt, y + \beta c\,dt, z + \gamma c\,dt) = W(x, y, z) + \frac{\partial W}{\partial x}\alpha c\,dt + \frac{\partial W}{\partial y}\beta c\,dt + \frac{\partial W}{\partial z}\gamma c\,dt \quad (2\text{-}41)$$

Equating (2-40) and (2-41) we have

$$\alpha\frac{\partial W}{\partial x} + \beta\frac{\partial W}{\partial y} + \gamma\frac{\partial W}{\partial z} = \frac{c_0}{c} = n \quad (2\text{-}42)$$

The direction numbers to a surface, $W =$ constant, are $\partial W/\partial x$, $\partial W/\partial y$, $\partial W/\partial z$. They will be proportional to the direction cosines so that we may write

$$\begin{aligned} \alpha &= b\,\frac{\partial W}{\partial x} \\ \beta &= b\,\frac{\partial W}{\partial y} \\ \alpha &= b\,\frac{\partial W}{\partial z} \end{aligned} \tag{2-43}$$

where the direction cosines also satisfy the subsidiary relation

$$\alpha^2 + \beta^2 + \gamma^2 = 1 \tag{2-44}$$

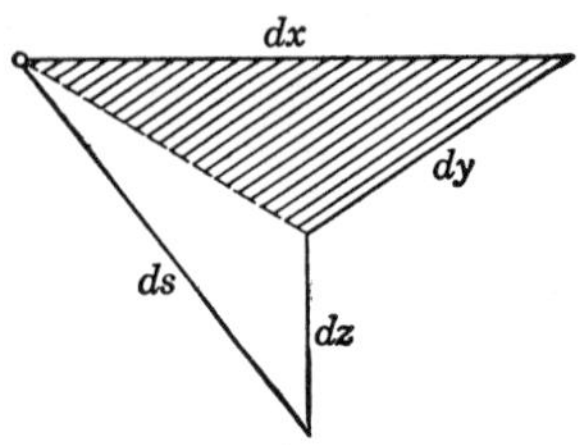

FIG. 2-3.

Multiplying the first of Eqs. (2-43) by α, the second by β, and the third by γ we obtain, using (2-42) and (2-44),

$$b = \frac{1}{n}$$

The direction cosines α, β, γ can be expressed in terms of the derivatives along the normals (Fig. 2-3) by

$$\begin{aligned} \alpha &= \frac{dx}{ds} \\ \beta &= \frac{dy}{ds} \\ \gamma &= \frac{dz}{ds} \end{aligned}$$

Substituting into (2-43) gives

$$\begin{aligned} n\,\frac{dx}{ds} &= \frac{\partial W}{\partial x} \\ n\,\frac{dy}{ds} &= \frac{\partial W}{\partial y} \\ n\,\frac{dz}{ds} &= \frac{\partial W}{\partial z} \end{aligned} \tag{2-45}$$

which are the same as Eqs. (2-27). They lead directly to the equations for the rays (2-28)

$$\begin{aligned} \frac{d}{ds}\left(n\,\frac{dx}{ds}\right) &= \frac{\partial n}{\partial x} \\ \frac{d}{ds}\left(n\,\frac{dy}{ds}\right) &= \frac{\partial n}{\partial y} \\ \frac{d}{ds}\left(n\,\frac{dz}{ds}\right) &= \frac{\partial n}{\partial z} \end{aligned} \tag{2-28}$$

or upon squaring and adding to the eikonal equation (2-8)

$$\left(\frac{\partial W}{\partial x}\right)^2 + \left(\frac{\partial W}{\partial y}\right)^2 + \left(\frac{\partial W}{\partial z}\right)^2 = 1 \tag{2-8}$$

Ray characteristics for velocity a function of one space coordinate only

For most physical cases of interest the velocity is a function of only one of the space coordinates, the depth. Equations (2-28) then reduce, for $n = n(z)$, to

$$\begin{aligned} n\frac{dx}{ds} &= \text{constant} \\ n\frac{dy}{ds} &= \text{constant} \\ \frac{d}{ds}\left(n\frac{dz}{ds}\right) &= \frac{dn}{dz} \end{aligned} \tag{2-46}$$

Combining the first and second gives

$$\frac{dx/ds}{dy/ds} = \frac{\alpha}{\beta} = \text{constant} \tag{2-47}$$

The restriction that the ratio of the direction cosines in the x and y directions be a constant implies that the ray path is confined to a plane normal to the xy plane. This can easily be seen from Fig. 2-3 where the light lines are similar triangles in the xy plane formed by dx, dy. The trace of ds in the xy plane necessarily satisfies (2-47). The trace of ds in the xy plane is confined to a line so that ds itself is confined to a plane containing this line and normal to the xy plane. For convenience and without loss of generality we shall choose this plane to coincide with the xz plane. Equations (2-46) become

$$n\frac{dx}{ds} = \text{constant} \tag{2-48}$$

$$\frac{d}{ds}\left(n\frac{dz}{ds}\right) = \frac{dn}{dz} \tag{2-49}$$

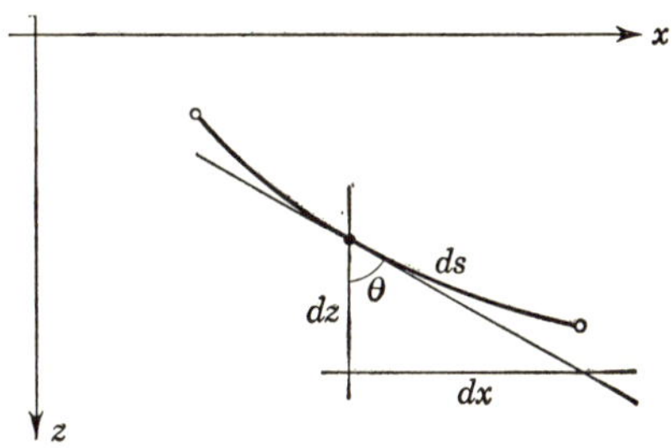

FIG. 2-4.

From Fig. 2-4 the direction cosines α, γ are given by

$$\alpha = \frac{dx}{ds} = \sin\theta$$

$$\beta = \frac{dz}{ds} = \cos\theta$$

where θ is the angle the ray makes with the z direction. Equation (2-48) then becomes, using the definition (2-9) that $n = c_0/c$ where c_0 is a constant,

$$\frac{\sin\theta}{c} = p \text{ (constant)} \tag{2-50}$$

Equation (2-43) is *Snell's law* and is the basic equation of ray acoustics. It states that the ratio of the sine of the angle of inclination of a ray at any given depth to the velocity at that depth is a constant along the ray path. The ray parameter p is a constant along a ray path but varies from one ray path to the next. It is sometimes useful to state this relation in terms of the angle of inclination and velocity of the ray at its source,

$$\frac{\sin\theta_0}{c_0} = \frac{\sin\theta}{c} \tag{2-51}$$

where c_0 is the velocity at the source and θ_0 the inclination of a particular ray at the source. Equation (2-49) reduces to

$$\begin{aligned} \frac{dn}{dz} = \frac{d}{ds}(n\cos\theta) &= -n\sin\theta\frac{d\theta}{ds} + \cos\theta\frac{dn}{dz}\frac{dz}{ds} \\ &= -n\sin\theta\frac{d\theta}{ds} + \cos^2\theta\frac{dn}{dz} \end{aligned}$$

and solving for $d\theta/ds$

$$\frac{d\theta}{ds} = -\frac{\sin\theta}{n}\frac{dn}{dz} \tag{2-52}$$

From the definition of n we have upon differentiating

$$\frac{dn}{n} = -\frac{dc}{c}$$

so that (2-52) becomes

$$\frac{d\theta}{ds} = \frac{\sin\theta}{c}\frac{dc}{dz}$$

or

$$\frac{d\theta}{ds} = p\frac{dc}{dz} \tag{2-53}$$

This is also a useful relation and states that the curvature of a ray $d\theta/ds$ is directly proportional to the velocity gradient dc/dz. For a region in which the velocity is increasing with depth, dc/dz positive, $d\theta/ds$ will be positive, and the ray will curve upward. Similarly for a region in which the velocity is decreasing the ray will curve downward. The rays always curve toward a region of minimum velocity; such a region will tend to form a wave guide for sound propagation. This result is to be expected from a qualitative consideration of the motion of the wave fronts. That portion of the wave front which is in a region of higher velocity will travel faster than that in a lower-velocity region and the wave front will be bent toward the lower-velocity region (Fig. 2-5).

The intensity I can be given by several equivalent relations. All refer back to the fact that the energy flow is defined by the ray paths and that the energy initially confined to a narrow bundle of rays near the source will continue to be confined within that bundle throughout its propagation. One such relation for the intensity is derived here. Consider a point source at the origin emitting P units of energy per unit time per unit

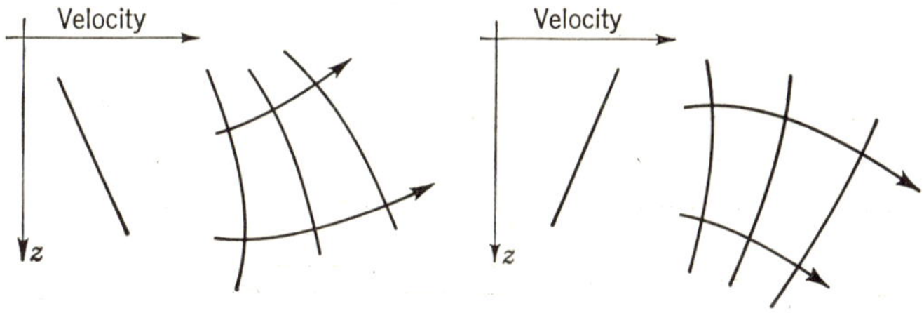

FIG. 2-5.

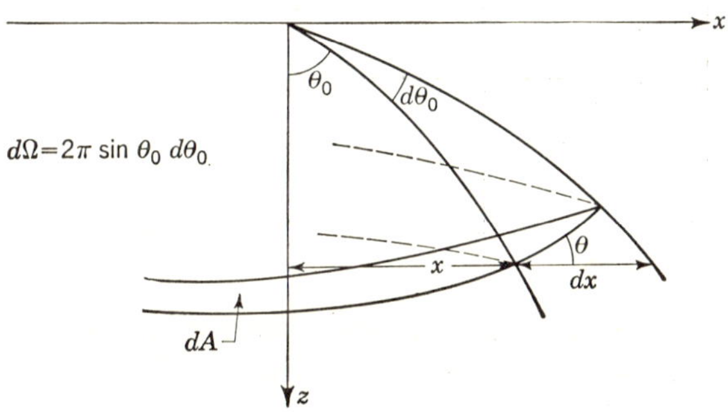

FIG. 2-6.

solid angle. Since the rays are confined to a plane with the z axis, there is cylindrical symmetry about this axis. We shall define the unit solid angle Ω with symmetry about the z axis so that, from a unit sphere about the origin in Fig. 2-6,

$$d\Omega = 2\pi \sin\theta_0\, d\theta_0$$

The intensity then at any other point along the ray path will be P multiplied by the ratio of $d\Omega$ to the area dA swept out by the wave surface, normal to the rays. From Fig. 2-6 this will be

$$I = P\frac{d\Omega}{dA} = P\frac{2\pi \sin\theta_0\, d\theta_0}{2\pi x \cos\theta\, dx} = P\frac{\sin\theta_0\, d\theta_0}{x\cos\theta\, dx}$$

The x coordinate of the ray can be expressed as a function of the initial angle θ_0 and the depth z through (2-50). At a given depth then

$$dx = \frac{\partial x}{\partial \theta_0} d\theta_0$$

or

$$I(x, \theta_0) = \frac{P \sin \theta_0}{x \cos \theta(\partial x/\partial \theta_0)} \tag{2-54}$$

If the source and receiver are at the same depth, $c = c_0$ and from (2-51), $\theta = \theta_0$ so that (2-54) becomes

$$I(X, \theta_0) = \frac{P \tan \theta_0}{X(\partial X/\partial \theta_0)} \tag{2-55}$$

The $1/x$ term is the loss due to cylindrical spreading out from the source, and the $\sin \theta_0/[\cos \theta(\partial x/\partial \theta_0)]$ term expresses the decrease or increase in intensity due to spreading or contracting of the ray bundle along its path.

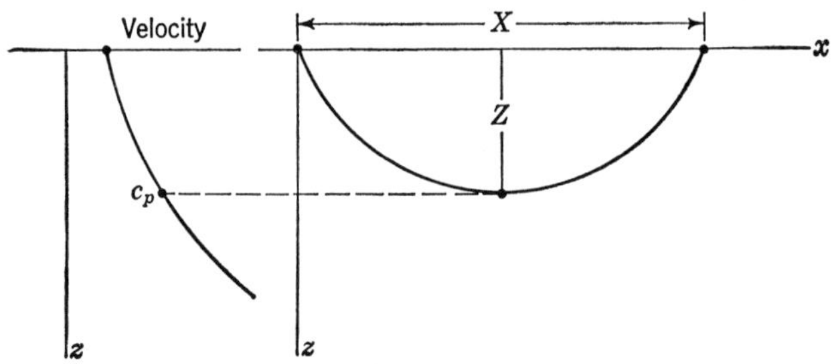

FIG. 2-7.

The travel time and horizontal range along a ray can be expressed in terms of the depth z or the angle of inclination θ by the following parametric equations. From Fig. 2-7

$$x = \int_0^z \tan \theta \, dz \tag{2-56}$$

and

$$t = \int_0^s \frac{ds}{c} = \int_0^z \frac{dz}{c \cos \theta} \tag{2-57}$$

where it is understood that θ is expressed as a function of z through (2-50). In terms of θ these integrals will be, taking differentials of (2-50) and substituting into (2-56) and (2-57)

$$x = \int_{\theta_0}^z \sin \theta \frac{dz}{\cos \theta} = \int_{\theta_0}^\theta \frac{\sin \theta \, d\theta}{pc'(z)} \tag{2-58}$$

and

$$t = \int_{\theta_0}^\theta \frac{d\theta}{c'(z) \sin \theta} \tag{2-59}$$

where it is understood that here the velocity gradient $c'(z)$ can be expressed in terms of θ through (2-50). To eliminate θ from (2-56) and (2-57) we have from (2-50) that

$$\tan\theta = \frac{pc}{\sqrt{1 - p^2c^2}}$$

and

$$\cos\theta = \sqrt{1 - p^2c^2}$$

so that

$$x = \int_0^z \frac{pc}{\sqrt{1 - p^2c^2}}\,dz \tag{2-60}$$

and

$$t = \int_0^z \frac{dz}{c\sqrt{1 - p^2c^2}} \tag{2-61}$$

It is convenient to define the parameter η

$$\eta \equiv \frac{1}{c} \tag{2-62}$$

Substituting into (2-60) and (2-61) we have

$$x = p\int_0^z \frac{dz}{(\eta^2 - p^2)^{1/2}} \tag{2-63}$$

and

$$t = \int_0^z \frac{\eta^2\,dz}{(\eta^2 - p^2)^{1/2}} \tag{2-64}$$

The travel time may also be expressed by

$$t = \int_0^z \left[\frac{p^2}{(\eta^2 - p^2)^{1/2}} + (\eta^2 - p^2)^{1/2}\right] dz$$

so that

$$t = px + \int_0^z (\eta^2 - p^2)^{1/2}\,dz \tag{2-65}$$

The horizontal range X and travel time T for the ray to reach the depth of the source again, i.e., if it does return to this depth, will be

$$X = 2p\int_0^Z \frac{dz}{(\eta^2 - p^2)^{1/2}} \tag{2-66}$$

and

$$T = 2\int_0^Z \frac{\eta^2\,dz}{(\eta^2 - p^2)^{1/2}} \tag{2-67}$$

where Z is the depth of the vertex, the maximum depth the ray penetrates.

We shall now derive a useful subsidiary relation between the ray parameter p, and the horizontal range x, and travel time t. Consider two adjacent rays defined by the ray parameters, p and $p + dp$, propagating out to a given depth z; and draw the wave surfaces normal to them. They will be separated by a horizontal range interval dx and a time interval dt. From the geometry of Fig. 2-8 we have

$$\sin \theta = \frac{c\,dt}{dx}$$

and from (2-50) this gives

$$p = \frac{dt}{dx} \tag{2-68}$$

Another relation for p in terms of the velocity c_p at the vertex of a ray can be given. At this point the angle of inclination θ is 90 deg so that $\sin \theta_p = 1$. Thus, from (2-50)

$$p = \frac{1}{c_p} \tag{2-69}$$

or from (2-62)

$$p = \eta_p \tag{2-70}$$

The velocity c_p at the vertex is the maximum velocity that the ray reaches and is attained at the greatest depth Z. If through a series of

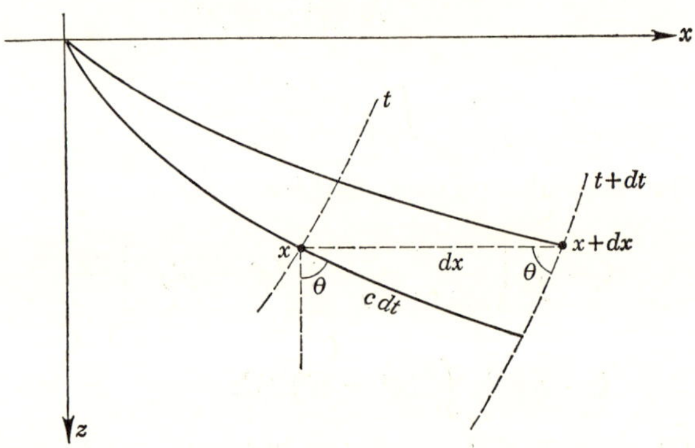

FIG. 2-8.

measurements from a source to a line of receivers at the surface, we should obtain a plot of T versus X, the inverse slope of this graph at any given range will be from (2-68) and (2-69) the velocity at the vertex for the ray emergent at this range (Fig. 2-9). The validity of this relation can be seen from the following geometrical construction. A ray emergent at the surface will be inclined at the same angle it had starting out from the source. The inverse slope at this point on the T versus X plot will be simply the trace velocity $\bar{c}$, with which the wave front associated with this ray moves along the surface $z = 0$. From the geometry of Fig. 2-10 this velocity will be

$$\bar{c} = \frac{c_0}{\sin \theta_0} \tag{2-71}$$

and from (2-51) remembering that $\sin \theta_p = 1$ at the vertex

$$\bar{c} = c_p \tag{2-72}$$

In many problems in geophysics the measurements give a plot of T versus X, and it is desired to find from this the velocity c as a function of depth z. From the above this reduces to the problem of finding the

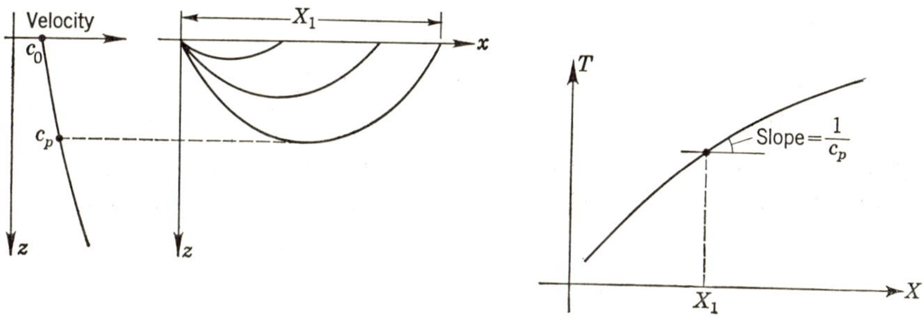

FIG. 2-9.

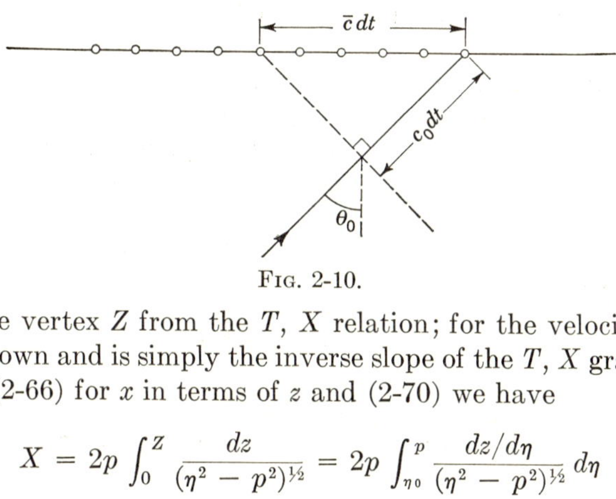

FIG. 2-10.

depth of the vertex Z from the T, X relation; for the velocity c_p at the vertex is known and is simply the inverse slope of the T, X graph. From expression (2-66) for x in terms of z and (2-70) we have

$$X = 2p \int_0^Z \frac{dz}{(\eta^2 - p^2)^{1/2}} = 2p \int_{\eta_0}^{p} \frac{dz/d\eta}{(\eta^2 - p^2)^{1/2}}\, d\eta \tag{2-73}$$

where the integration is taken along the ray path. Let us apply the operation $\int_{p_0}^{p_1} (p^2 - \eta_1{}^2)^{-1/2}\, dp$ to both sides of (2-73). This is an integration across the rays from the ray at zero range down to the ray whose vertex is at a depth where the velocity is $c_1 = 1/\eta_1$. In this operation the limits of integration have the relations

$$p_0 = \eta_{p0} = \eta_0 = \frac{1}{c_0} \qquad \text{and} \qquad p_1 = \eta_{p1} = \eta_1 = \frac{1}{c_1}$$

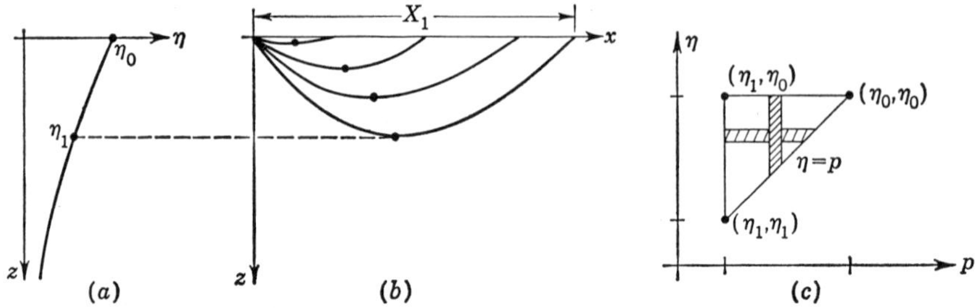

FIG. 2-11.

Expression (2-73) becomes

$$\int_{\eta_0}^{\eta_1} \frac{X}{(p^2 - \eta_1^2)^{1/2}}\, dp = \int_{\eta_0}^{\eta_1} dp \int_{\eta_0}^{\eta_p} \frac{2p(dz/d\eta)}{(p^2 - \eta_1^2)^{1/2}(\eta^2 - p^2)^{1/2}}\, d\eta \qquad (2\text{-}74)$$

From Fig. 2-11c we see that the order and limits of integration for the double integral on the right-hand side can be changed as follows:

$$\int_{\eta_0}^{\eta_1} \frac{X}{(p^2 - \eta_1^2)^{1/2}}\, dp = \int_{\eta_0}^{\eta_1} d\eta \int_{p}^{\eta_1} \frac{dz/d\eta}{[-p^4 + (\eta^2 + \eta_1^2)p^2 - \eta_1^2\eta^2]^{1/2}}\, 2p\, dp$$

Integrating by parts on the left-hand side and carrying out the first integration on the right-hand side we obtain

$$\left[X \cosh^{-1}\left(\frac{p}{\eta_1}\right)\right]_{\eta_0}^{\eta_1} - \int_{\eta_0}^{\eta_1} \frac{dX}{dp} \cosh^{-1}\left(\frac{p}{\eta_1}\right) dp$$
$$= \int_{\eta_0}^{\eta_1} \frac{dz}{d\eta} \left\{\sin^{-1}\left[\frac{2p^2 - (\eta^2 + \eta_1^2)}{(\eta^2 - \eta_1^2)}\right]\right\}_{p=\eta}^{\eta_1} d\eta$$

The first term on the left-hand side is zero since $X = 0$ when $p = \eta_0$. We then have

$$\int_{\eta_0}^{\eta_1} \frac{dX}{dp} \cosh^{-1}\left(\frac{p}{\eta_1}\right) dp = \pi \int_{\eta_0}^{\eta_1} \frac{dz}{d\eta}\, d\eta$$

or

$$\int_0^{X_1} \cosh^{-1}\left(\frac{p}{\eta_1}\right) dX = \pi \int_0^{Z_1} dz = \pi Z_1$$

and substituting for η_1 and p finally

$$Z_1 = \frac{1}{\pi} \int_0^{X_1} \cosh^{-1}\left(\frac{c_1}{c_p}\right) dX = \frac{1}{\pi} \int_0^{X_1} \cosh^{-1}\left(\frac{c_1}{\bar{c}}\right) dX \qquad (2\text{-}75)$$

This is an expression for Z_1, in terms of T and X, the quantities $1/c_1$ and $1/c_p$ given by $(dT/dX)_1$ and (dT/dX). Generally the integration is carried out numerically from the T versus X plot. The quantity $\bar{c}$ is the inverse slope to this graph and is determined point by point along the graph from $X = 0$ to $X = X_1$. The quantity c_1 is the inverse slope at $X = X_1$; the integration can be carried out to determine the value of Z_1 at which the velocity c_1 is reached. The velocity-depth relation, $c = c(z)$, can be determined by carrying out successive integrations of (2-75) for $X = X_1, X_2, X_3, \ldots$. In this derivation it has been implicitly assumed that η was a continuously decreasing function of depth (Fig. 2-11a). For an increasing section, $d\eta/dz$ would change signs and the integral (2-74) would become indeterminate. Expression (2-75) is valid only for those cases where the velocity is a continuously increasing function of depth.

Before going into specific velocity functions in the later chapters there are several somewhat more general relations which we can obtain here. First we wish to derive an expression for dX/dp. From (2-66) this will be

$$\frac{dX}{dp} = 2\int_0^Z \frac{dz}{(\eta^2 - p^2)^{1/2}} + 2p\,\frac{d}{dp}\int_0^Z \frac{dz}{(\eta^2 - p^2)^{1/2}} \tag{2-76}$$

Before carrying out the indicated differentiation on the last term we shall integrate by parts

$$\begin{aligned}\int_0^Z \frac{dz}{(\eta^2 - p^2)^{1/2}} &= \int_0^Z \frac{dz}{d\eta}\,\frac{d\eta/dz}{(\eta^2 - p^2)^{1/2}}\,dz = \int_0^Z f(\eta)\,\frac{d\eta/dz}{(\eta^2 - p^2)^{1/2}}\,dz \\ &= \left[f(\eta)\cosh^{-1}\left(\frac{\eta}{p}\right)\right]_{\eta_0}^{p} - \int_0^Z f'(\eta)\cosh^{-1}\left(\frac{\eta}{p}\right)\frac{d\eta}{dz}\,dz \\ &= -f(\eta_0)\cosh^{-1}\left(\frac{\eta_0}{p}\right) - \int_0^Z f'(\eta)\cosh^{-1}\left(\frac{\eta}{p}\right)\frac{d\eta}{dz}\,dz\end{aligned}$$

so that (2-66) will be

$$\begin{aligned}\frac{dX}{dp} &= 2\int_0^Z \frac{dz}{(\eta^2 - p^2)^{1/2}} + 2p\left[\frac{\eta_0 f(\eta_0)}{p(\eta_0^2 - p^2)^{1/2}} + \int_0^Z \frac{\eta f'(\eta)}{p(\eta^2 - p^2)^{1/2}}\,\frac{d\eta}{dz}\,dz\right] \\ &= 2\,\frac{\eta_0 f(\eta_0)}{(\eta_0^2 - p^2)^{1/2}} + 2\int_0^Z [f(\eta) + \eta f'(\eta)]\,\frac{d\eta/dz}{(\eta^2 - p^2)^{1/2}}\,dz\end{aligned} \tag{2-77}$$

Let us now make the substitution, $\zeta = \zeta(z)$, where

$$\zeta(z) \equiv \frac{1}{c}\frac{dc}{dz} \tag{2-78}$$

Then, we shall have

$$\eta f(\eta) = \eta\,\frac{dz}{d\eta} = -c\,\frac{dz}{dc} = -\frac{1}{\zeta}$$

remembering from (2-62) that $d\eta/\eta = -(dc/c)$. Taking derivatives with respect to z will give

$$[f(\eta) + \eta f'(\eta)]\,\frac{d\eta}{dz} = \frac{1}{\zeta^2}\frac{d\zeta}{dz}$$

Substituting into (2-77) gives finally

$$\frac{dX}{dp} = -\frac{2}{\zeta_0(\eta_0^2 - p^2)^{1/2}} + 2\int_0^z \frac{d\zeta/dz}{\zeta^2(\eta^2 - p^2)^{1/2}}\,dz \qquad (2\text{-}79)$$

In many geophysical problems the velocity increases gradually with depth so that ζ will be a moderate positive quantity and $d\zeta/dz$ small. Starting out from $z = 0$, the integrated term of (2-79) will be large compared with the integral so that dX/dp will be negative. For increasing depth, $p = \eta_p$ will decrease and dX/dp will decrease. From Fig. 2-11*d* we can then obtain Fig. 2-11*e*; and since $p = dT/dX$, from Fig. 2-11*e*, we

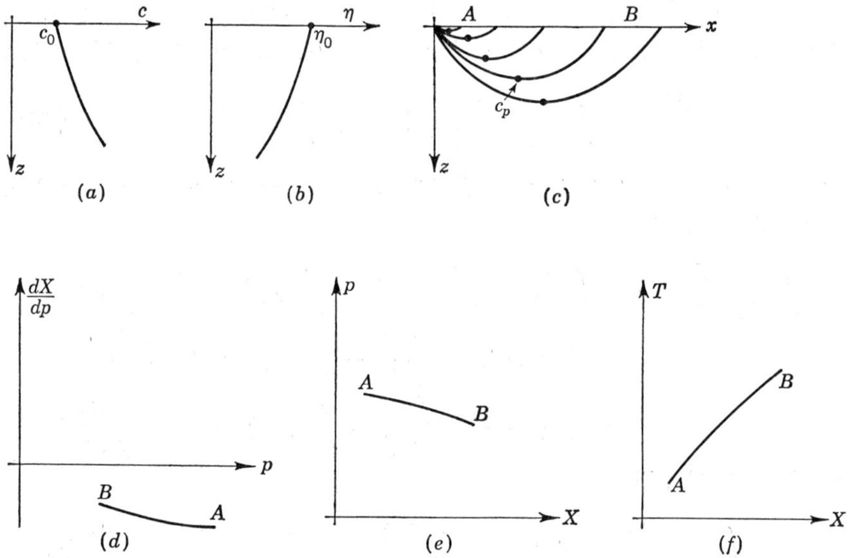

FIG. 2-12.

can obtain the graph of T versus X. It is seen, as expected, that T increases gradually with X and that the inverse slope, $c_p = 1/(dT/dX)$, increases with X corresponding to penetration of the deeper, higher-velocity media. Furthermore, the intensity, from (2-55), is proportional to $\tan \theta_0/X(dX/d\theta_0)$. Since $p = \sin \theta_0/c$, the ordinate of Fig. 2-11*d* or the slope of Fig. 2-11*e* can be considered to be a measure of $\partial X/\partial\theta_0$. Thus we see that the intensity will decrease gradually with range, the rate of decrease being smaller the larger the velocity gradient.

Let us assume now that at some depth there is a marked increase in the rate of increase in velocity and then a return to the conditions of a gradual increase in velocity. Then ζ will become large and $d\zeta/dz$ will be at first a moderate positive number and then a moderate negative number. If this rate of increase is sufficiently large, the integral of (2-79) can become larger than the integrated term. dX/dp will become positive and then return to its normal condition with increasing depth. As before, from the graph of dX/dp versus p (Fig. 2-13*d*) the graphs of p versus X and T versus X can be obtained. We see here that there is a reversed segment from B to C in the travel-time curve. Within this range interval there is triplication in the travel-time curve, three arrivals being received at the same range. The normal segment from A to B has a lower velocity, corresponding to the medium above the depth of the rapid increase in velocity, than the normal segment from C to D, corresponding to penetration of the higher-velocity media below the depth of rapid increase. At

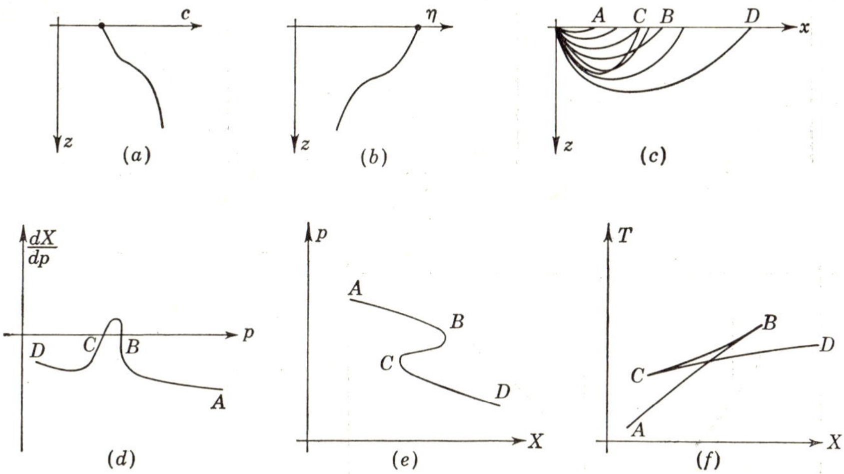

FIG. 2-13.

the points B and C $1/(\partial X/\partial\theta_0)$ will be zero; and these points will be bright spots of intensity. Along the reversed segment from B to C the intensity will be low.

If, on the other hand, the rate of increase is not sufficiently large for the value of dX/dp to pass through zero, the resulting ray situation will be as illustrated in Fig. 2-14. There is no triplication in the travel-time curve. The travel-time curve changes slope rapidly in the vicinity of B, and in this vicinity the intensity will be high.

Let us now consider the case in which at some depth there is a decrease in velocity followed at a greater depth by a return to the more normal condition of a gradual increase of velocity with depth. At the depth where the decrease in velocity begins, ζ will be zero; and the integral of (2-79) will become divergent. At depths in the region of decreasing velocity, since $p = \eta_p$, the value of p should increase. However, this is an impossibility, for a ray which is inclined at an angle θ_1 at a depth of higher velocity c_1 will from Snell's law (2-50) be inclined at a steeper

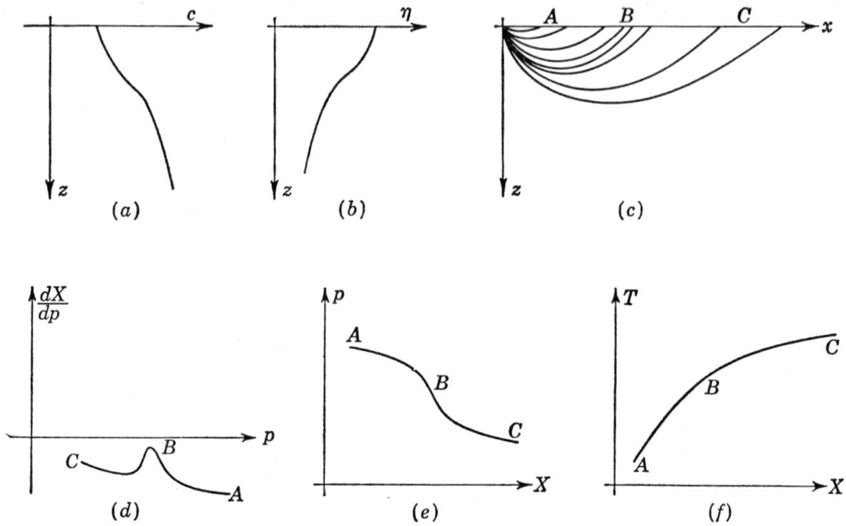

FIG. 2-14.

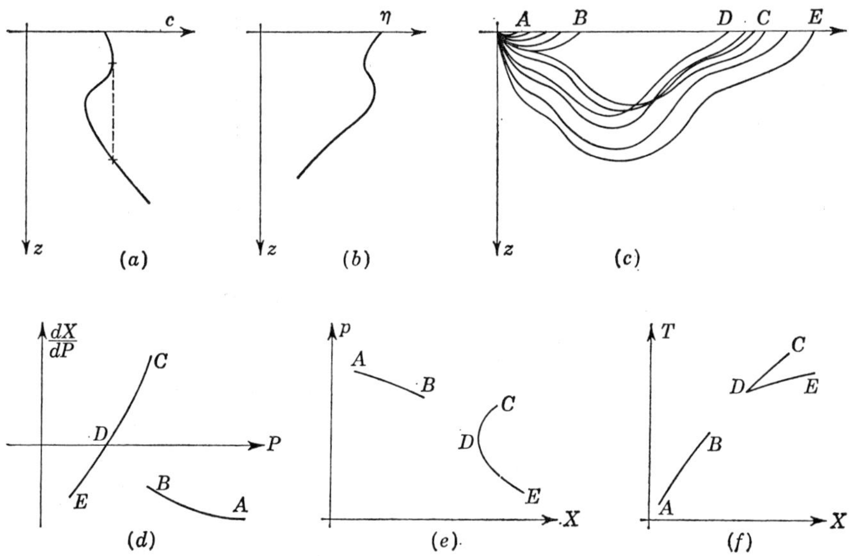

FIG. 2-15.

angle θ_2 at a greater depth of lower velocity c_2. It cannot have a vertex until the depth is reached at which the velocity is equal to that at the depth where it first began to decrease (Fig. 2-15*a*). In this interval of forbidden depths the quantity $(\eta^2 - p^2)^{1/2}$ in the integral of (2-79) is indeterminate.

Considered in another way, we want to obtain the value of dX/dp for all rays; starting from grazing incidence at the source, $\theta_0 = 90$ deg, and moving around to normal incidence, $\theta_0 = 0$ deg, all rays will be covered. Since $p = \sin \theta_0/c_0$, p will be a continually decreasing quantity in this analysis so that dX/dp will be a single-valued function of p. As a result, there will be a discontinuity in Z and a corresponding discontinuity in X and T.

We can get an approximation to dX/dp for those rays whose vertex is below the negative-gradient section from the case of a discontinuous decrease in velocity with a normal gradual increase in velocity above and below. The integrals above and below the discontinuity will be small because $d\zeta/dz$ is small; and the integrated term becomes

$$\frac{dX}{dp} \approx -\frac{2}{\zeta_0(\eta_0{}^2 - p^2)^{1/2}} + \frac{2}{\zeta_a(\eta_a{}^2 - p^2)^{1/2}} - \frac{2}{\zeta_b(\eta_b{}^2 - p^2)^{1/2}}$$

where η_a and η_b are the values of η just above and just below the discontinuity. For the first ray to be returned from below the discontinuity (ray C of Fig. 2-15) p will be nearly equal to η_a so that dX/dp will change from a moderate negative value at B to a large positive value at C. For increasing depth (increasing p) the importance of the η_a term will diminish and dX/dp will approach its normal value. These circumstances are illustrated in Fig. 2-15*d* from which Figs. 2-15*e* and 2-15*f* have been derived. There is a discontinuity in the travel-time curve (Fig. 2-15*f*) such that over the range interval from B to D no rays are received at the surface. This is known as a *shadow zone*. Beyond the shadow zone two rays are received out to the range of C. From Fig. 2-15*d* or 2-15*e* the intensity in this region will be higher than if there had been no negative gradient section. Physically this is understandable in that the energy which has been removed from the shadow zone must appear somewhere else. Further, in the immediate vicinity of D, $\partial X/\partial\theta_0$ is very small; this will be a focus point of intensity. It is a characteristic of a velocity structure with a negative gradient section that there will be a shadow zone followed by a focus point and a zone of moderately high intensity.

Ray characteristics for a constant velocity gradient

The simplest type of velocity function is that in which the velocity is a linear function of the depth. This is also a most useful function, for many physical problems can be approximated by a constant velocity gradient. Those problems which do not meet this condition can often be divided up into a number of depth intervals in each of which the velocity can be approximated by a constant gradient. For a constant gradient we have

$$c = c_0 + gz \tag{2-80}$$

where g is a constant. From Eq. (2-53) we shall have

$$\frac{d\theta}{ds} = p\,\frac{dc}{dz} = pg \tag{2-81}$$

This states that the curvature along any given ray is a constant; the ray path is an arc of a circle. The radius of curvature of the arc is the reciprocal of (2-81), and from (2-50) and (2-51) this will be

$$\text{Radius of curvature} = \frac{1}{d\theta/ds} = \frac{c_0}{g\sin\theta_0} \tag{2-82}$$

From the geometrical construction of Fig. 2-16 it can be seen that the centers of all the circular arcs lie on a line c_0/g units of distance above the surface. If the gradient changes at some greater depth, the centers of the circles for the ray paths through the depth interval of the new gradient

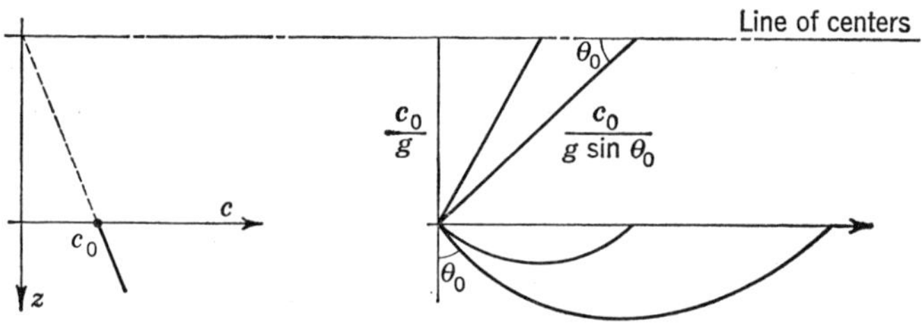

FIG. 2-16.

will lie along a different line. If the gradient changes from positive to negative, the line of centers will shift from above the ray to below the ray.

The equations for the horizontal range x, travel time t, and depth z, can be given conveniently parametrically in terms of the angle θ. From (2-80) we shall have

$$\frac{dc}{dz} = c' = g$$

and from (2-50) and (2-51)

$$p = \frac{\sin\theta}{c} = \frac{\sin\theta_0}{c_0}$$

so that (2-58) will be

$$x = \int_{\theta_0}^{\theta} \frac{\sin\theta\,d\theta}{pc'} = \frac{c_0}{g\sin\theta_0}\int_{\theta_0}^{\theta}\sin\theta\,d\theta = \frac{c_0}{g}\left(\frac{\cos\theta_0 - \cos\theta}{\sin\theta_0}\right) \tag{2-83}$$

and (2-59) will become

$$t = \int_{\theta_0}^{\theta} \frac{d\theta}{c'\sin\theta} = \frac{1}{g}\int_{\theta_0}^{\theta}\frac{d\theta}{\sin\theta} = \frac{1}{g}\log\frac{\tan\theta/2}{\tan\theta_0/2} \tag{2-84}$$

and (2-80) will become

$$z = \frac{c - c_0}{g} = \frac{c_0}{g}\left(\frac{\sin\theta}{\sin\theta_0} - 1\right) \tag{2-85}$$

An expression for the ray intensity I of Eq. (2-54) under the conditions of a constant gradient can be derived as follows. We must first obtain an evaluation of the quantity $\partial x/\partial\theta_0$. From (2-83) this will be

$$\begin{aligned}\frac{\partial x}{\partial\theta_0} &= \frac{c_0}{g\sin^2\theta_0}\left[\sin\theta_0\left(-\sin\theta_0 + \sin\theta\frac{\partial\theta}{\partial\theta_0}\right) - (\cos\theta_0 - \cos\theta)\cos\theta_0\right] \\ &= \frac{c_0}{g\sin^2\theta_0}\left[-1 + \cos\theta\cos\theta_0 + \sin\theta\sin\theta_0\frac{\partial\theta}{\partial\theta_0}\right] \end{aligned} \tag{2-86}$$

The quantity $\partial\theta/\partial\theta_0$ can be determined by differentiating (2-51) with respect to θ_0, giving

$$\begin{aligned}\cos\theta\frac{\partial\theta}{\partial\theta_0} &= \frac{c}{c_0}\cos\theta_0 \\ \frac{\partial\theta}{\partial\theta_0} &= \frac{\sin\theta}{\sin\theta_0}\frac{\cos\theta_0}{\cos\theta}\end{aligned} \tag{2-87}$$

Substitution of (2-87) into (2-86) gives finally

$$\begin{aligned}\frac{\partial x}{\partial\theta_0} &= \frac{c_0}{g\sin^2\theta_0}\left(-1 + \cos\theta\cos\theta_0 + \sin\theta\sin\theta_0\frac{\sin\theta\cos\theta_0}{\cos\theta\sin\theta_0}\right) \\ &= \frac{c_0}{g\sin^2\theta_0}\left(\frac{-\cos\theta + \cos^2\theta\cos\theta_0 + \sin^2\theta\cos\theta_0}{\cos\theta}\right) \\ &= \frac{c_0}{g\sin^2\theta_0}\left(\frac{\cos\theta_0 - \cos\theta}{\cos\theta}\right) \\ &= \frac{x}{\cos\theta\sin\theta_0}\end{aligned} \tag{2-88}$$

so that the intensity I of Eq. (2-54) becomes upon substitution of (2-88)

$$\begin{aligned}I &= P\frac{\sin\theta_0}{x\cos\theta\,\partial x/\partial\theta_0} \\ &= P\frac{\sin^2\theta_0}{x^2}\end{aligned} \tag{2-89}$$

The intensity decreases proportionally to the square of the horizontal range. This result is similar to that for a spherical ray spreading in a constant velocity medium (1-39) in that the intensity decrease is proportional to the square of a distance, but here the distance is the ratio of the horizontal range to the sine of the initial angle of inclination, not the ray path distance itself. Graphically this relation is shown in Fig. 2-17.

Development of solution in terms of normal modes

From the solution (1-36) we can take sine or cosine function for the arbitrary function defined there so that

$$\psi = \frac{\sin}{\cos}\left[k(x - ct) + \eta\right] \tag{2-90}$$

or

$$\psi = e^{i[k(x-ct)+\eta]}$$

where k and η are arbitrary constants. Motion of this type is known as *simple harmonic motion.* If we look at the motion at a particular instant

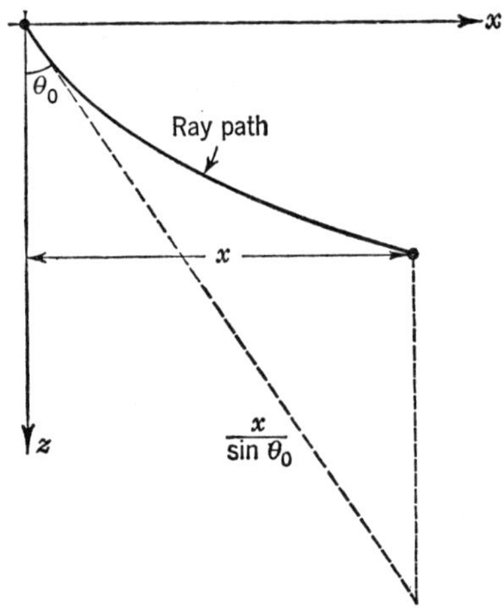

FIG. 2-17.

of time, it will be represented by an oscillation back and forth across the x axis. The distance between two successively repeated points, such as the distance between two crests, as indicated in Fig. 2-18, is the *wavelength* λ. It is that distance which is traversed for one complete cycle of the sine function. Thus, from Eq. (2-90)

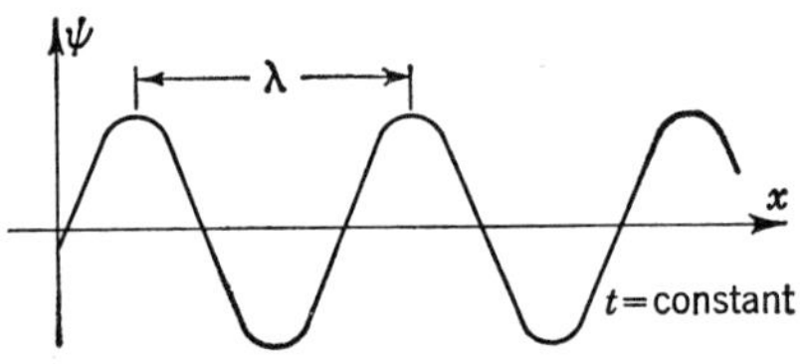

FIG. 2-18.

$$kx = k\lambda = 2\pi$$
$$\lambda = \frac{2\pi}{k} \tag{2-91}$$

Similarly, if we stand at one point in space, and observe the motion as a function of time, it will be a series of oscillations. The time for one complete cycle is the *period* τ, and from Eq. (2-90)

$$kct = kc\tau = 2\pi$$
$$\tau = \frac{2\pi}{kc} = \frac{\lambda}{c} \tag{2-92}$$

The *frequency* f is defined as the number of oscillations which take place in one second and is simply the reciprocal of the period:

$$f = \frac{1}{\tau} = \frac{c}{\lambda} \tag{2-93}$$

This leads to the familiar relation

$$c = f\lambda \tag{2-94}$$

which states that the velocity of propagation of simple harmonic motion is equal to the product of frequency and wavelength. This is physically understandable in that, if we are at a fixed point in space and observe the number of cycles that pass in a second, the velocity of propagation of the

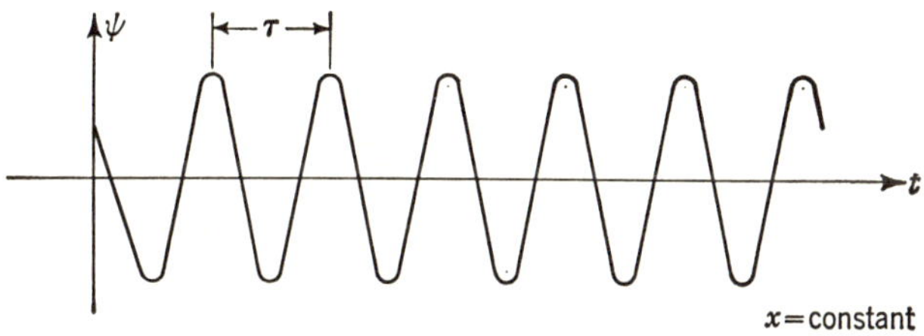

FIG. 2-19.

wave motion will be the product of this number by the length of one cycle. This is simply Eq. (2-94).

Equation (2-90) may now be written in the form

$$\psi = \frac{\sin}{\cos}\left[2\pi\left(\frac{x}{\lambda} - ft\right) + \eta\right] \tag{2-95}$$

In some problems it is more convenient to write Eq. (2-90) in the form

$$\psi = \frac{\sin}{\cos}\left[(kx - wt) + \eta\right] \tag{2-96}$$

where ω is the *circular frequency* and is defined as the number of radians per second,

$$\omega = 2\pi f = ck \tag{2-97}$$

and k is the original constant of Eq. (2-90), the *wave number.* The constant η is referred to as the initial phase.

Equation (2-96) may be expanded into product terms of sines and cosines so that the solution may be written

$$\psi = \frac{\sin}{\cos}\left(kx\right)\frac{\sin}{\cos}\left(\omega t\right) \tag{2-98}$$

When such a solution is generalized in the form

$$\psi = \sum_{n=1}^{\infty} A_n \frac{\sin}{\cos}\left(knx\right) \frac{\sin}{\cos}\left(\omega_n t\right)$$

or

$$\psi = \sum_{n=1}^{\infty} A_n \frac{\sin}{\cos}\left(knx\right) e^{i\omega_n t} \tag{2-99}$$

it is referred to as a *normal-mode solution.* In Eq. (2-99) ω_n is given by

$$\omega_n = nck \tag{2-100}$$

from Eq. (2-97). Such a solution states that the motion is composed of a *set of discrete frequencies* $\omega_1, \omega_2, \omega_3, \ldots$ whose amplitude varies in a sinusoidal manner.

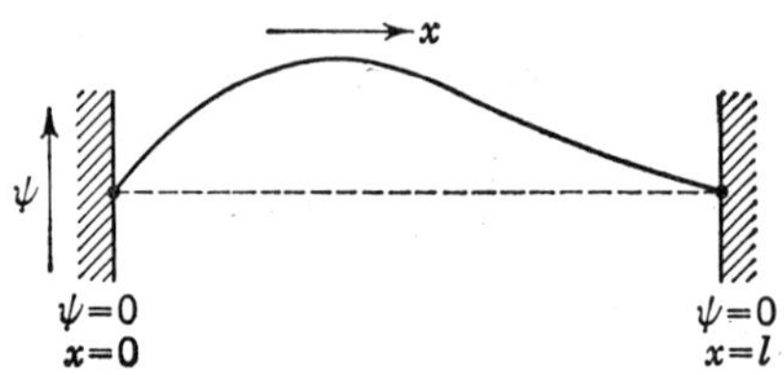

FIG. 2-20.

This type of solution is extremely useful in problems where fixed boundary conditions have to be met. As an example, let us take the simple case of the motion of a flexible string stretched between two rigid supports, ψ representing the displacement of the string. [Its motion is governed by Eq. (1-35).] Then, since $\psi = 0$ at $x = 0$ for all time, the cos (knx) term in (2-99) must drop out leaving

$$\psi = \sum_{n=1}^{\infty} A_n \sin\,(knx) \frac{\sin}{\cos}\left(\omega_n t\right)$$

Then, since $\psi = 0$ at $x = l$ also for all time, we must satisfy the condition that

$$\sin\,(knl) = \sin\left(2\pi n \frac{l}{\lambda}\right) = 0$$

This is met if

$$\lambda = 2l \tag{2-101}$$

The final solution becomes

$$\psi(x, t) = \sum_{n=1}^{\infty} a_n \sin\left(\frac{n\pi x}{l}\right) \cos\,(2\pi f_n t) + b_n \sin\left(\frac{n\pi x}{l}\right) \sin\,(2\pi f_n t) \tag{2-102}$$

where

$$f_n = \frac{nc}{2l} \tag{2-103}$$

Equation (2-102) satisfies both boundary conditions. From (2-102) the motion of the string can be thought of as composed of a discrete set of simple harmonic vibrations such that the distance between the supports is an integral number of half wavelengths as shown in Fig. 2-21.

The arbitrary constants of (2-102) are determined by the initial conditions, i.e., by the initial displacement, $\psi(x, 0)$, and initial velocity,

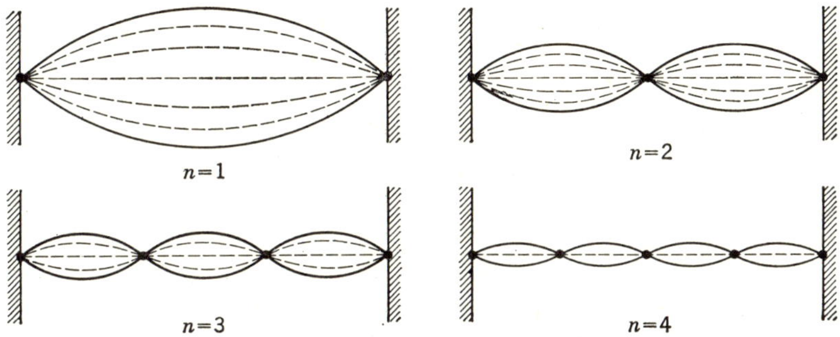

FIG. 2-21.

$(\partial\psi/\partial t)(x, 0)$, to the given string. The easiest method of doing this is through a *Fourier synthesis*. From the given initial conditions and (2-102) we have

$$\psi(x, 0) = \sum_{n=1}^{\infty} a_n \sin\left(\frac{n\pi x}{l}\right)$$

and

$$\frac{\partial\psi}{\partial t}(x, 0) = 2\pi \sum_{n=1}^{\infty} b_n f_n \sin\left(\frac{n\pi x}{l}\right)$$

Then, performing the integration

$$\int_0^l \psi(x, 0) \sin\left(\frac{m\pi x}{l}\right) dx = \sum_{n=1}^{\infty} a_n \int_0^l \sin\left(\frac{n\pi x}{l}\right) \sin\left(\frac{m\pi x}{l}\right) dx$$

we obtain

$$a_m = \frac{2}{l} \int_0^l \psi(x, 0) \sin\left(\frac{m\pi x}{l}\right) dx \tag{2-104}$$

since

$$\begin{aligned}\int_0^\pi \sin(n\xi)\sin(m\xi)\,d\xi &= 0 \qquad \text{for} \qquad m \neq n \\ &= \frac{\pi}{2} \qquad \text{for} \qquad m = n\end{aligned}$$

Similarly, we obtain

$$b_m = \frac{1}{\pi f_n l} \int_0^l \frac{\partial\psi}{\partial t}(x, 0) \sin\left(\frac{m\pi x}{l}\right) dx \tag{2-105}$$

The solution of the more complicated normal-mode problems that follow in the later chapters is quite similar to this simple case with the exception that the waves progress outward in a second space coordinate. The normal-mode patterns exist in the depth coordinate z because of the plane parallel boundaries perpendicular to z. The propagation is outward in a radial direction.

Normal-mode solution in problems with more complicated wave equations

In many problems the wave equation is more complicated than that of (1-35) either through the variation of the velocity c with one or more of the space coordinates, usually the depth, or through the introduction of additional terms to the equation, or both. In such cases it cannot be assumed that Eq. (2-90) will be a solution. However, let us try a solution of the form

$$\psi = Ae^{i[k(Vt-x)]} = Ae^{i(\omega t-kx)} \tag{2-106}$$

where

$$\omega = Vk \tag{2-107}$$

We cannot assume a priori that the velocity of wave propagation V is equal to c as was true in the simple case of constant coefficients. In general, (2-106) will be a solution to the more complex wave equations; but upon substitution a conditional equation between k and V results which also must be satisfied. This conditional equation is referred to as the *characteristic equation* (not to be confused with the equation of characteristics).

$$G(k, V) = 0 \tag{2-108}$$

Equation (2-108) states that the velocity of wave propagation V, also referred to as the *phase velocity*, is a function of the wave number or through (2-107) the frequency. This phenomenon is known as *dispersion.* Each frequency will travel with a different velocity. An initial disturbance will gradually become a long train of waves whose frequency will vary in a regular manner, those frequencies associated with the higher velocities being near the head of the train and those associated with lower velocities near the rear.

Let us now generalize this solution to arbitrary initial conditions. One of the easier methods is to follow an analysis similar to that for the generalization of the vibrating-string solution. However, in this case, at present, there are no boundary conditions so that an integration over all wave numbers (or frequencies) rather than a set of discrete frequencies is necessary. This is achieved through the *Fourier integral transforms*

$$\begin{aligned} f(x) &= \frac{1}{2\pi}\int_{-\infty}^{\infty} e^{ikx}g(k)\,dk \\ g(k) &= \int_{-\infty}^{\infty} e^{-ik\chi}f(\chi)\,d\chi \end{aligned} \tag{2-109}$$

for the initial conditions, $f(\chi)$, given as a particular distribution of pressure, velocity, and/or displacement at time $t = 0$ or

$$\begin{aligned} f(t) &= \frac{1}{2\pi} \int_{-\infty}^{\infty} e^{i\omega t} g(\omega)\, d\omega \\ g(\omega) &= \int_{-\infty}^{\infty} e^{-i\omega\tau} f(\tau)\, d\tau \end{aligned} \tag{2-110}$$

for the initial conditions, $f(\tau)$, given as a particular pressure, velocity, and/or displacement time function at $x = 0$. In (2-109) $g(k)$ is in general a complex number and represents the amplitude and phase of the *spectrum* of wave numbers which when added together give the desired initial displacement $f(x)$. Similarly $g(\omega)$ represents the amplitude and phase of the spectrum of frequencies which are required to produce the initial disturbance $f(t)$. More simply stated $g(\omega)$ gives the spectral amplitude and phase of $f(t)$.

From this formulation we can write down the complete solution for the given initial conditions. Using the initial conditions of (2-110) for a given source function, $f(t)$, acting at the origin, $x = 0$, the solution will be

$$\Psi(x, t) = \frac{1}{2\pi} \int_{-\infty}^{\infty} g(\omega) e^{i[\omega t - k(\omega)x]}\, d\omega \tag{2-111}$$

For a given source distribution, $f(x)$, acting at $t = 0$, the solution will be

$$\Psi(x, t) = \frac{1}{2\pi} \int_{-\infty}^{\infty} g(k) e^{i[kx - \omega(k)t]}\, dk \tag{2-112}$$

We can be sure that this is the correct solution; for in general an integration of the simple harmonic solution to the wave equation over a parameter not entering in the original wave will also be a solution; and at $x = 0$, Eq. (2-111) reduces to

$$\Psi(x, 0) = \frac{1}{2\pi} \int_{-\infty}^{\infty} g(\omega) e^{i\omega t}\, d\omega = f(t)$$

the specified initial conditions. The problem reduces to the determination of the Fourier transform $g(\omega)$ of the initial conditions $f(t)$, and the evaluation of the integral (2-111). In most problems certain boundary conditions also have to be met which often lead to an additional integration generally over the parameter k, such an integration being taken prior to the evaluation (2-111).

Evaluation of an integral such as (2-111) is generally facilitated through the approximation of the method of stationary phase. If we have an integral of the form

$$u = \int_a^b \varphi(z) e^{if(z)}\, dz \tag{2-113}$$

it represents integration of a rapidly oscillating function $e^{if(z)}$ of amplitude

$\varphi(z)$. If $\varphi(z)$ is a slowly changing function with respect to one oscillation of $e^{if(z)}$, the various contributions to the integral will cancel out by annulling plus and minus values except in the neighborhood of those values of z, if any, for which $f(z)$ is stationary. Here and only here will a nonoscillating contribution exist. Let us designate by α the value of z at a stationary value of $f(z)$; then

$$z = \alpha + \xi$$

and

$$f'(\alpha) = 0$$

so that in the neighborhood of α, by a series expansion,

$$f(x) = f(\alpha) + \tfrac{1}{2}\xi^2 f''(\alpha) + \cdots$$

Since the integral has an appreciable value in this neighborhood only, the limits of integration may be extended to $-\infty$ and ∞

$$u = \varphi(\alpha)e^{if(\alpha)} \int_{-\infty}^{\infty} e^{i\frac{1}{2}\xi^2 f''(\alpha)}\, d\xi \tag{2-114}$$

The value of this integral is known.[1] It is

$$u = \frac{\pi^{\frac{1}{2}}\varphi(\alpha)}{|\frac{1}{2}f''(\alpha)|^{\frac{1}{2}}} e^{i[f(\alpha)\pm(\pi/4)]} \tag{2-115}$$

where | | indicates the absolute value and the plus or minus sign in the exponential is taken as $f''(\alpha)$ is negative or positive. Only the second term of the series expansion for $f(\alpha)$ has been used. The evaluation (2-115) is valid then only under the condition that the third term is much smaller than the second,

$$\frac{|\frac{1}{6}\xi^3 f'''(\alpha)|}{|\frac{1}{2}\xi^2 f''(\alpha)|} = \frac{|\xi f'''(\alpha)|}{|3f''(\alpha)|} \ll 1$$

There are some examples for which this inequality does not hold, in particular where $f''(\alpha)$ also equals zero. This requires a further evaluation.

The solution for (2-112) [or (2-111)] in terms of (2-115) when the above condition is met is

$$\Psi(x, t) = \frac{\varphi(k)}{|2\pi t(d^2\omega/dk^2)|^{\frac{1}{2}}} e^{i[kx-\omega t\pm(\pi/4)]} \tag{2-116}$$

where

$$\varphi(k) \equiv g(k)$$

and

$$f(k) = kx - \omega(k)t$$

[1] B. O. Pierce, "A Short Table of Integrals," p. 63, Sec. 492, Ginn & Company, Boston, 1929.

We then have for the value of k at the point of stationary phase

$$f'(k) = x - \frac{d\omega}{dk} t = 0 \tag{2-117}$$

The particular value of k that is used, and also ω by (2-108), is determined as a function of x and t through Eq. (2-117). Thus, at a particular value of x and t the dominant energy contribution will have a particular wave number k and frequency ω. Rewriting (2-117) in the form

$$U \equiv \frac{x}{t} = \frac{d\omega}{dk} \tag{2-118}$$

we see that a particular value of wave number k and frequency ω found at a space-time location (x_1, t_1) will be found again at (x_2, t_2) such that

$$\frac{x_1}{t_1} = \frac{x_2}{t_2}$$

It will have traveled with a velocity $(d\omega/dk)_{k_1}$, given by (2-118). Thus the energy associated with a particular frequency group will travel with a velocity U known as the *group velocity*, which is different from the velocity of wave propagation or phase velocity V.

The existence and significance of these two velocities can be demonstrated physically through a consideration of the mechanisms that occur in such propagation. First, however, it is desirable to write down some other forms of (2-118).

$$U = \frac{d\omega}{dk} = \frac{d(Vk)}{dk} = V + k\frac{dV}{dk} \tag{2-119}$$

Since $k = 2\pi/\lambda$, $dk = -2\pi\lambda^{-2}\,d\lambda$,

$$U = V - \lambda\frac{dV}{d\lambda} \tag{2-120}$$

Since $\omega = kV$, $U = d\omega/[d(\omega/V)]$,

$$\frac{1}{U} = \frac{1}{V} - \frac{\omega}{V^2}\frac{dV}{d\omega} \tag{2-121}$$

Consider a dispersive wave train in which the phase velocity V increases with frequency. Then, at a given distance the wave train may resemble Fig. 2-22 as a function of time. At some slightly greater distance A will arrive at a later time A', having traveled with a velocity V_A. B travels more slowly than A so that the time between B and B' will be greater than between A and A' and relative separation $A - B$ will increase. The same is true in a corresponding manner for every other point on the wave train. The wave train is lengthened as it progresses out, and the crest between AB which was associated with a frequency $1/[2(B - A)]$ at the near distance is associated with a lower frequency $1/[2(B' - A')]$ at a greater distance. The frequency associated with a particular phase of the motion, such as $A - A'$, is variable with distance. In the example of Fig. 2-22 the increase in distance was chosen such that the frequency of the crest BC at the nearer distance is found at the trough $A'B'$ at the greater distance. The group velocity being that of a particular frequency group is the velocity from BC to $A'B'$ whereas the phase velocity is that

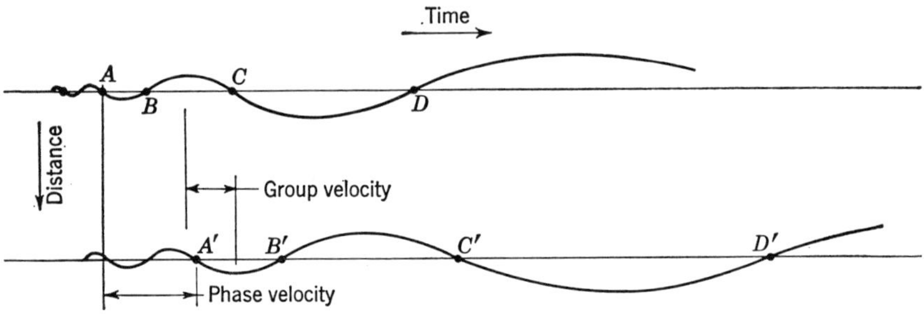

Fig. 2-22.

from AB to $A'B'$. In this example the group velocity is greater than the phase velocity.

The expression for the group velocity in terms of the phase velocity can be derived from a simple diagram similar to that shown above. For simplicity in arithmetic the coordinates of Fig. 2-22 will be reversed, and we shall consider the ground motion at a specific instant of time and then at some slightly later time. Also in this case we shall assume that the

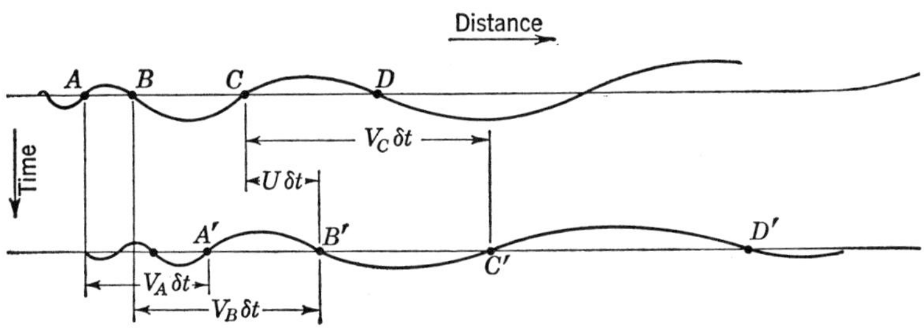

Fig. 2-23.

phase velocity V decreases with frequency. In this case it is assumed that the time interval δt between the two examinations is such that the wavelength originally associated with C is found at B' at the later time. Then

$$U = \frac{B' - C}{\delta t} = \frac{C' - C}{\delta t} - \frac{C' - B'}{\delta t} = V_C - \frac{1}{2}\frac{\lambda_{B'C'}}{\delta t} \qquad (2\text{-}122)$$

Now δt has been chosen such that $C' - B' = D - C$. $C - B$ becomes

$C' - B'$ in this time interval so that δt is given by

$$\delta t = \frac{\frac{1}{2}(\lambda_{CD} - \lambda_{BC})}{V_C - V_B} \tag{2-123}$$

where $\frac{1}{2}(\lambda_{CD} - \lambda_{BC})$ is the difference in distance between CD and BC expressed in wavelengths. It is the relative distance that BC must increase in passing to $B'C'$, and $V_C - V_B$ is the relative velocity at which C is spreading away from B. Substituting (2-123) into (2-122) gives

$$U = V_C - \lambda_{CD} \frac{V_C - V_B}{\lambda_{CD} - \lambda_{BC}}$$

or in the limit

$$U = V - \lambda \frac{dV}{d\lambda} \tag{2-124}$$

which is simply Eq. (2-120).

The phase velocity is determined from the physical characteristics of the medium. The group velocity is determined from the net effect of adding up all the little elements of different frequency in the wave train traveling with different velocities. In the case where the dispersion is brought about through the variation of velocity with one of the space coordinates, generally depth, either in a continuous manner or discontinuously across boundaries, it is referred to as *geometrical dispersion.* In the case where the dispersion is brought about by a dependence on frequency of the physical parameters determining the wave velocity it is referred to as *material dispersion.*

Another useful concept in visualizing the mechanism of dispersive wave propagation is to consider a group of records that might be obtained at a number of receiving stations away from the source. In Fig. 2-24 the records are aligned at time $t = 0$. Each frequency of the wave train will propagate through the wave train with its appropriate group velocity, and this group velocity will be constant for a given frequency and different for different frequencies. The location, then, of a given frequency in the wave trains from successive receiving stations will plot on a straight line passing through the origin. The slope of this line will be the group velocity. Such lines have been drawn for three different frequencies, f_1, f_2, and f_3 in Fig. 2-24. If we had chosen to follow a particular phase of the wave train from record to record, such as the crest A, it would plot on a curve passing through the origin. The phase A is associated with a different frequency from one record to the next. In Fig. 2-24 the frequency decreases with distance for a particular phase of the wave train. The slope of this curve at any point will be the phase velocity for the particular frequency associated with that point. Such curves have been drawn for the crests A, B, C, and D in Fig. 2-24.

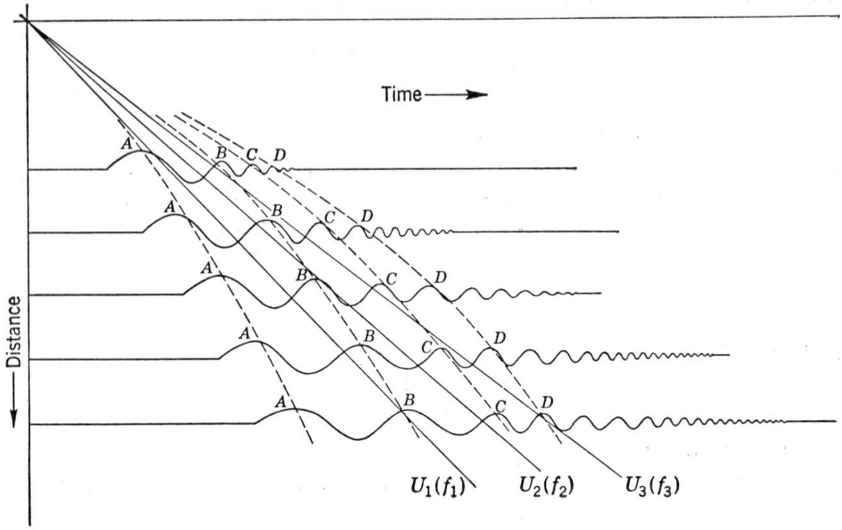

FIG. 2-24.

As an example of dispersion we may take the previous example of the vibrating string which we shall now subject to an additional restoring force proportional to the displacement, the equation of motion becoming

$$\frac{\partial^2\psi}{\partial t^2} = c^2 \frac{\partial^2\psi}{\partial x^2} - a^2\psi \tag{2-125}$$

Substituting (2-106) into (2-125) gives

$$k^2V^2 = c^2k^2 + a^2 \tag{2-126}$$

corresponding to (2-108). From (2-126) we obtain for the phase and group velocities

$$V = \left(c^2 + \frac{a^2}{k^2}\right)^{1/2}$$

or (2-127)

$$V = c\left(\frac{\omega^2}{\omega^2 - a^2}\right)^{1/2}$$

and

$$U = \frac{c^2}{V} \tag{2-128}$$

For small wave numbers (low frequencies greater than $\omega = a$) the phase velocity approaches ∞ and the group velocity approaches zero. For large wave numbers (high frequencies) both the phase and the group velocities approach the value c, the solution for the simpler wave equation (1-35).

Refraction of a dispersive wave train

Let us examine here the refraction of a dispersive wave train across a boundary, the medium on one side characterized by phase and group velocities, V and U, respectively, and the medium on the other side by V_1 and U_1. In particular we are interested in determining the law of refraction and the travel time from a point in the first medium to a point in the second. The dispersive wave train in each medium can be represented by a Fourier integral with respect to the frequency ω following the developments given in the previous section on dispersion. The contribution at O_1, from a source at O, can be given by an integral over the surface, S, of the excitation initiated there from the source O. The resulting disturbance at O_1 will then be

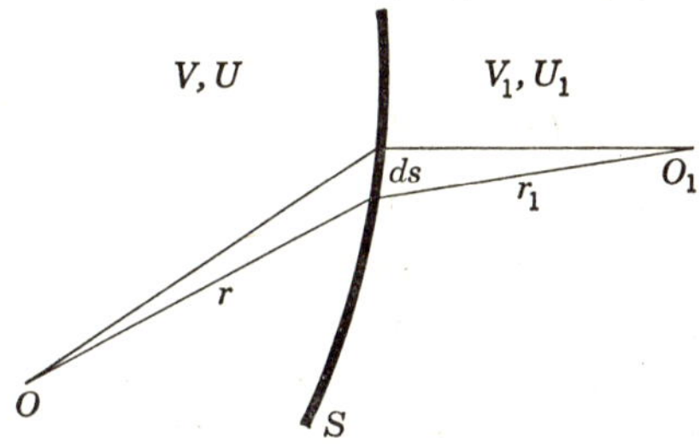

FIG. 2-25.

$$\iint f(\omega)g(s) \exp\left[i\omega\left(\frac{r}{V} + \frac{r_1}{V_1} - t\right)\right] d\omega\, ds \tag{2-129}$$

As before the amplitude functions $f(\omega)$ and $g(s)$ will be slowly varying with respect to the oscillations of the exponential. The integral will have a dominant contribution at stationary values of the exponential with respect to both ω and s. This condition is met if

$$\frac{1}{V}\frac{dr}{ds} + \frac{1}{V_1}\frac{dr_1}{ds} = 0 \tag{2-130}$$

and

$$r\,\frac{d(\omega/V)}{d\omega} + r_1\,\frac{d(\omega/V_1)}{d\omega} - t = 0 \tag{2-131}$$

The space derivatives dr/ds and $-dr_1/ds$ are equal to the sines of the angles of incidence and refraction, respectively. Equation (2-130) reduces to

$$\sin\theta/V = \sin\theta_1/V_1 \tag{2-132}$$

which is simply Snell's law and states that a dispersive wave train is refracted across a boundary according to Snell's law for the respective phase velocities. From (2-120) and (2-121) Eq. (2-131) reduces to

$$t = \frac{r}{U} + \frac{r}{U_1} \tag{2-133}$$

which states that the travel time from O to O_1 is determined by the group velocities in the respective media, the distances r and r_1 being determined by (2-132). These results are understandable physically in that the particle motions in each medium are determined by the phase velocities so that one might expect that refraction across a boundary should also be determined by these velocities. On the other hand, the velocity of propagation of a particular frequency is obtained from the summation of the individual contributions in the wave train and will be the respective group velocity in each medium.

Boundary conditions

In almost all acoustic problems boundary conditions of one sort or another have to be met. Let us examine one of the simplest cases, a simple harmonic plane wave incident on a plane boundary between two fluids across which there is a change in velocity and density. We shall orient the coordinate system such that the y axis is along the boundary, the x axis normal to the boundary, and the z axis parallel to the incident plane wave. A plane wave is one whose wave front is plane and one for which the motion of any given wave front is everywhere the same at any given instant of time. The motion of a plane wave in three dimensions is described by one space coordinate, that measured along the ray direction. In this case the incident plane wave is a function of x and y only. For a plane wave the wave equation becomes in terms of r, the distance along the ray,

$$\frac{\partial^2 \varphi}{\partial t^2} = c^2 \frac{\partial^2 \varphi}{\partial r^2} \tag{2-134}$$

whose simple harmonic solution is of the form (2-96) for a wave propagating outward along r,

$$\varphi = Ae^{i(\omega t - kr + \eta)} \tag{2-135}$$

In terms of a coordinate system where the z axis is normal to r this becomes, from Fig. 2-26,

$$\begin{aligned} \varphi &= Ae^{i(\omega t - kx \cos\theta - ky \sin\theta + \eta)} \\ &= Ae^{i(\omega t - ax - by + \eta)} \end{aligned} \tag{2-136}$$

where

$$\begin{aligned} a &\equiv k \cos\theta \\ b &\equiv k \sin\theta \end{aligned} \tag{2-137}$$

The quantities a and b may be considered as the wave numbers measured along the x and y directions, respectively, corresponding to the wave number measured along the ray direction. The wave numbers measured in the x and y coordinate directions are always less than those along the ray direction. Equation (2-136) may be rewritten in a slightly different form as

$$\varphi = A \exp \left\{ i \left[\omega \left(t - \frac{x}{V_a} - \frac{y}{V_b} \right) + \eta \right] \right\} \tag{2-138}$$

where

$$V_a \equiv \frac{c}{\cos \theta} \qquad \text{and} \qquad V_b \equiv \frac{c}{\sin \theta} \tag{2-139}$$

The quantities V_a and V_b may be considered as the velocities of propagation that would be measured if one observed along the x or y directions, respectively, corresponding to the velocity of propagation of the medium c, which is measured along the ray direction. The velocities V_a and V_b are always greater than c, as can be seen from Eqs. (2-139) or Fig. 2-27.

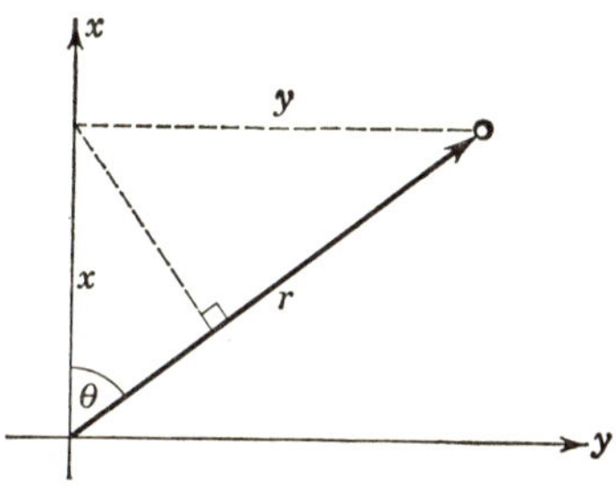

FIG. 2-26.

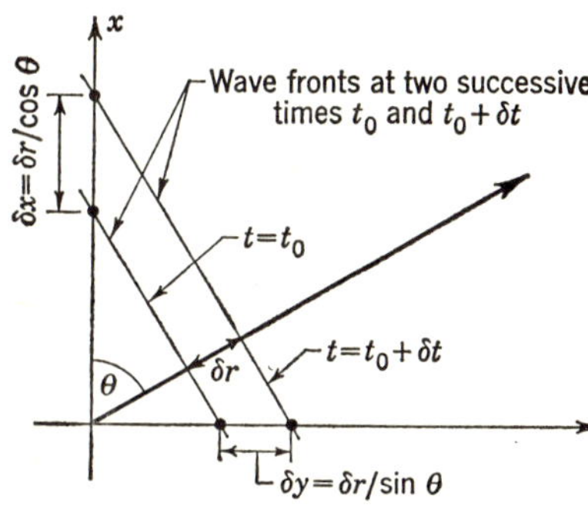

FIG. 2-27.

In general when a plane wave, φ_{in}, is incident on a plane boundary between two media we might expect two contributions to result, one, φ_{re}, *reflected* back into the first medium from the boundary and one, φ_2, *refracted* into the second medium. From (2-136) and Fig. 2-28 their solutions will be of the form

$$\begin{aligned} \varphi_{\text{in}} &= Ae^{i(\omega_{\text{in}}t + a_{\text{in}}x + b_{\text{in}}y)} \\ \varphi_{\text{re}} &= Be^{i(\omega_{\text{re}}t - a_{\text{re}}x + b_{\text{re}}y)} \end{aligned} \tag{2-140}$$

and

$$\varphi_2 = Ce^{i(\omega_2 t + a_2 x + b_2 y)}$$

where the initial phase has been omitted for convenience. The boundary conditions that must be met by the acoustic vibrations in the two media at the interface between the two in order to preserve the continuity of the interface are that (1) there be continuity of vertical particle velocity and (2) there be continuity of pressure. These boundary conditions must be met everywhere along the interface and at all times. From Fig. 2-28 and Eqs. (1-83) and (1-85) the boundary conditions are, in terms of the velocity potentials,

$$\frac{\partial \varphi_1}{\partial x} = \frac{\partial \varphi_2}{\partial x} \tag{2-141}$$

and at $x = 0$

$$\rho_1 \frac{\partial \varphi_1}{\partial t} = \rho_2 \frac{\partial \varphi_2}{\partial t} \tag{2-142}$$

In order to assure that these conditions will be met at all times and at all positions along the boundary the coefficients of the three exponentials in

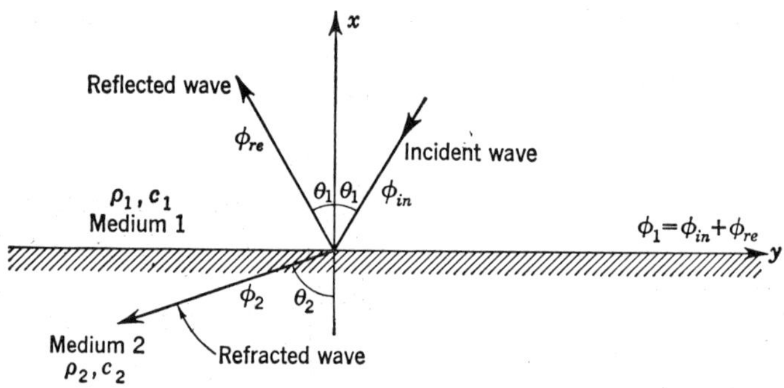

FIG. 2-28.

t of Eqs. (2-140) must be equal and the coefficients of the three exponentials in y must be equal. This gives

$$\omega_{\text{in}} = \omega_{\text{re}} = \omega_2 \equiv \omega \tag{2-143}$$

and

$$b_{\text{in}} = b_{\text{re}} = b_2 \equiv b$$
$$\frac{\omega}{c_1} \sin \theta_{\text{in}} = \frac{\omega}{c_1} \sin \theta_{\text{re}} = \frac{\omega}{c_2} \sin \theta_2 \tag{2-144}$$

or simply

$$\sin \theta_{\text{in}} = \sin \theta_{\text{re}} \equiv \sin \theta_1 \tag{2-145}$$

and

$$\frac{\sin \theta_1}{c_1} = \frac{\sin \theta_2}{c_2} \tag{2-146}$$

The first condition (2-143) demonstrates that the frequency of the reflected wave and the frequency of the refracted wave are the same as that of the incident wave. The second condition (2-144) redemonstrates Snell's law. The angle of the reflected ray is equal to the angle of the incident ray, and the angle of the refracted ray is determined from the angle of the incident ray through Snell's law. The coefficients of x become

$$a_{\text{in}} = a_{\text{re}} = \frac{\omega}{c_1} \cos \theta_1 = k_1 \cos \theta_1 = k_1 \sqrt{1 - \sin^2 \theta_1} \equiv a_1 \tag{2-147}$$

and

$$a_2 = \frac{\omega}{c_2} \cos \theta_2 = k_2 \cos \theta_2 = k_1 \sin \theta_1 \cot \theta_2 = k_1 \sqrt{\frac{c_1^2}{c_2^2} - \sin^2 \theta_1} \qquad \text{(2-148)}$$

Equations (2-140) become

$$\begin{aligned} \varphi_{\text{in}} &= Ae^{i(\omega t + a_1 x + by)} &&= Ae^{i\omega[t+(x \cos \theta_1 + y \sin \theta_1)/c_1]} \\ \varphi_{\text{re}} &= Be^{i(\omega t - a_1 x + by)} &&= Be^{i\omega[t+(-x \cos \theta_1 + y \sin \theta_1)/c_1]} \\ \varphi_2 &= Ce^{i(\omega t + a_2 x + by)} &&= Ce^{i\omega[t+(x \cos \theta_2 + y \sin \theta_2)/c_2]} \end{aligned} \qquad \text{(2-149)}$$

To determine the ratio of the amplitude of the reflected and refracted waves to the incident wave, Eqs. (2-149) are substituted into the boundary conditions (2-141) and (2-142), remembering φ_1 represents the sum of φ_{in} and φ_{re},

$$a_1(A - B) = a_2 C \qquad \text{(2-150)}$$

and

$$\rho_1(A + B) = \rho_2 C$$

from which we obtain

$$\frac{B}{A} = \frac{a_1\rho_2 - a_2\rho_1}{a_1\rho_2 + a_2\rho_1}$$

and (2-151)

$$\frac{C}{A} = \frac{2a_1\rho_1}{a_1\rho_2 + a_2\rho_1}$$

Upon substitution for a_1 and a_2 from (2-147) and (2-148), we obtain

$$\frac{B}{A} = \frac{\rho_2 \cot \theta_1 - \rho_1 \cot \theta_2}{\rho_2 \cot \theta_1 + \rho_1 \cot \theta_2} \qquad \text{(2-152)}$$

and

$$\frac{C}{A} = \frac{2\rho_1 \cot \theta_1}{\rho_2 \cot \theta_1 + \rho_1 \cot \theta_2} \qquad \text{(2-153)}$$

These ratios may be written in a slightly different form in terms of the velocities and densities of the two media and the sine of the angle of incidence,

$$\frac{B}{A} = \frac{\dfrac{\rho_2}{\rho_1} - \dfrac{\sqrt{(c_1^2/c_2^2) - \sin^2 \theta_1}}{\sqrt{1 - \sin^2 \theta_1}}}{\dfrac{\rho_2}{\rho_1} + \dfrac{\sqrt{(c_1^2/c_2^2) - \sin^2 \theta_1}}{\sqrt{1 - \sin^2 \theta_1}}} \qquad \text{(2-154)}$$

and

$$\frac{C}{A} = \frac{2}{\dfrac{\rho_2}{\rho_1} + \dfrac{\sqrt{(c_1^2/c_2^2) - \sin^2 \theta_1}}{\sqrt{1 - \sin^2 \theta_1}}} \qquad \text{(2-155)}$$

For normal incidence Eqs. (2-154) and (2-155) reduce to

$$\frac{B}{A} = \frac{\rho_2 c_2 - \rho_1 c_1}{\rho_2 c_2 + \rho_1 c_1} \tag{2-156}$$

and at $\theta_1 = 0$

$$\frac{C}{A} = \frac{2\rho_1 c_2}{\rho_2 c_2 + \rho_1 c_1} \tag{2-157}$$

The quantity (2-154) is often referred to as the *Rayleigh reflection coefficient.*

From (2-154) it is apparent that if c_2 is less than c_1 the ratio B/A, i.e., the reflectivity, will decrease as the angle decreases from normal incidence, $\theta_1 = 0$, since the radical in the numerator of the second term decreases less rapidly than the radical in the denominator. The reflectivity will decrease to zero at an angle of incidence such that

$$\frac{\rho_2}{\rho_1} = \frac{\sqrt{(c_1^2/c_2^2) - \sin^2\theta_1}}{\sqrt{1 - \sin^2\theta_1}} \tag{2-158}$$

Beyond this angle of incidence the reflectivity will increase to a value of unity at grazing incidence, $\theta_1 = 90$ deg. The sign of the reflectivity in this region will be negative indicating that the reflected wave has been reflected from the boundary with a phase change of 180 deg with respect to the incident wave.

For the case when c_2 is greater than c_1, the reflectivity will increase from normal incidence to a value of unity at an intermediate angle such that

$$\sin\theta_1 = \frac{c_1}{c_2} \tag{2-159}$$

This angle is referred to as the *critical angle.* Beyond this angle of incidence the radical in the numerator of the second term of (2-154) becomes imaginary. The imaginary radical can be expressed by $\pm i\sqrt{\sin^2\theta_1 - (c_1^2/c_2^2)}$ where $\sqrt{\sin^2\theta_1 - (c_1^2/c_2^2)}$ is a real quantity. The quantity $-i\sqrt{\sin^2\theta_1 - (c_1^2/c_2^2)}$ is the correct representation for this physical problem, giving for the refracted wave (2-166) an amplitude which decreases exponentially away from the boundary into the lower medium and not the physically inconsistent picture of an amplitude which increases continuously toward infinity in the lower medium. Thus, Eq. (2-154) becomes, for incidence more grazing than the critical angle,

$$\frac{B}{A} = \frac{\dfrac{\rho_2}{\rho_1} + i\dfrac{\sqrt{\sin^2\theta_1 - (c_1^2/c_2^2)}}{\sqrt{1 - \sin^2\theta_1}}}{\dfrac{\rho_2}{\rho_1} - i\dfrac{\sqrt{\sin^2\theta_1 - (c_1^2/c_2^2)}}{\sqrt{1 - \sin^2\theta_1}}} \tag{2-160}$$

which can be written in the following form, after a small amount of algebra,

$$\frac{B}{A} = e^{i\epsilon} \tag{2-161}$$

where

$$\frac{\epsilon}{2} = \tan^{-1} \frac{\rho_1 \sqrt{\sin^2 \theta_1 - (c_1^2/c_2^2)}}{\rho_2 \sqrt{1 - \sin^2 \theta_1}} \tag{2-162}$$

Thus, at angles more grazing than the critical angle the amplitude of the reflected wave remains unity, but there is a phase change of ϵ between the reflected and incident waves. This phase change is 0 deg at the critical angle and increases to 180 deg at grazing incidence. The expression (2-155) becomes

$$\frac{C}{A} = \frac{2}{\left[\dfrac{\rho_2^2}{\rho_1^2} + \dfrac{\sin^2 \theta_1 - (c_1^2/c_2^2)}{1 - \sin^2 \theta_1}\right]^{1/2}} e^{i(\epsilon/2)} \tag{2-163}$$

Substituting (2-161) into (2-140) gives for the reflected wave

$$\varphi_{\text{re}} = Ae^{i(\omega t + by - a_1 x + \epsilon)} \tag{2-164}$$

For the refracted wave the quantity a_2 of (2-148) becomes imaginary and, as has been discussed above, is replaced by

$$a_2 = k_1 \sqrt{\frac{c_1^2}{c_2^2} - \sin^2 \theta_1} = -ik_1 \sqrt{\sin^2 \theta_1 - \frac{c_1^2}{c_2^2}} = -ia_2' \tag{2-165}$$

Substituting (2-163) and (2-165) into (2-140) gives for the refracted wave

$$\varphi_2 = A \frac{2}{\left[\dfrac{\rho_2^2}{\rho_1^2} + \dfrac{\sin^2 \theta_1 - (c_1^2/c_2^2)}{1 - \sin^2 \theta_1}\right]^{1/2}} e^{a_2' x} e^{i[\omega t + by + (\epsilon/2)]} \tag{2-166}$$

Equation (2-164) states, as we have said before, that the amplitude of the reflected wave is equal to that of the incident wave with a phase change of ϵ. There is *total reflection* and no transmission of energy into the lower medium. Equation (2-166) represents a somewhat different wave. It is *bound* to the discontinuity surface and propagates parallel to this surface with a velocity less than that of the lower medium except at the critical angle reflection for which the bound wave velocity is equal to that of the lower medium. There is no transmission of energy into the lower medium, and the amplitude decreases exponentially away from the surface into the lower medium. This wave is in no sense a free wave; it can exist only as a consequence of reflection at angles of incidence more grazing than critical and is constrained to move along the boundary with the trace velocity of the reflection along the boundary. It is a diffraction effect across the interface associated with and bound to the reflected wave.

For comparison the reflectivity and phase angle as a function of the angle of incidence are sketched in Fig. 2-29*a*, *b*, *c*, *d*, and *e* from Eqs. (2-154) and (2-162) for various velocity and acoustic resistance contrasts. Figure 2-29*a*, *b*, and *c* is plotted for the more usual cases found in nature of

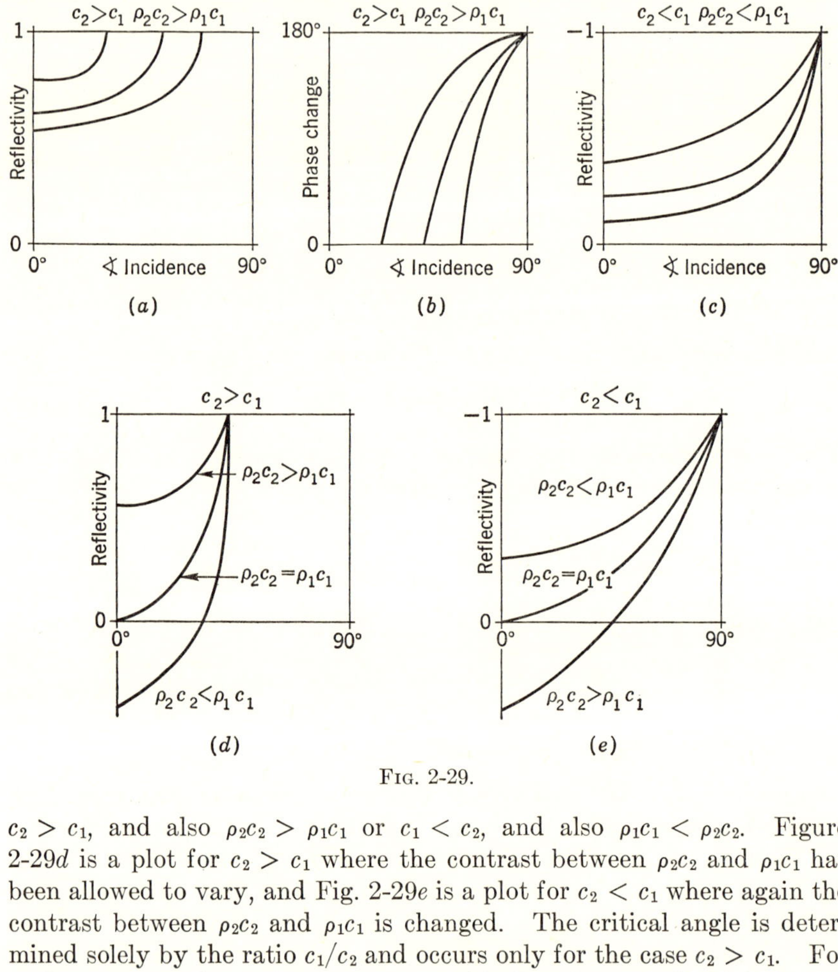

FIG. 2-29.

$c_2 > c_1$, and also $\rho_2 c_2 > \rho_1 c_1$ or $c_1 < c_2$, and also $\rho_1 c_1 < \rho_2 c_2$. Figure 2-29*d* is a plot for $c_2 > c_1$ where the contrast between $\rho_2 c_2$ and $\rho_1 c_1$ has been allowed to vary, and Fig. 2-29*e* is a plot for $c_2 < c_1$ where again the contrast between $\rho_2 c_2$ and $\rho_1 c_1$ is changed. The critical angle is determined solely by the ratio c_1/c_2 and occurs only for the case $c_2 > c_1$. For such a contrast there will always be a critical angle beyond which there is total reflection with a change in phase of the reflected wave with respect to the incident wave. For $c_2 < c_1$ total reflection occurs only at grazing incidence. The phenomenon of no reflected wave can occur only at an intermediate angle and only for those cases in which the velocity contrast is the opposite of that of the acoustic contrast. In such a case there is a phase change of 180 deg of the reflected wave with respect to the incident wave across the angle of no reflection.

If we had chosen the direction for the x axis as down rather than up in Fig. 2-28, the amplitude reflection and refraction coefficient of (2-154) and (2-155) would have remained the same; but in (2-165) it would have been necessary to take the positive sign for the imaginary so that the refracted wave (2-166) would decrease toward infinity into the lower medium. Therefore the phase changes of (2-161) and (2-163) would be the negative of that given

$$\frac{B}{A} = e^{-i\epsilon} \qquad \text{and} \qquad \frac{C}{A} = \frac{2}{\left(\dfrac{\rho_2^2}{\rho_1^2} + \dfrac{\sin^2\theta_1 - (c_1^2/c_2^2)}{1 - \sin^2\theta_1}\right)^{1/2}} e^{-i(\epsilon/2)}$$

and (2-164) and (2-166) would become

$$\varphi_{\text{re}} = Ae^{i(\omega t + by + a_1 x - \epsilon)}$$

$$\varphi_2 = A\frac{2}{\left(\dfrac{\rho_2^2}{\rho_1^2} + \dfrac{\sin^2\theta_1 - (c_1^2/c_2^2)}{1 - \sin^2\theta_1}\right)^{1/2}} e^{-a_2 x} e^{i[\omega t + by - (\epsilon/2)]}$$

At first glance it might seem inconsistent that simply by a change in the sense of the x coordinate axis the phase change upon reflection from the

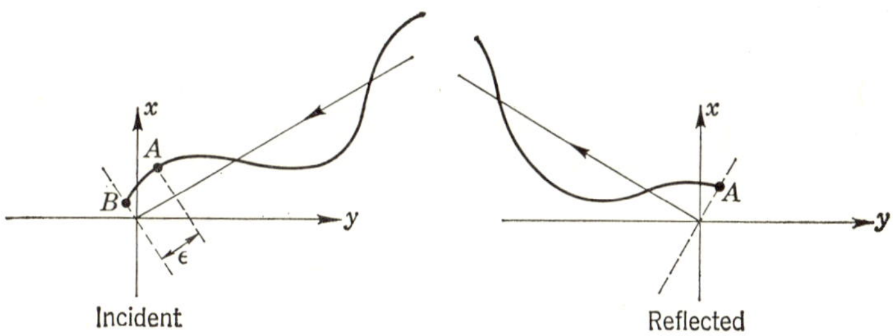

Fig. 2-30.

same boundary in the same physical problem should change from ϵ to $-\epsilon$. However, this is not the case; and its resolution brings out a rather important fact to be remembered when considering a phase change incurred upon reflection from a boundary. In the first case the incident and reflected waves are given by

$$e^{i(\omega t + by + a_1 x)} \qquad \text{and} \qquad e^{i(\omega t + by - a_1 x + \epsilon)}$$

(In the figures below the amplitude is measured along the normal to the ray.) Let us look at the motion at some particular instant of time. The incident wave is propagating in the negative x and negative y directions at an angle determined by a and b. The reflected wave is propagating in the negative y and positive x directions at an angle equal to that of the incident wave across the x axis. If there were no phase change, the reflected wave would simply continue the motion of the incident wave in a different direction. However, with the phase change, the reflected wave starts its motion with the amplitude A rather than B (Fig. 2-30). The phase of the reflected wave is *advanced* with respect to the incident wave. In the second case, the incident and reflected waves are given by

$$e^{i(\omega t+by-a_1x)} \qquad \text{and} \qquad e^{i(\omega t+by+a_1x-\epsilon)}$$

In this case also, as seen in Fig. 2-31, the beginning of the reflected wave is referred back to the point C along the incident wave since here the sense

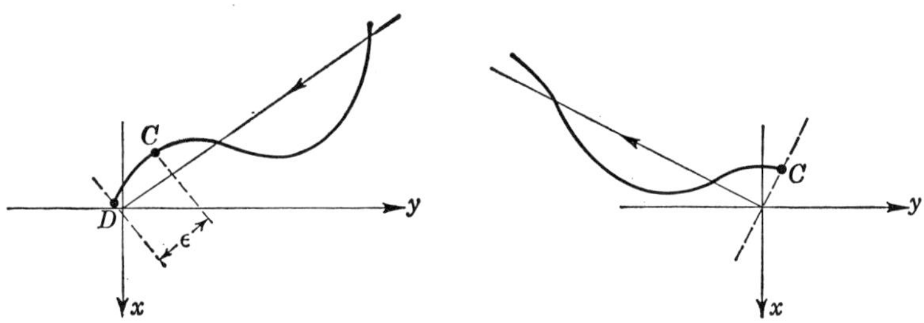

FIG. 2-31.

of x has been reversed and a phase change of $-\epsilon$ refers to a position back along the x axis. The result is the same as before. The important thing to remember is that the phase of the reflected wave is advanced with respect to the incident wave; the phase difference between the reflected and incident waves represents a negative phase change.

BIBLIOGRAPHY

Baker, B. B., and E. T. Copson: "The Mathematical Theory of Huygens' Principle," Oxford University Press, New York, 1939.

Bateman, H.: "Partial Differential Equations of Mathematical Physics," Cambridge University Press, New York, 1932, and Dover Publications, New York, 1944.

Bullen, K. E.: "Introduction to the Theory of Seismology," Cambridge University Press, New York, 1947.

Dobrin, M. B.: Dispersion in Seismic Surface Waves, *Geophysics*, vol. 16, pp. 63–80, 1951.

Eckart, C.: Approximate Solutions of One-dimensional Wave Equations, *Rev. Mod. Phys.*, vol. 20, pp. 399–417, 1948.

Havelock, T. H.: "The Propagation of Disturbances in Dispersive Media," Cambridge Tracts in Mathematics and Mathematical Physics No. 17, Cambridge University Press, New York, 1914.

Jeffreys, H.: "The Earth," Cambridge University Press, New York, 1952.

Kaufman, H.: Velocity Functions in Seismic Prospecting, *Geophysics*, vol. 18, pp. 289–297, 1953.

Lamb, H.: "Hydrodynamics," Cambridge University Press, New York, 1932, and Dover Publications, New York, 1945.

Luneberg, R. K.: "Mathematical Theory of Optics," Brown University Series in Applied Mathematics, Providence, 1944.

Slichter, L. B.: Theory of the Interpretation of Seismic Travel-time Curves in Horizontal Structures, *Physics*, vol. 3, pp. 273–295, 1932.

Spitzer, L., P. G. Bergmann, and A. Yaspan (eds.): Physics of Sound in the Sea, Div. 6, vol. 8, *Natl. Defense Research Council Summary Tech. Rept.*, Washington, 1945.

Strutt, J. W. (Lord Rayleigh): "The Theory of Sound," The Macmillan Company, London, 1894, and Dover Publications, New York, 1945.

3. *Transmission in Shallow Water*

The transmission of sound in the ocean is dependent on several environmental factors, the most important of which are the depth and configuration of the bottom, the physical properties of the bottom material, the sound velocity structure within the ocean, the distribution and character of sound scatterers within the ocean, and the shape of the ocean surface. A complete formulation with all possible variations of the above environmental conditions would be exceedingly difficult to obtain. Instead we shall consider the effects of each or groups of each on sound transmission. In some cases we shall use only simple theoretical models, the simplicity of which is not always a good approximation to the actual physical case.

Because of the peculiar distribution of the depths in the ocean, the solution of underwater-sound problems divides into two categories. For one the depths of water are less than 300 feet, and for the other the depths of water are around 10,000 to 20,000 feet. This distribution is illustrated in Fig. 3-1, a bathymetric (submarine topographic) map of the North Atlantic Ocean. The major portion of the ocean on this map consists of a broad basin whose depths are greater than 2,500 fathoms,[1] the *ocean basins.* The next most extensive area is that from land's end out to about 50 fathoms. This second area is known as the *continental shelf.* Its size can vary considerably with geography from nonexistent, as off the Bahama Islands in the lower left-hand corner of Fig. 3-1, to rather large widths of the order of 100 nautical miles,[2] as off the northeastern coast of the United States and Canada. From 50 fathoms the depth decreases rapidly down to a depth of about 1,500 fathoms. This region is referred to as the *continental slope.* From here the depth decreases more gradually to the ocean basins, this region being referred to as the *continental rise.* The area with depths between 50 and 1,500 fathoms is a small portion of the total oceanic area.

In this chapter we shall discuss sound transmission through the ocean in the shallow-water area of the continental shelf. As mentioned, the

[1] 1 fathom = 6 feet.

[2] 1 nautical mile = 1 degree of latitude = 6,080 feet.

transmission is dependent on the state of the environment. It is not the purpose here to describe in anything more than a cursory manner what these environmental conditions are, or to go into a discussion of the origin of them or their dependence on other variables. This is more properly a part of the sciences of physical oceanography, marine geology, marine geophysics, and marine biology. We shall merely state what the conditions are and describe the transmission of sound under these conditions.

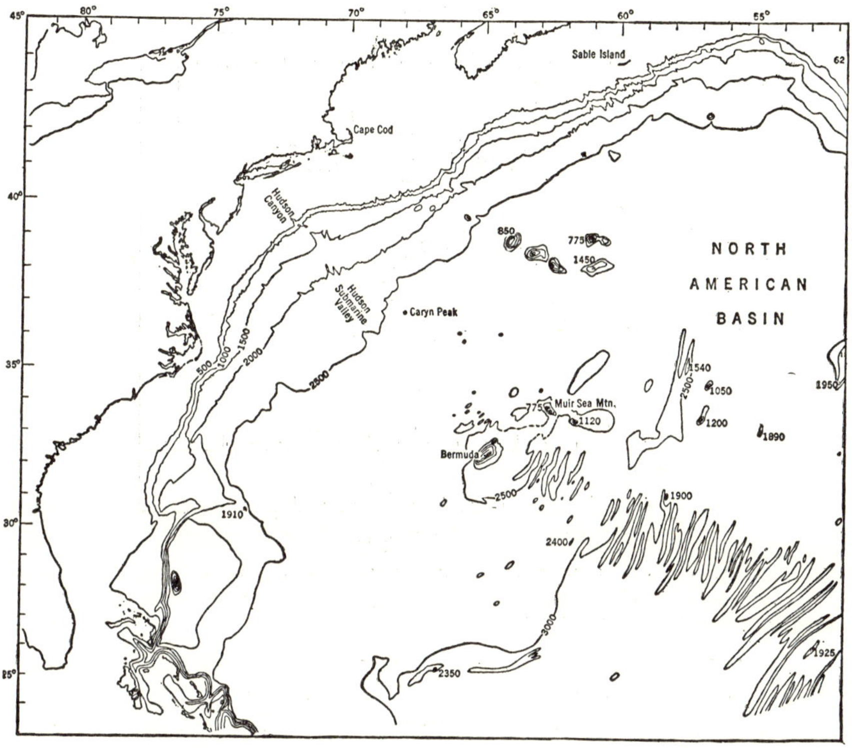

FIG. 3-1. Bathymetric sketch map of the North Atlantic. Contour interval 500 fathoms. (*From I. Tolstoy, Bull. Geol. Soc. Amer., vol.* 62, 1951.)

The velocity of sound in sea water is determined by the temperature, salinity, and pressure (depth). The variations in salinity are generally of the least importance in determining sound velocity except in those areas where there is a substantial efflux of fresh water from the continent out over the more saline ocean water. In this case a sharp salinity gradient with depth is often developed. The distribution of temperature with depth varies from one shallow-water area to the next and from one season to the next because of the efflux of fresh water, currents, atmospheric variations, and other causes. The resultant variations in sound velocity are generally greater and more variable in shallow water than over the same depth range in deep water. A common temperature, salinity, and density profile for shallow water is shown in Fig. 3-2. The density is determined from the temperature and salinity, an increase in density being proportional to an increase in salinity or a decrease in temperature. The region of steep density gradient is the more stable region and less subject to change. The regions above and below this are less stable and more subject to change through convective overturn or other mixing

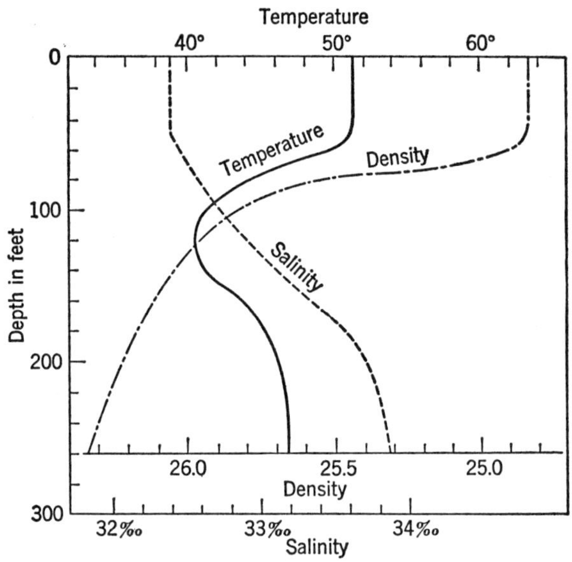

FIG. 3-2. Temperature is in degrees Fahrenheit. Salinity is in parts per thousand. Density is parts per thousand above 1 g/cm³. *(From C. O'D. Iselin, Natl. Defense Research Council, Div. 6, vol. 6A, 1946.)*

process induced by external forces. This overturn when complete produces a thorough admixture of the water resulting in a constant distribution of temperature and salinity with depth over that region, as shown for the upper 50 feet of Fig. 3-2. This situation is more prevalent in deep water than in shallow water. With such a condition of isothermal water the sound velocity will then increase with depth because of the effect of increased pressure alone and form a sound channel with the ocean surface. As illustration of the seasonal variation of temperature and sound velocity with depth in a particular shallow-water area, Figs. 3-3 and 3-4 are shown.

The bottom properties pertinent to sound transmission in shallow water can also vary considerably from one area to the next. One can find places where the ocean is bounded by hard crystalline rock or rock covered by a thin veneer of sediment; reef or reef debris; or gravel, sand, sand and mud, or mud sediments. The more common types are sand,

sand and mud, or mud. The areas of mud-size particles can vary in compactness from those of a hard clay to a loose suspension into which objects laid on the bottom can sink to an appreciable depth. Little is known of the bottom materials beyond their geologic properties; in particular little is known of the acoustic properties of the immediate bottom

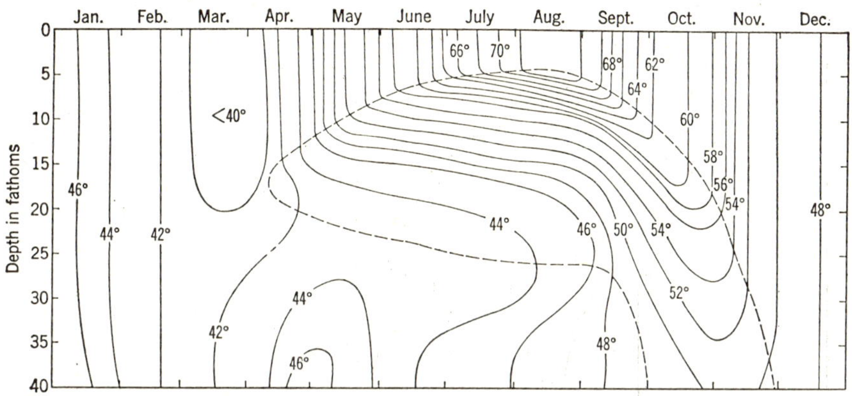

FIG. 3-3. Seasonal cycle of temperature in coastal waters off Montauk Point. Contour interval 2° F. (*From Woods Hole Oceanographic Institution, Natl. Defense Research Council*, 1941.)

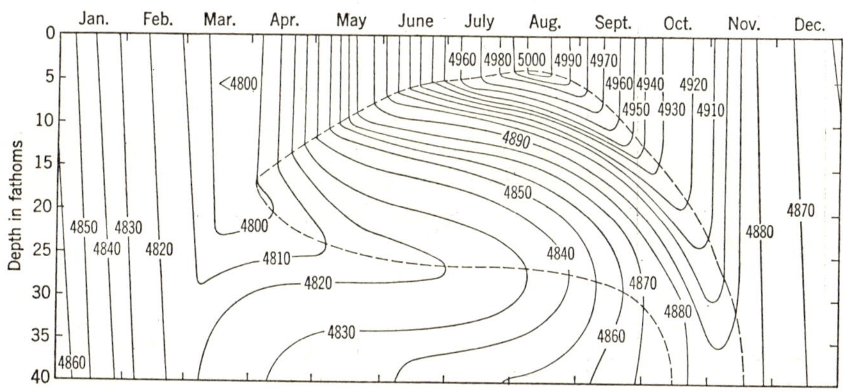

FIG. 3-4. Seasonal cycle of speed of sound in coastal waters off Montauk Point. Contour interval 10 ft/sec. (*From Woods Hole Oceanographic Institution, Natl. Defense Research Council*, 1941.)

materials and the variation of these properties as a function of depth into the bottom. In several areas the sound velocity through the bottom has been measured but this is generally not sufficient to determine its acoustic properties for underwater sound transmission. As illustration of this variability, a bottom-sediment chart for a portion of the continental shelf off San Francisco is shown in Fig. 3-5.

These three, depth and configuration of the bottom, sound velocity distribution in the ocean, and bottom and subbottom properties, are generally the most important environmental factors for the determination of underwater sound transmission. Their effects and those of other environment conditions will be discussed in this and later chapters.

For a ray solution to be valid the condition (2-22) must be satisfied. Strictly the condition (2-22) is not satisfied at any frequency for the discontinuous velocity change at the ocean surface and ocean bottom. We

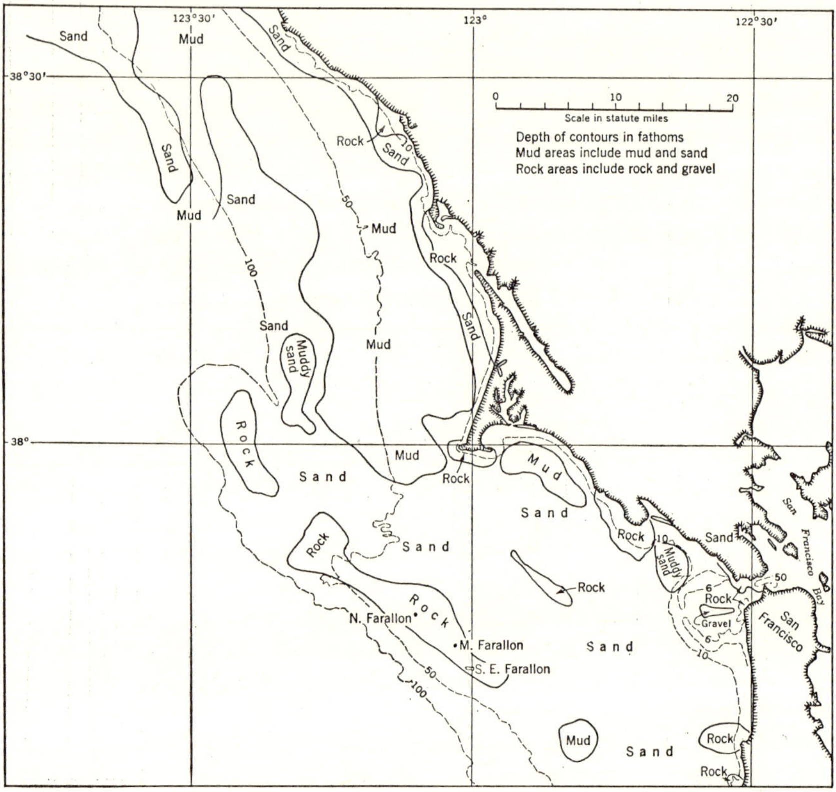

FIG. 3-5. (*From F. P. Shepherd, Submarine Geology, Harper & Brothers, New York,* 1948.)

may, however, treat this as a separate problem, as we have done in Chap. 2, and add its solution for the boundary effect to the more usual ray solution up to and away from the boundary. We, then, need to be concerned only with the sound velocity changes in the ocean. For the condition on the space rate of change of amplitude the change for sound of wavelength of the order of the depth of water will not be small over the distance of a wavelength. For smaller wavelengths, although each ray path either directly through the ocean or via some combination of bottom and surface reflections may satisfy this condition individually, their combined interference may not. At some still shorter wavelength individual arrivals will be separated and appear as distinct pulses corresponding to the possible ray paths from source to receiver; these wavelengths are within the range of validity of the ray solution. A safe limit for the validity of the ray solution is $\lambda = \frac{1}{10}h$ where h is the depth of water. (This limit is also dependent on the range.) For an average shallow-water depth of 50 feet the ray solution will be valid for frequencies above $f = 5{,}000 \times 10/50 = 1{,}000$ cycles per sec.

We shall learn later that a normal-mode solution can be interpreted and computed conveniently for wavelengths of the order of half the depth of water or greater. For shorter wavelengths the complication of additional terms, but not a sufficient number so that an integral approximation can be given for the series, generally makes the interpretation obscure. For the same water depth of 50 feet the normal-mode solution can be evaluated for frequencies below $f = 5{,}000 \times 2/50 = 200$ cycles per sec. In this frequency region the multiply reflected arrivals from the ocean surface and bottom do not arrive as discrete pulses but interfere to produce a dispersive wave train, their characteristics being described by the normal-mode solution.

The order of procedure in this chapter will be to give a discussion of shallow-water transmission in terms of rays, first. Second, the reflection of a pulse from a boundary across which there is a change in acoustic properties is given; there is a change in shape of the reflected pulse with respect to the incident pulse when the phase change upon reflection is not zero. In Chap. 2, the reflection for a simple harmonic source was given; here the more general solution for a pulse is given. Third, the interference effect of a source with its acoustic image from the ocean surface is given; this illustrates for a very simple case the manner in which a ray solution leads over to a normal-mode solution. Fourth, the interference effects developed for plane-wave incidence normal to a single boundary and to two boundaries are discussed; this section is included as an elementary example of some of the features encountered in normal-mode propagation. Finally, fifth, the general normal-mode solution for shallow-water transmission is derived, and the physical characteristics of its solution are discussed.

Ray solution

The ray solution is fairly simple and straightforward. The principles developed in the preceding chapter are applied to the particular sound velocity structure in the ocean in order to determine the ray paths out from a given source location, the changes of intensity along the ray path, and the travel time.

When a ray is reflected from the bottom or surface of the ocean, account must be taken of the resultant changes in amplitude and phase and the change in ray path direction. It can be demonstrated that over the domain for which the ray solution is valid the change in amplitude and phase for the reflected ray will be given by the Rayleigh reflection coefficient derived in Chap. 2 for plane waves (2-154) and (2-161). The derivation of this property is moderately complex and will not be given here. Instead in a later chapter the general form of the reflection coefficient for spherical waves will be derived; and it will be shown to reduce to the Rayleigh reflection coefficient at a modest distance from the source corresponding to the domain for which the ray solution criterion that the fractional change in the space rate of change of the amplitude be small (2-19) becomes valid.

For the ocean the usually assumed lower boundary condition is that there is a discontinuous increase in velocity across the ocean bottom from a lower velocity in the ocean to a higher velocity in the sediment. The amplitude of the reflected ray in this case will increase from a finite value at normal incidence to a value of unity (i.e., perfect reflection) at the critical angle and remain unity out to grazing incidence (Fig. 2-29). Furthermore the most rapid rate of increase is immediately below the critical angle. The phase change is 0 deg from normal incidence out to the critical angle and from there increases in a regular manner to 180 deg at grazing incidence. The phase change is independent of frequency. The usually assumed upper boundary condition is that there is a discontinuous change to a massless medium, vacuum. This is generally a valid assumption because of the large contrasts in acoustical properties, particularly density, between water and air. The resultant pressure on the upper surface must then be zero; otherwise there would be an infinite acceleration of the boundary. The right-hand side of Eq. (2-142) will be zero and from (2-150) and (2-154) the amplitude of the reflected wave will be equal to that of the incident wave and 180 deg out of phase with it. For all cases, both reflection from the ocean bottom and surface, the angle of the reflected ray is equal to the angle of the incident ray as measured across the normal to the reflecting surface.

It is important to note here that the phase angle discussed above refers to the velocity potential from the solution of Chap. 2. From (1-83) and (1-85) and (2-140) the pressure will be in the same phase as the velocity potential, but the component of particle velocity normal to the reflecting surface will be 180 deg out of phase. Thus, for the ocean surface the pressure of the incident and reflected waves will be 180 deg out of phase. The total pressure on the surface, being the sum of the two, will be zero. However, the vertical component of the particle velocity of the two waves will be in phase and the total vertical particle velocity will be a maximum. The condition of zero disturbance for any of the physically measurable parameters such as pressure, particle velocity, or displacement in a wave-propagation problem is called a *node* and the condition of maximum disturbance an *antinode*. The ocean surface will be a node for pressure and an antinode for the vertical component of particle velocity.

In this section we shall be concerned primarily with the changes of intensity and direction of the reflected rays. In a later section we shall consider the change in shape of a pulse reflected from bottom caused by the phase change at angles of incidence more grazing than the critical angle.

Let us consider first an isovelocity shallow-water area, i.e., one in which there is no change in sound velocity either vertically or laterally. Then, only the effects from the repeated reflections from the ocean surface and bottom need be considered. All ray paths will be straight lines which will be changed in direction upon reflection by an angle across the normal to the reflecting surface equal to the angle of incidence. The solution then resolves itself simply into accounting for the effects of all the rays from a given source to a given receiver. From expression (1-39) the wave surfaces emanating from a point source will be spherical with radius equal to the total ray path distance and, from the definition of wave surfaces, normal to the rays. From (1-39) and (1-78) the intensity along any ray will decrease inversely as the square of the total ray path distance from the source. The time of arrival at the receiver is given here simply by the ray path distance divided by the velocity of sound in water.

The sequence of ray paths is illustrated in Fig. 3-6 for a source and receiver at the surface. First to arrive is the direct ray D. Next is the ray which has undergone one bottom reflection, followed successively by the rays which have undergone two, three, four, etc., bottom reflections. The times of arrival will then be, from Fig. 3-6,

$$
\begin{aligned}
T_D &= \frac{X}{c} = \frac{[X^2 + (0Z)^2]^{1/2}}{c} \\
T_{B1} &= 2\,\frac{[(X/2)^2 + Z^2]^{1/2}}{c} = \frac{[X^2 + (2Z)^2]^{1/2}}{c} \\
T_{B2} &= \frac{[X^2 + (4Z)^2]^{1/2}}{c} \\
T_{B3} &= \frac{[X^2 + (6Z)^2]^{1/2}}{c} \\
T_{B4} &= \frac{[X^2 + (8Z)^2]^{1/2}}{c} \\
&\cdot \\
&\cdot \\
&\cdot \\
T_{Bn} &= \frac{[X^2 + (2nZ)^2]^{1/2}}{c}
\end{aligned}
\qquad (3\text{-}1)
$$

Each successive higher-order bottom reflection arrives at a later time. (The expression *order of reflection* is used here to designate the number of reflections a given ray takes from a particular reflecting horizon in traveling from source to receiver; a ray which has undergone one reflection from the bottom ($B1$ of Fig. 3-6) is referred to as a first-order bottom reflection, one that has undergone two reflections from the bottom ($B2$ of Fig. 3-6) as a second-order bottom reflection, etc.) Further, from Eq.

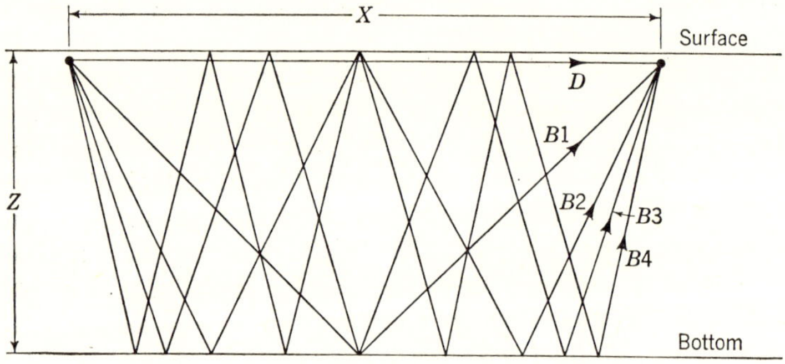

FIG. 3-6.

(3-1) it can be seen that the time difference between successive orders increases with increasing order.

The ray intensity will decrease as the square of the numerator of Eq. (3-1); it will be lower the higher the order. There will be in addition a decrease in intensity for those rays which are reflected from the bottom at angles steeper than the critical angle. There will be no bottom-reflection loss for those rays which are reflected from the bottom at angles of incidence more grazing than the critical angle, but the loss will increase at an increasing rate once the order has been reached at which the rays

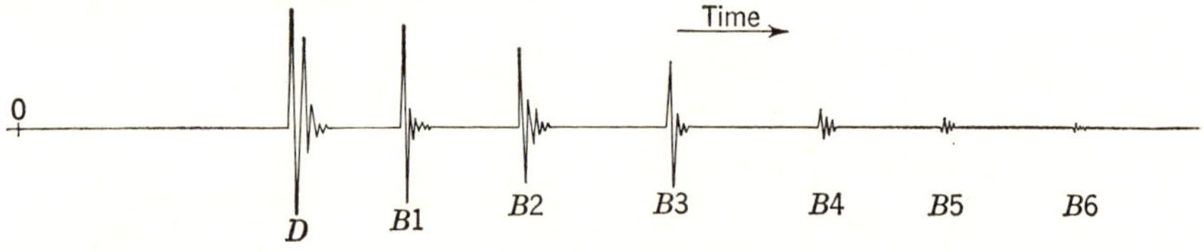

FIG. 3-7.

are incident on the bottom at angles steeper than critical because (1) each higher order has to undergo one more bottom reflection with consequent loss on that reflection and (2) the loss per reflection increases with decreasing angle of incidence, i.e., increasing order of reflection. Diagrammatically the sequence of arrivals will be as shown in Fig. 3-7 where this figure represents a record of pressure, displacement, or particle velocity amplitude as a function of time at a fixed receiver location.

The equations for the times of arrival as a function of distance [Eqs. (3-1)] may be rewritten as

$$
\begin{aligned}
T_D - \frac{X}{c} &= 0 \\
T_{B1}^2 - \frac{X^2}{c^2} &= \left(\frac{2Z}{c}\right)^2 \\
T_{B2}^2 - \frac{X^2}{c^2} &= \left(\frac{4Z}{c}\right)^2 \\
T_{B3}^2 - \frac{X^2}{c^2} &= \left(\frac{6Z}{c}\right)^2 \\
&\cdot \\
&\cdot \\
&\cdot \\
T_{Bn}^2 - \frac{X^2}{c^2} &= \left(\frac{2nZ}{c}\right)^2
\end{aligned}
\tag{3-2}
$$

Regarding T_{Bn} as a function of X, they are the equations of hyperbolas. Their intercepts at $X = 0$ are successively the round-trip time for a vertical reflection from the bottom and twice, three times, etc., that time, as can also be seen from Fig. 3-6. The equation for T_D as a function of X is a straight line whose inverse slope is the velocity of sound in water. At great range the equations for T_{Bn} approach the line for T_D. Figure 3-8 is a plot of T versus X. Equations (3-2) considered as a function of

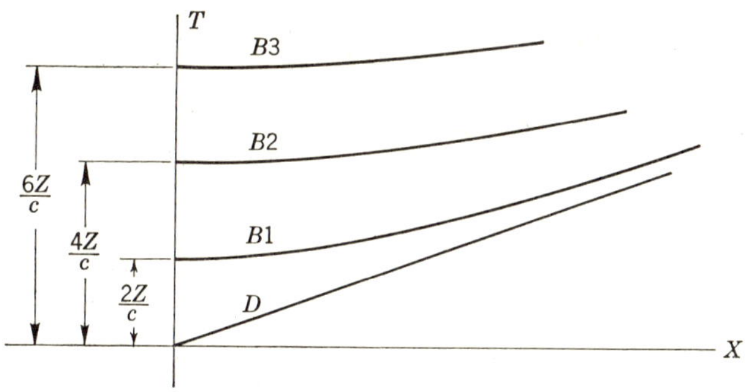

FIG. 3-8.

T^2 versus X^2 are parallel straight lines whose inverse slopes are equal to the velocity of sound in water squared. Figure 3-9 is a plot of T^2 versus X^2.

Let us consider for a moment a shallow-water channel bounded both top and bottom by reflectors whose reflectivity is unity for all angles of incidence. Then we know that the energy will be allowed to spread out in one direction, x, only so that the total received energy from all ray paths will have to decrease on the average in inverse proportion to the distance, x, i.e., cylindrical spreading. This is contrasted with the variation of intensity along any given ray, which as we have said before will decrease in inverse proportion to the distance squared, i.e., spherical spreading. For our case in which the bottom has a critical angle the total received energy near the source will decrease as x^{-2} since D will have the shortest path and be the predominant signal and all the bottom reflections will suffer a loss by refraction of energy into the bottom upon each reflection. With increasing range more and more bottom-reflection paths will be reflected from the bottom at angles less grazing than critical. The total received energy will approach the condition of decreasing proportionally to x^{-1}. Schematically these relations are illustrated in Fig. 3-10.

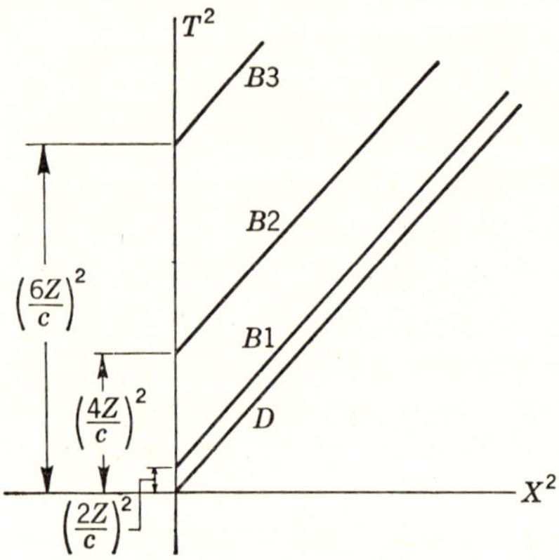

Fig. 3-9.

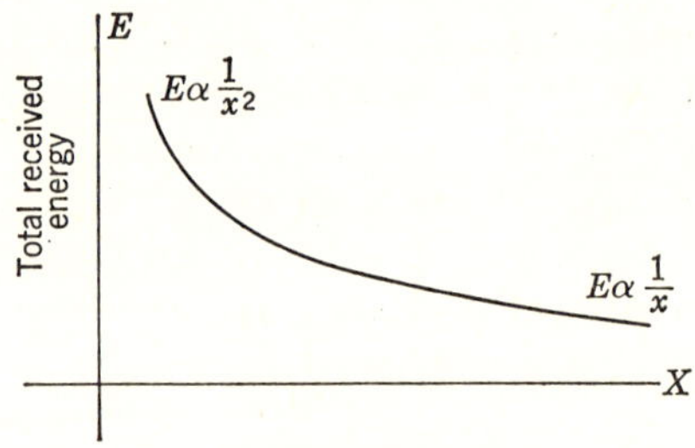

Fig. 3-10.

In Fig. 3-6 we considered the source and receiver at the surface. In general they will both be at some depth below the surface. In this case instead of having only one arrival which has undergone one bottom reflection, there will be four. The same will be true for all orders of bottom reflection. Each order of bottom reflection will consist of four arrivals. If Bn is the nth bottom-reflection order, all four arrivals will have undergone n bottom reflections; two will have undergone n surface reflections, one $n - 1$ surface reflections, and one $n + 1$ surface reflections. The two arrivals which have taken n surface reflections will not arrive at the same time unless the source and receiver depths are equal and the bottom horizontal. In addition there will be a surface-reflected arrival, S, associated with D which has undergone no bottom reflections. In the system outlined this may be considered as the zero-order bottom-reflection arrival which has undergone $0 + 1$ surface reflections. The direct and surface-reflected arrivals and the four sets of arrivals associated with the first- and second-order bottom reflections are shown in

Fig. 3-11. Subscripts have been used to distinguish the four arrivals, subscript 1 for the arrival which has taken $n - 1$ surface reflections, subscript 2 for the arrival which has taken n surface reflections and is outgoing from the source in the same direction as 1, subscript 3 for the other arrival which has taken n surface reflections and is outgoing from the source in the same direction as 4, and subscript 4 for the arrival which has taken $n + 1$ surface reflections. The arrival sequence will now be as shown in Fig. 3-12. An equally satisfactory system could be obtained from referring the arrivals to a constant order of surface reflection. The

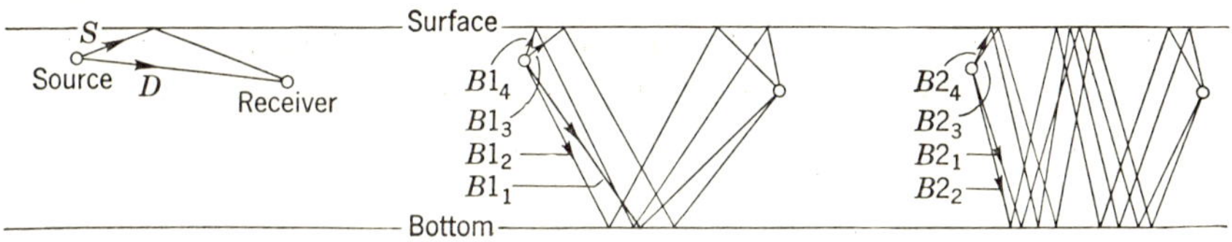

FIG. 3-11.

system chosen generally depends on the relative distances of the source and receiver from the ocean surface and bottom. Usually the one discussed here of referring to a constant order of bottom reflection is more convenient.

To take a still more general solution we should consider the variations of sound velocity with depth in the ocean. As has been discussed, the velocity of sound in a shallow-water environment can be and usually is quite variable both geographically and temporally. No particular sound velocity profile can be taken as typical. We shall discuss transmission under four conditions which include in a general way the properties that

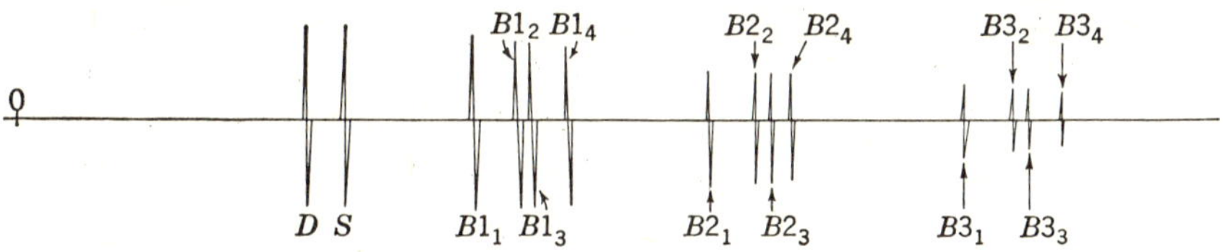

FIG. 3-12.

might be expected for any particular sound velocity profile found in nature. These are a negative velocity gradient from the surface to the bottom, a positive gradient, a positive followed by a negative gradient, and a negative followed by a positive gradient.

For the condition of a negative velocity gradient from the surface to the bottom there will be downward refraction of the rays. This produces a limited range of reception for each order and for each arrival in each order, as contrasted with a theoretically infinite range of reception for each arrival for the isovelocity condition of Fig. 3-11. Another effect is that a portion of the energy initially upgoing from the source is turned down without undergoing a surface reflection. The ray diagrams corresponding to the D, S, and the $B1$ arrivals are given in Fig. 3-13. Depending on the particular sound velocity structure there may be one or more zones of triple-valued functions, as discussed in Chap. 2, so that, for instance, instead of receiving just D and S three arrivals may be received. The time of arrival must now be given by the line integral $\int ds/c(z)$ along the ray path, and the intensity calculated by similar relations from Chap. 2. Often where an explicit ray solution for arrival time and intensity of the rays as a function of depth and range cannot be obtained, which is usually the case, it is convenient to approximate the actual velocity profile by a number of constant gradients and use the appropriate equations of Chap. 2 for a constant gradient as an accumulative solution for numerical calculations.

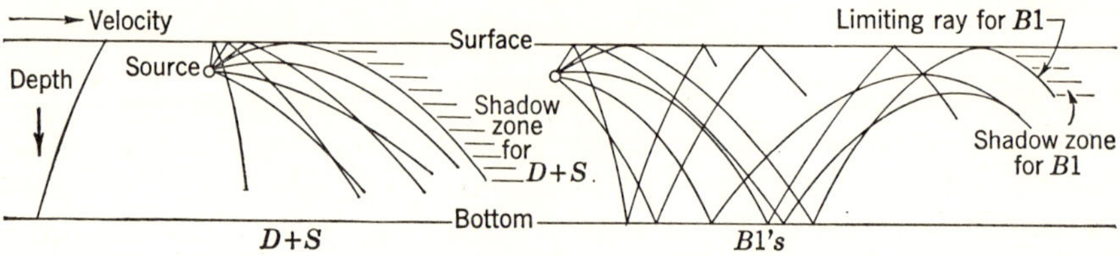

Fig. 3-13.

The condition of a limited range for any given ray type is a characteristic of the ray solution for any problem in which the transmitting medium is bounded top and bottom and in which the sound velocity is not constant. This should be obvious; it is due to the existence of the boundaries themselves. It refers only to a given ray-path type, such as the rays which have taken only one bottom reflection, and does not refer to the accumulative effect from all ray paths. For instance, as one follows a particular arrival from one recording station to the next with increasing range from the source, that arrival will eventually disappear and no longer be present on recordings at greater ranges. The region in which there is no reception for a particular type of ray is referred to as the *shadow zone* for that type. The boundary between the regions of reception and no reception is called the *shadow boundary.* If the shadow boundary is defined by a single ray path, such as in Fig. 3-13, this ray is called the *limiting ray.* If the shadow boundary is defined by the envelope of a sequence of rays, this envelope is a *caustic* and is referred to as such. At the other end of the range scale there is a limit defined by critical angle reflection from the bottom. At ranges less than the range for the critical angle reflection there will be a decrease in the intensity of the received signal due to the loss of energy by refraction into the bottom. Thus, a limited range interval for *best reception* by a particular order of bottom reflection is delimited. At the long-range end it is defined by the limiting ray beyond which no ray paths of the given type can be followed; at the short-range end it is defined by the *critical angle ray* beyond which there is perfect reflectivity and inside which there is lossy reflectivity from the bottom. The definition of this range interval at the lower end becomes better with increasing order of reflection. The appropriate range intervals are shown in Fig. 3-14 for a source and receiver at the surface and a horizontal bottom and in Fig. 3-15 for a sloping bottom. For Fig. 3-14

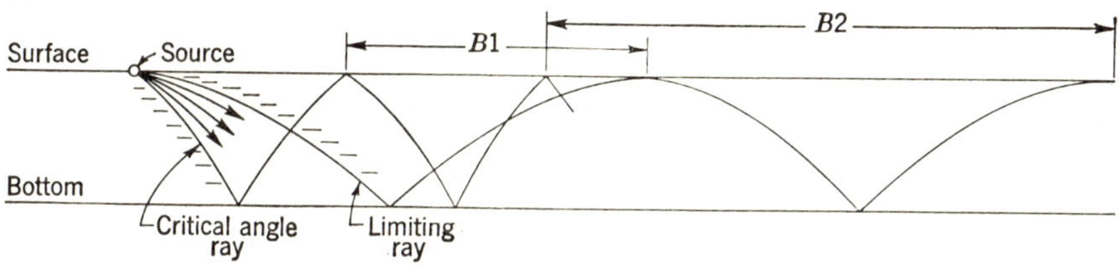

FIG. 3-14.

the rays are repetitive. The range interval for $B2$ is twice that for $B1$; the interval for $B3$ would be three times $B1$ and in general the interval for a given order n would be n times the range interval for $B1$. The same is true for the beginning and ending ranges of the zone of best reception. For the upward-sloping bottom shown in Fig. 3-15 the intervals will be less than those for the horizontal bottom. The concept of the zone of best reception for each order of bottom reflection is a useful theoretical tool as a first approximation to the transmission in a complex physical

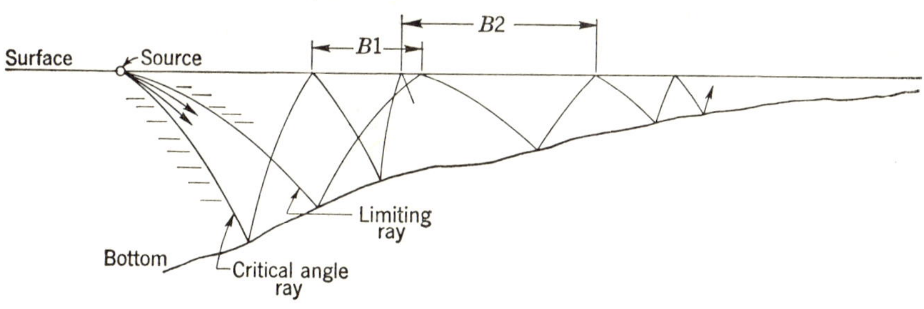

FIG. 3-15.

problem of a variable sound velocity profile and changing bottom topography. Often it is helpful to plot the range interval of best reception for the various orders versus the range; such a graph is called a *transmission diagram*. An example is shown in Fig. 3-16. More will be said about them in Chap. 4.

For the condition of a positive velocity gradient from the surface to the bottom there will be upward refraction of the rays. Again a limiting ray or caustic will be defined. Here it will be a ray which is grazing or nearly grazing to the bottom depending on the actual velocity conditions. Because of the upward bending of the rays a larger portion of the energy will be reflected from the bottom at angles of incidence beyond critical than in the isovelocity or negative gradient cases; there will be a smaller loss of energy out of the shallow-water channel, and this case should show in general better transmission characteristics. Some of the rays will be returned upward without a bottom reflection. They are continuous with their appropriate order of bottom reflection, the whole sequence being considered as one family which has taken n bottom reflections or inversions at depth or combination of bottom reflections and inversions. The abbreviation RSR, standing for refracted surface reflected, is used to designate that portion of the family which has been returned by an inversion at depth rather than a bottom reflection. In this case the first order grades continuously into the direct, and no distinction is made between the two. As we shall see in Chap. 5, there are mechanisms other than the primary one discussed here related to the critical angle for the loss of energy upon bottom reflection so that none of the B arrivals may be as good as their associated RSR arrivals, again indicating better transmission characteristics for this case than those for the isovelocity or negative gradient cases. The limiting ray for each order carries that order out to longer ranges than for the corresponding negative-gradient case. This is particularly noticeable for a shallow source and shallow receiver; compare

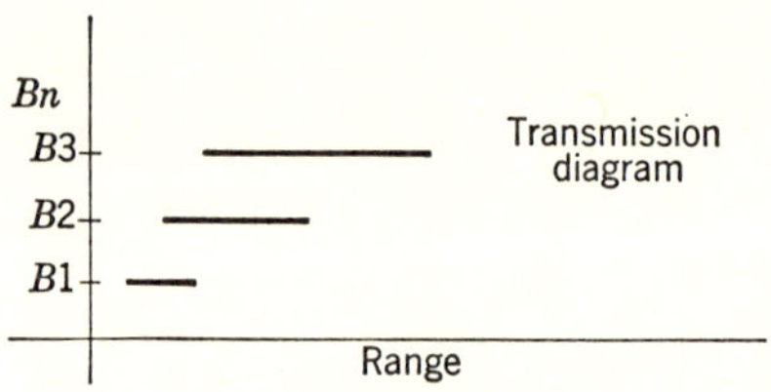

FIG. 3-16.

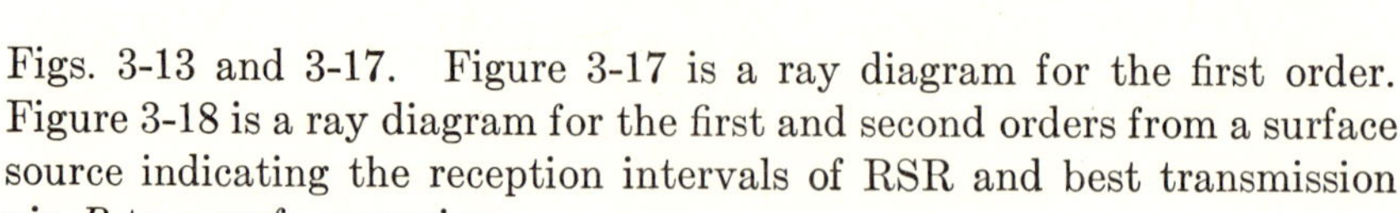

FIG. 3-17.

Figs. 3-13 and 3-17. Figure 3-17 is a ray diagram for the first order. Figure 3-18 is a ray diagram for the first and second orders from a surface source indicating the reception intervals of RSR and best transmission via B to a surface receiver.

For the condition of a positive followed by a negative velocity gradient there will be upward refraction of the rays in the upper portion of the water, positive gradient, and downward refraction in the lower portion, negative gradient. An interesting effect is introduced here. A ray which becomes horizontal at the depth of maximum velocity can be refracted upward or downward; there is a divergence of energy away from this depth. Such a ray is referred to as a *split-beam ray*. In this case the split-beam ray is also the limiting ray, as shown in Fig. 3-19. As before there will be RSR rays; only here they will be confined to the

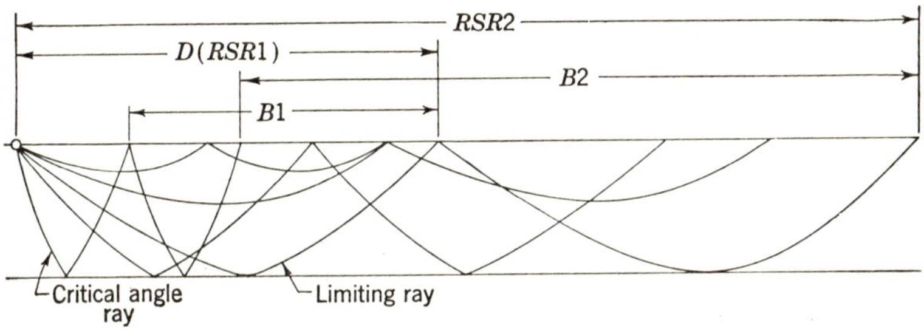

FIG. 3-18.

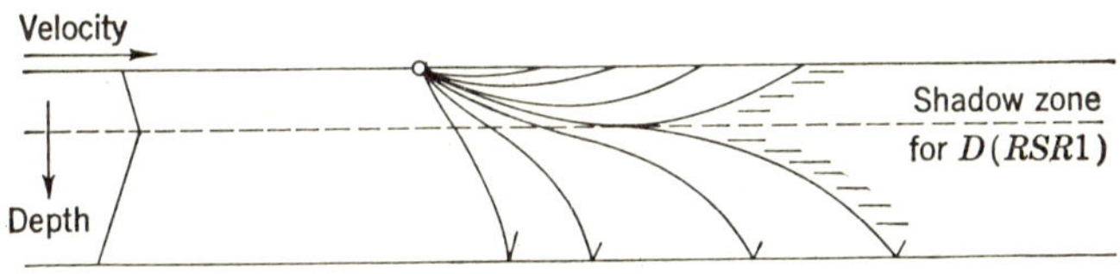

FIG. 3-19.

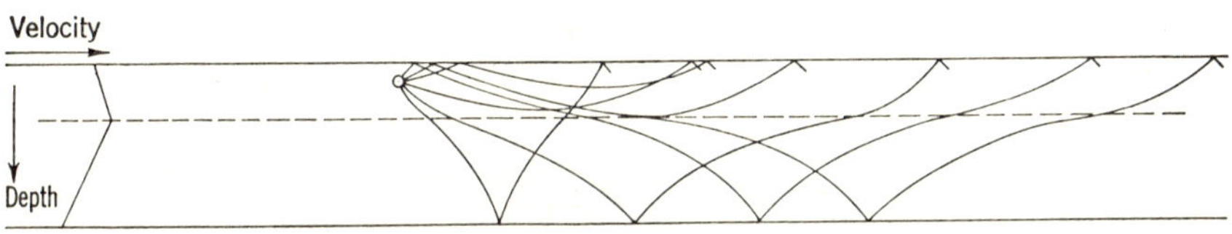

FIG. 3-20.

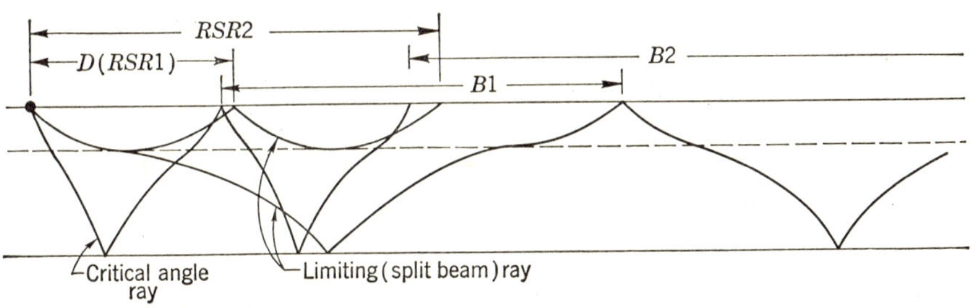

FIG. 3-21.

upper portion of the water, that of the positive gradient, and will not form a continuous group with their associated bottom reflections. Figure 3-20 is a ray diagram for the RSR1 (*D* and *S*) and *B*1 rays. Figure 3-21 is a diagram for the range intervals of reception for *D* and RSR2 and the range interval of best reception for *B*1 and *B*2 from a surface source to a surface receiver.

For the condition of a negative followed by a positive velocity gradient there will be downward refraction of the rays in the upper portion of the water, negative gradient, and upward refraction in the lower portion, positive gradient. Here there will be a convergence of energy toward the depth of minimum velocity. A new type of ray is introduced here that is reflected from neither the bottom nor the surface; it is refracted upward before reaching the bottom and is refracted downward before reaching the surface. This type of ray, which is inverted both upward and downward before reaching the surface or bottom, will be referred to as a SOFAR ray from its original discovery and proposed applications in deep-water transmission for sound fixing and ranging. The existence and extent of the SOFAR transmission depends on the depths of the source and receiver. If the source is located at a depth for which the sound velocity is greater than that at either the upper or lower boundary, there

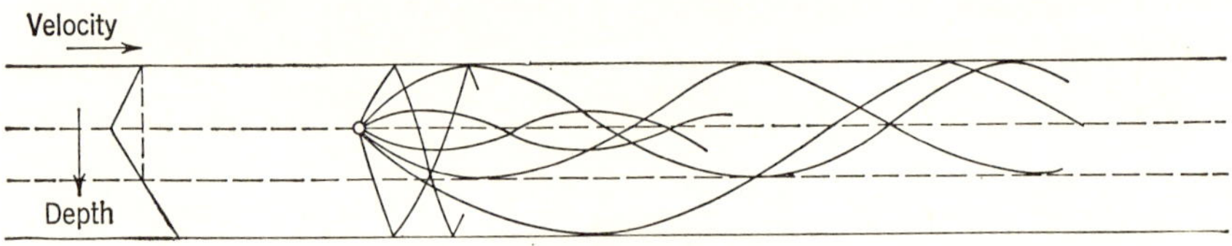

FIG. 3-22.

will be no SOFAR rays. This conclusion follows directly from the discussion of Chap. 2. From Eq. (2-51) we have that

$$\frac{\sin \theta_0}{c_0} = \frac{\sin \theta}{c} \tag{2-51}$$

Thus, for a ray which is horizontal at the source, $\theta_0 = 0$, the angle will not become zero again following its path toward the reflecting surface if $c_0 > c_b$ where c_b is the velocity at the boundary. All other rays from the source, $\theta_0 \neq 0$, will be steeper along their paths than this ray. If the velocity at the boundary is greater than that at the source, the horizontal ray from the source will become horizontal again at the depth where $c = c_0$ and be refracted away from the boundary without undergoing a reflection. The cone of rays defined by angles θ_0' and 90 deg + θ_0' where $\sin \theta_0' = c_0/c_b$ will all be refracted away from the boundary without a reflection. In this case the B, RSR, and SOFAR rays all form a continuous group for each order of bottom reflection or inversion at depth. Figure 3-22 is a ray diagram for a source at the depth of minimum velocity. Figure 3-23 is a ray diagram of the first order for each type from a shallower source. Figure 3-24 is a ray diagram of the range intervals of reception of the first and second orders of RSR and the range interval of best reception for $B1$ and $B2$ from a surface source to a surface receiver; in this case there will be no SOFAR rays.

The velocity structure of a negative followed by a positive velocity gradient is the same as the last one discussed in the general ray section of Chap. 2. We expect then to observe a shadow zone followed by a focus peak and a zone of moderately high intensity for the RSR and SOFAR sequence of each order. Immediately following the focus peak the RSR sequence is double-valued.

All the cases considered here consist of guided or channel propagation. The energy is constrained to spread laterally in two dimensions in contrast to the unguided case of spreading outward in all (three) space dimensions. The various orders of *B* which are reflected from the bottom at angles more grazing than critical, the RSR, and the SOFAR rays all give guided transmission. The *B* rays beyond the critical angle are con-

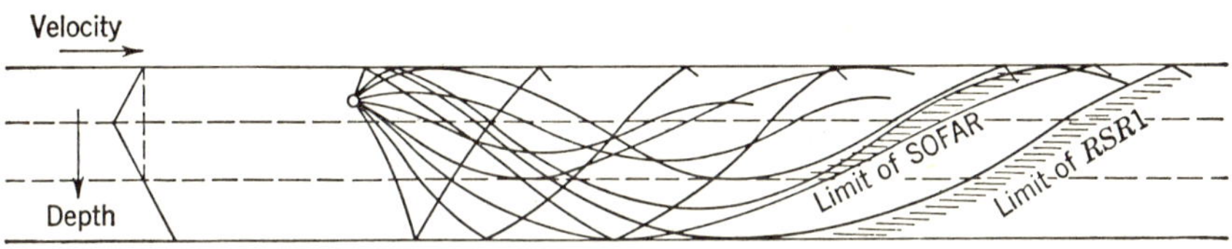

FIG. 3-23.

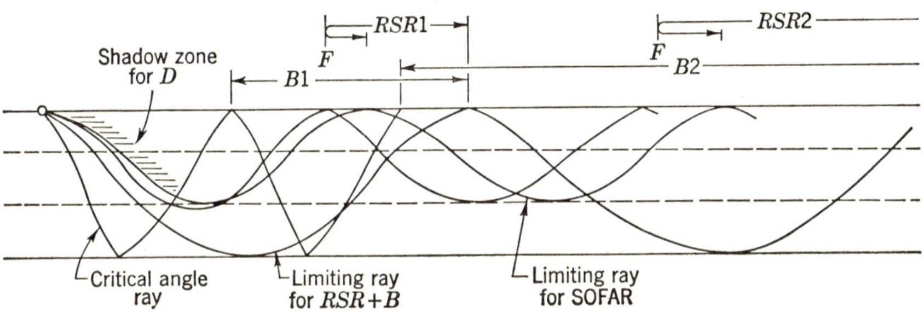

FIG. 3-24.

strained to spread in two dimensions by reflection from the surface and bottom without loss upon either reflection. The RSR rays are confined by a surface reflection and an inversion at depth without reflection. The SOFAR rays are confined by an inversion upward at depth and an inversion downward at elevation from the mean position. The *B* type will exist for any sound velocity structure in the ocean. Both the RSR and SOFAR types are dependent on the velocity structure and the depths of the source and receiver. A positive velocity gradient must exist for RSR, and a negative followed by a positive gradient must exist for SOFAR.

Reflection of a pulse

The problem that we shall next consider is the reflection of a pulse from a boundary across which there is a change in acoustical properties. In

Chap. 2 the solution was given for the reflection of a simple harmonic plane wave. The results were that for reflection from a higher to lower velocity medium at all angles of incidence and for reflection from a lower to higher velocity medium at angles of incidence steeper than the critical angle the ratio of the amplitude of the reflected to incident wave was

$$\frac{B}{A} = \frac{\dfrac{\rho_2}{\rho_1} - \dfrac{\sqrt{(c_1^2/c_2^2) - \sin^2\theta_1}}{\sqrt{1 - \sin^2\theta_1}}}{\dfrac{\rho_2}{\rho_1} + \dfrac{\sqrt{(c_1^2/c_2^2) - \sin^2\theta_1}}{\sqrt{1 - \sin^2\theta_1}}} \tag{2-154}$$

with no phase change. However, for reflection from a lower to higher velocity medium at angles of incidence more grazing than the critical angle the amplitude ratio was unity, and there was a change in phase of the reflected wave with respect to the incident wave which increased from 0 deg at the critical angle to 180 deg at grazing incidence. For this case,

$$\frac{B}{A} = e^{i\epsilon} \tag{2-161}$$

where

$$\begin{aligned} \frac{\epsilon}{2} &= \tan^{-1}\left[\frac{\rho_1\sqrt{\sin^2\theta_1 - (c_1^2/c_2^2)}}{\rho_2\sqrt{1 - \sin^2\theta_1}}\right] \\ &= \tan^{-1}\left[\frac{\rho_1\sqrt{\sin^2\theta_1 - (c_1^2/c_2^2)}}{\rho_2\cos\theta_1}\right] \\ &= \tan^{-1}\left[\frac{\rho_1}{\rho_2}\sqrt{\tan^2\theta_1 - \frac{c_1^2}{c_2^2}\sec^2\theta_1}\right] \end{aligned} \tag{2-162}$$

It might be supposed at first glance that having obtained all the features for the reflection of a simple harmonic wave nothing new would result from consideration of the reflection of a pulse. Such, however, is not the case. A pulse incident on a boundary at angles more grazing than critical will be reflected with a change in shape.

Physically we can obtain an understanding of this phenomenon through the following reasoning. We may consider any pulse as consisting of the summation of a continuous spectrum of frequencies with specified amplitudes and phases. This we express mathematically through the Fourier integral theorem, (2-110). The condition (2-161) states that the spectrum is shifted by a constant phase angle ϵ independent of frequency. Let us represent the original pulse by a Fourier integral of the form

$$\int_0^\infty g(\omega)e^{i\omega t}\,d\omega \tag{3-3}$$

or

$$\int_0^\infty \gamma(\omega)e^{i[\omega t + \varphi(\omega)]}\,d\omega \tag{3-4}$$

where $\gamma(\omega)$ and $\varphi(\omega)$ are the amplitude and argument of $g(\omega)$. The reflected pulse will then be given by an integral of the form

$$\int_0^\infty \gamma(\omega)e^{i[\omega t+\varphi(\omega)+\epsilon]}\,d\omega \tag{3-5}$$

The integral (3-5) is not equal to a constant multiplied by integral (3-4). Let us consider the manner in which the change of shape occurs. For a low frequency and one somewhat higher, a graph of $\gamma(\omega)e^{i[\omega t+\varphi(\omega)]}$

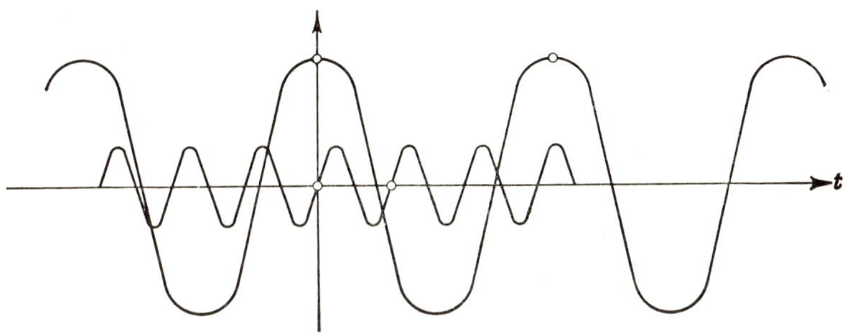

FIG. 3-25.

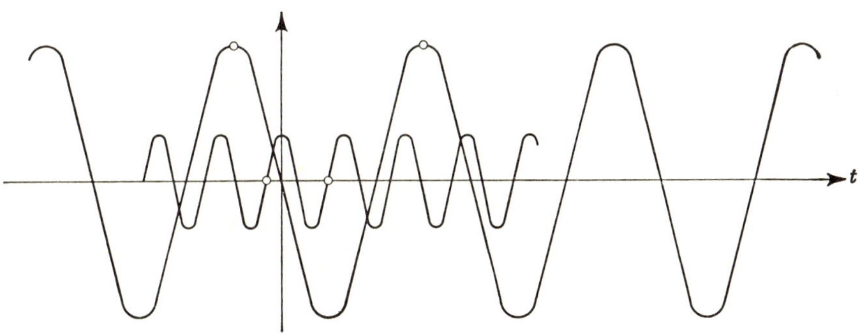

FIG. 3-26.

versus t might be as shown above in Fig. 3-25. If the phase change is 45 deg, $\gamma(\omega)e^{i[\omega t+\varphi(\omega)+\epsilon]}$ will be that shown in Fig. 3-26 for the same frequencies. The two frequencies have been shifted a different relative amount with respect to the time axis. Their summation will not be the same as that for Fig. 3-25. By the same reasoning the integral (3-5) will result in a pulse shape different from (3-4). We see that in order to reproduce the same pulse it is necessary that the relative phase change for each frequency be such that the shift along the time axis is the same for all frequencies. From the definition of the period τ and from Fig. 3-25 or 3-26 the time shift for a phase change ϵ is

$$\Delta t = \epsilon\tau \tag{3-6}$$

or from (2-93) and (2-97)

$$\Delta t = \frac{2\pi\epsilon}{\omega} \tag{3-7}$$

For Δt of (3-7) to be a constant, ϵ must be proportional to ω,

$$\Delta t = \frac{2\pi a\omega}{\omega} = 2\pi a \tag{3-8}$$

where a is a constant and the phase change is

$$\epsilon = a\omega \tag{3-9}$$

In order to reproduce the original pulse shape the phase change must be proportional to the frequency. An analogy can be made with electric or acoustic filters; for the filter to reproduce the portion of the spectrum of interest without phase distortion the phase change through the filter must be proportional to frequency.

The general solution for the shape of the reflected pulse may be formulated through the use of the Fourier integral theorem. From (2-110) we have

$$f(t) = \frac{1}{2\pi}\int_{-\infty}^{\infty} d\omega \int_{-\infty}^{\infty} f(\tau)e^{i\omega(t-\tau)}\, d\tau \tag{2-110}$$

In all the problems considered here the function $f(\tau)$ will always be a real quantity so that we may simplify (2-110) to the following in terms of real variables

$$f(t) = \frac{1}{2\pi}\int_{-\infty}^{\infty} d\omega \int_{-\infty}^{\infty} f(\tau)\cos\,[\omega(t-\tau)]\, d\tau \tag{3-10}$$

or since the cosine is an even function of ω,

$$f(t) = \frac{1}{\pi}\int_{0}^{\infty} d\omega \int_{-\infty}^{\infty} f(\tau)\cos\,[\omega(t-\tau)]\, d\tau \tag{3-11}$$

Expanding the cosine term, we may rewrite (3-11) in another form useful for calculation,

$$f(t) = \frac{1}{\pi}\int_{0}^{\infty} p(\omega)\cos\,(\omega t)\, d\omega + \frac{1}{\pi}\int_{0}^{\infty} q(\omega)\sin\,(\omega t)\, d\omega \tag{3-12}$$

where from (3-11) and (3-12) $p(\omega)$ and $q(\omega)$ are defined by

$$p(\omega) \equiv \int_{-\infty}^{\infty} f(\tau)\cos\,(\omega\tau)\, d\tau$$

and (3-13)

$$q(\omega) \equiv \int_{-\infty}^{\infty} f(\tau)\sin\,(\omega\tau)\, d\tau$$

In general if $f(t)$ represents the source condition, it will be initiated at some particular time. This time we shall designate as time zero so that $f(t)$ has a value for t greater than zero and is zero for t less than zero. Under this condition Eqs. (3-13) reduce to

$$p(\omega) = \int_0^\infty f(\tau) \cos(\omega\tau)\, d\tau$$

and (3-14)

$$q(\omega) = \int_0^\infty f(\tau) \sin(\omega\tau)\, d\tau$$

If instead of a vibration, $f(t)$, we wished to consider a plane wave propagating outward in space of the form $f(t - lx + my)$, Eqs. (3-10), (3-11), and (3-12) would be of the same form, substituting the variable

$$T = t - lx + my$$

for t, where $f(t - lx + my)$ now expresses the motion following the ray path.

The simple harmonic solution for a plane wave reflected from a boundary at angles more grazing than the critical angle will then be from (2-164) simply

$$\Phi_{re} = \varphi_a e^{i[\omega(t-lx+my)+\epsilon]} \tag{3-15}$$

where

$$l = \frac{a_1}{\omega} = \frac{k_1 \cos\theta_1}{\omega} = \frac{\cos\theta_1}{c_1}$$

and (3-16)

$$m = \frac{b}{\omega} = \frac{k_1 \sin\theta_1}{\omega} = \frac{\sin\theta_1}{c_1}$$

and ϵ is the constant phase change upon reflection given by (2-162). Let us consider an incident pulse, $f(t + lx + my)$. If it is reflected from the boundary without change in phase or amplitude, it will be $f(t - lx + my)$. Then if $f(t - lx + my) = f(T)$ is defined for T greater than zero and is zero for T less than zero, the reflected pulse will be of the form (3-15) in (3-12)

$$y(T) = \frac{1}{\pi}\int_0^\infty p(\omega)\cos(\omega T + \epsilon)\, d\omega + \frac{1}{\pi}\int_0^\infty q(\omega)\sin(\omega T + \epsilon)\, d\omega \tag{3-17}$$

where the value of the phase, $T = 0$, corresponds to the arrival beginning that would exist along the reflected ray path for a disturbance at the source at $t = 0$ if there were no changes in wave shape upon reflection. We may be sure that this integral satisfies the conditions of our problems, for the integrand is of the form (3-15) of the simple harmonic solution and Eq. (3-17) reduces to Eq. (3-12), the initial pulse $f(T)$, for no phase change. Expanding the cosine and sine terms of (3-17), we have

$$\begin{aligned} y(T) = \cos\epsilon &\left[\frac{1}{\pi}\int_0^\infty p(\omega)\cos(\omega T)\, d\omega + \frac{1}{\pi}\int_0^\infty q(\omega)\sin(\omega T)\, d\omega\right] \\ &+ \sin\epsilon\left[\frac{1}{\pi}\int_0^\infty q(\omega)\cos(\omega T)\, d\omega + \frac{-1}{\pi}\int_0^\infty p(\omega)\sin(\omega T)\, d\omega\right] \end{aligned} \tag{3-18}$$

The first bracket is simply $f(T)$. The second bracket is of the same form as (3-12) and may be considered as defining a function described from (3-13) by the relations

$$q(\omega) = \int_{-\infty}^{\infty} F(\tau) \cos(\omega\tau)\, d\tau$$

and (3-19)

$$-p(\omega) = \int_{-\infty}^{\infty} F(\tau) \sin(\omega\tau)\, d\tau$$

Equation (3-18) may then be written as

$$y(T) = \cos\epsilon\, f(T) + \sin\epsilon\, F(T) \tag{3-20}$$

The reflected pulse is equal to the sum of two functions, the initial pulse diminished in amplitude by the factor $\cos\epsilon$ and a second pulse defined by the relations (3-19).

It is of interest to calculate the change in shape for a given initial pulse $f(T)$ as a function of the phase change ϵ. For illustration we shall take an exponential pulse

$$\begin{aligned} f(T) &= e^{-\lambda T} \qquad && \text{for } T \geqslant 0 \\ &= 0 && T < 0 \end{aligned} \tag{3-21}$$

where λ is the decay constant of the pulse. Equation (3-21) is a reasonably good representation for an underwater explosion. From (3-14) p and q will then be

$$p(\omega) = \int_0^{\infty} e^{-\lambda\tau} \cos(\omega\tau)\, d\tau = \frac{\lambda}{\lambda^2 + \omega^2}$$

and (3-22)

$$q(\omega) = \int_0^{\infty} e^{-\lambda\tau} \sin(\omega\tau)\, d\tau = \frac{\omega}{\lambda^2 + \omega^2}$$

From (3-18) we shall then have for F

$$\begin{aligned} F(T) &= \frac{1}{\pi}\left[\int_0^{\infty} \frac{\omega\cos(\omega T)}{\lambda^2+\omega^2}\, d\omega - \int_0^{\infty} \frac{\lambda \sin(\omega T)}{\lambda^2+\omega^2}\, d\omega\right] \\ &= \frac{1}{\pi}\int_0^{\infty} \frac{\omega\cos(\omega T) - \lambda\sin(\omega T)}{\lambda^2+\omega^2}\, d\omega \end{aligned} \tag{3-23}$$

The integral (3-23) is a form of the exponential integral,[1] a tabulated function. Equation (3-23) then reduces, in terms of the exponential integral, to

$$\begin{aligned} F(T) &= \frac{-1}{\pi} e^{-\lambda T}\overline{Ei}(\lambda T) \qquad && \text{for } T \geqslant 0 \\ &= \frac{-1}{\pi} e^{\lambda T} Ei(-\lambda T) && T \leqslant 0 \end{aligned} \tag{3-24}$$

[1] E. Jahnke and F. Emde, "Tables of Functions," pp. 1–5, Dover Publications, New York, 1945.

and the reflected pulse y can be written as

$$y(T) = \cos \epsilon \, e^{-\lambda T} - \sin \epsilon \, e^{-\lambda T} \frac{\overline{Ei}(\lambda T)}{\pi}$$

$$= \left[\cos \epsilon - \sin \epsilon \frac{\overline{Ei}(\lambda T)}{\pi} \right] e^{-\lambda T} \qquad \text{for } T \geqslant 0 \qquad (3\text{-}25)$$

and

$$y(T) = -\sin \epsilon \frac{Ei(-\lambda T)}{\pi} e^{\lambda T} \qquad \text{for } T < 0 \qquad (3\text{-}26)$$

The reflected pulse can now be calculated easily as a function of T. Figure 3-27 is a plot of y versus T for various values of the phase change. The reflected pulse is the same as the original exponential pulse at critical

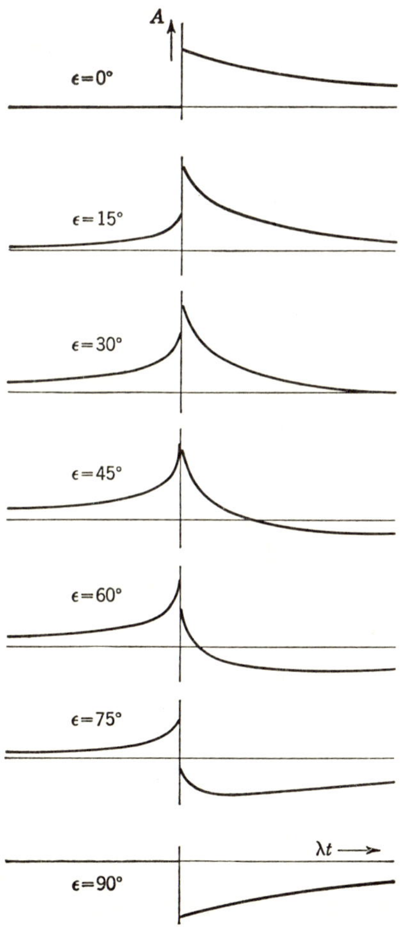

FIG. 3-27. (*From A. B. Arons and D. R. Yennie, J. Acoust. Soc. Amer., vol.* 22, 1950.)

angle reflection; here the phase change is zero. It gradually develops into a sharp spike about $T = 0$ with a forerunner for $T < 0$ and finally reduces to a negative exponential pulse at grazing incidence.

For reflection at angles of incidence for which there is no phase change and only a constant amplitude change it is immediately apparent from (3-11) or Figs. 3-25 and 3-26 that the original pulse shape is retained with a reduction in amplitude. Also from (3-11) or Figs. 3-25 and 3-26, it is apparent that for a phase change of $\pm\pi$ the resultant pulse is the negative of the original pulse.

There are one or two moderately subtle points connected with the formulation presented which are worth mentioning briefly. If we had used the Fourier integral formulation in the form (2-110) and imposed a phase change ϵ constant over all values of ω, we should have obtained

$$\begin{aligned} y(T) &= \frac{1}{2\pi} \int_{-\infty}^{\infty} d\omega \int_{-\infty}^{\infty} f(\tau) e^{i\omega(T-\tau)} e^{i\epsilon}\, d\tau \\ &= \frac{e^{i\epsilon}}{\pi} \int_{-\infty}^{\infty} d\omega \int_{-\infty}^{\infty} f(\tau) e^{i\omega(T-\tau)}\, d\tau = e^{i\epsilon} f(T) \end{aligned}$$

The same result would have been obtained, of course, if we had used the real-value expression of (2-110), (3-10),

$$\begin{aligned} y(T) &= \frac{1}{2\pi} \int_{-\infty}^{\infty} d\omega \int_{-\infty}^{\infty} f(\tau) \cos\,[\omega(T-\tau)+\epsilon]\, d\tau \\ &= \frac{\cos\epsilon}{2\pi} \int_{-\infty}^{\infty} d\omega \int_{-\infty}^{\infty} f(\tau) \cos\,[\omega(T-\tau)]\, d\tau \\ &\qquad - \frac{\sin\epsilon}{2\pi} \int_{-\infty}^{\infty} d\omega \int_{-\infty}^{\infty} f(\tau) \sin\,[\omega(T-\tau)]\, d\tau \\ &= \frac{\cos\epsilon}{2\pi} \int_{0}^{\infty} d\omega \int_{-\infty}^{\infty} f(\tau) \cos\,[\omega(T-\tau)]\, d\tau = \cos\epsilon f(T) \end{aligned}$$

since the integrand of the second double integral is an odd function of ω so that an integral $\int_{-\infty}^{\infty} d\omega$ will be zero. However, if we look back to our problem to find what the value of the phase change will be for negative values of ω, we see that we must choose the positive sign for the imaginary radical so that Eq. (2-166) will still be a wave which diminishes in amplitude toward infinity into the lower medium, Eq. (2-165) becoming for $\omega = -w$, w being a positive number,

$$a_1 = k_1 \sqrt{\frac{c_1^2}{c_2^2} - \sin^2\theta_1} = i\frac{\omega}{c_1}\sqrt{\sin^2\theta_1 - \frac{c_1^2}{c_2^2}} = -i\frac{w}{c_1}\sqrt{\sin^2\theta_1 - \frac{c_1^2}{c_2^2}}$$

The phase change, (2-162), then becomes the negative of what it was for positive ω.

$$\begin{aligned} \text{Phase change upon reflection} &= \epsilon \qquad \text{for } \omega > 0 \\ &= -\epsilon \qquad \omega < 0 \end{aligned}$$

Under these conditions the integral (3-10) becomes

$$y(T) = \frac{\cos \epsilon}{2\pi} \int_{-\infty}^{\infty} d\omega \int_{-\infty}^{\infty} f(\tau) \cos [\omega(T - \tau)]\, d\tau$$
$$- \frac{\sin \epsilon}{2\pi} \int_{0}^{\infty} d\omega \int_{-\infty}^{\infty} f(\tau) \sin [\omega(T - \tau)]\, d\tau$$
$$+ \frac{\sin \epsilon}{2\pi} \int_{-\infty}^{0} d\omega \int_{-\infty}^{\infty} f(\tau) \sin [\omega(T - \tau)]\, d\tau$$
$$= \frac{\cos \epsilon}{2\pi} \int_{0}^{\infty} d\omega \int_{-\infty}^{\infty} f(\tau) \cos [\omega(T - \tau)]\, d\tau$$
$$- \frac{\sin \epsilon}{2\pi} \int_{0}^{\infty} d\omega \int_{-\infty}^{\infty} f(\tau) \sin [\omega(T - \tau)]\, d\tau$$

which is the same as (3-18) upon substitution of the relations (3-13) for $f(\tau)$.

A second point is that the quantity $F(T)$ obtained from the integration (3-24), and consequently $y(T)$, will in general have a value for T less than zero. We can check that this is indeed the correct solution by applying the law of conservation of energy. From (1-71), (1-77), and (1-85) the total energy of the incident pulse will be proportional to $\int_{-\infty}^{\infty} f^2(T)\, dT$ and a similar expression for the reflected pulse. Since there is total reflection at angles of incidence more grazing than the critical angle and no energy lost into the second medium, the energy of the reflected wave must be equal to that of the incident wave

$$\int_{-\infty}^{\infty} f^2(T)\, dT = \int_{-\infty}^{\infty} y^2(T)\, dT$$

From (3-20) we have for $\int_{-\infty}^{\infty} y^2(T)\, dT$

$$\int_{-\infty}^{\infty} y^2(T)\, dT = \cos^2 \epsilon \int_{-\infty}^{\infty} f^2(T)\, dT + 2 \sin \epsilon \cos \epsilon \int_{-\infty}^{\infty} f(T)F(T)\, dT$$
$$+ \sin^2 \epsilon \int_{-\infty}^{\infty} F^2(T)\, dT$$

These integrals may be evaluated from (3-12), (3-13), (3-14), (3-18), and (3-19)

$$\int_{-\infty}^{\infty} f^2(T)\, dT = \int_{0}^{\infty} f(T)f(T)\, dT = \frac{1}{\pi} \int_{0}^{\infty} f(T)\, dT \int_{0}^{\infty} [p(\omega) \cos (\omega T)$$
$$+ q(\omega) \sin (\omega T)]\, d\omega$$
$$= \frac{1}{\pi} \int_{0}^{\infty} d\omega \Big[p(\omega) \int_{0}^{\infty} f(T) \cos (\omega T)\, dT$$
$$+ q(\omega) \int_{0}^{\infty} f(T) \sin (\omega T)\, dT \Big]$$
$$= \frac{1}{\pi} \int_{0}^{\infty} [p^2(\omega) + q^2(\omega)]\, d\omega$$

and similarly

$$\int_{-\infty}^{\infty} f(T)F(T)\,dT = \int_{0}^{\infty} f(T)F(T)\,dT = 0$$

$$\int_{-\infty}^{\infty} F^2(T)\,dT = \frac{1}{\pi}\int_{0}^{\infty} [q^2(\omega) + p^2(\omega)]\,d\omega$$

so that

$$\int_{-\infty}^{\infty} y^2(T)\,dT = \frac{1}{\pi}(\cos^2 \epsilon + \sin^2 \epsilon)\int_{0}^{\infty} [p^2(\omega) + q^2(\omega)]\,d\omega = \int_{0}^{\infty} f^2(T)\,dT$$

the desired result.

The fact that $y(T)$ has a value for $T < 0$ can be explained in physical terms. From the simple harmonic solution of Chap. 2 we obtained the result that the phase change ϵ was an advance in phase of the reflected wave with respect to the incident wave. Thus, if we chose to follow a

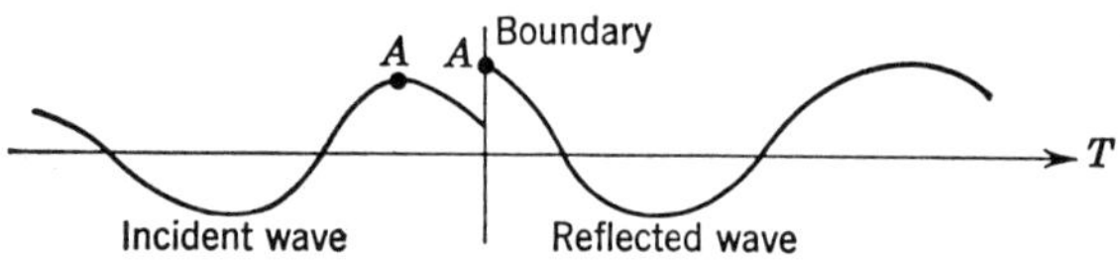

FIG. 3-28.

particular phase A of the incident wave along to the reflected wave we would see that, because of the phase advance upon reflection, this phase appeared in the reflected wave before the same phase had reached the boundary in the incident wave along the incident ray path. The time shortening of the arrival of phase A along the reflected ray over the simple ray path time would be $\Delta t = 2\pi\epsilon/\omega$. For high frequencies the time advance will be small, and for low frequencies the time advance will be large. The physical mechanism underlying the phase retardation is not so easily explained. It is certainly related to the boundary effect itself and to the fact that the underlying medium has a higher velocity than the overlying medium. In the vicinity of the reflection at the boundary the particle motion of the lower medium will be such as to preserve the boundary conditions (2-141) and (2-142). The particle motion in the lower medium will take place at a higher velocity c_2 than that for the upper medium c_1. Its net effect on the resultant reflected wave may be such as to allow the phase A to appear before that time calculated for the ray path with the velocity c_1 in the upper medium.

Surface image interference

In the first section we discussed the general characteristics of the ray solution for a variable velocity medium bounded by a free surface above and by a surface across which there was a change in acoustic properties below. In the previous section we examined the change in shape of a pulse caused by a phase change upon reflection from a boundary. Here we should like to consider the interference effects that can exist between the direct wave from source to receiver and the wave reflected from the free surface. Such an effect will arise, particularly in cases where the source is near the surface, when the surface reflected wave begins to arrive before the termination of the direct wave.

Let us consider a point source in a constant velocity medium bounded by a free surface (Fig. 3-29). From (1-39) the solution for the direct wave in terms of the velocity potential will be

$$\psi_D = \frac{1}{r} f\left(t - \frac{r}{c}\right)$$

where $f(t)$ is the source function and r is the distance between source and receiver. From the discussion in the first part of this chapter and the

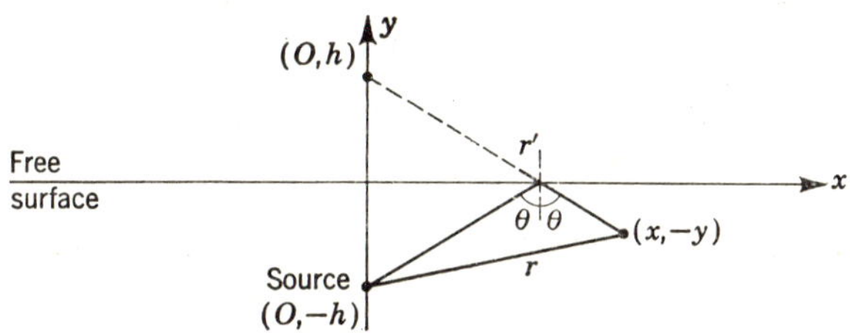

FIG. 3-29.

latter part of the second chapter we know that the angle of the reflected wave is equal to the angle of the incident wave as measured across the normal to the reflecting surface and that the amplitude of the reflected wave is equal to that of the incident wave with a phase change of 180 deg. The reflected wave is the negative of the incident wave. The velocity potential of the reflected wave will then be

$$\psi_{\text{re}} = -\frac{1}{r'} f\left(t - \frac{r'}{c}\right)$$

where r' is the distance between source and receiver via the reflection path. The total received signal will then be simply

$$\psi = \psi_D + \psi_{\text{re}} = \frac{1}{r} f\left(t - \frac{r}{c}\right) - \frac{1}{r'} f\left(t - \frac{r'}{c}\right) \tag{3-27}$$

where from the geometry of Fig. 3-29 we have that

$$r = \sqrt{x^2 + (y - h)^2}$$

and

$$r' = \sqrt{x^2 + (y + h)^2} \tag{3-28}$$

From Eqs. (3-27) we see that we might consider our problem as the interference effect between two sources, one our original source located a distance $-h$ below the reflecting surface and the other the negative of the original source located a distance h above the surface. The effect of the free surface can be replaced by an *image source* located an equal distance above the surface. The geometry of Fig. 3-30 attests to this reasoning.

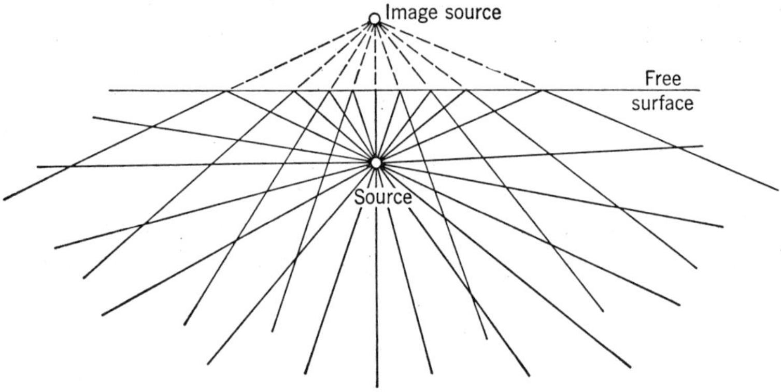

FIG. 3-30.

For a simple harmonic source, $f(t) = a \cos(2\pi ft)$, the steady-state solution will be

$$\psi = a\left\{\frac{1}{r}\cos\left[2\pi f\left(t - \frac{r}{c}\right)\right] - \frac{1}{r'}\cos\left[2\pi f\left(t - \frac{r'}{c}\right)\right]\right\} \tag{3-29}$$

We wish to consider the case where y and h are small with respect to x. Then, r and r' may be approximated by the following

$$r = x\left[1 + \frac{(y-h)^2}{x^2}\right]^{1/2} \approx x\left[1 + \frac{1}{2}\frac{(y-h)^2}{x^2}\right]$$

and similarly (3-30)

$$r' \approx x\left[1 + \frac{1}{2}\frac{(y+h)^2}{x^2}\right]$$

We may then write successively for ψ from (3-29)

$$\begin{aligned}\psi &\approx \frac{a}{x}\left\{\cos\left[2\pi f\left(t - \frac{r}{c}\right)\right] - \cos\left[2\pi f\left(t - \frac{r'}{c}\right)\right]\right\} \\ &= \frac{a}{x}\left\{-2\sin\left[2\pi f\left(t - \frac{r+r'}{2c}\right)\right]\sin\left[2\pi f\left(\frac{r-r'}{2c}\right)\right]\right. \\ &\approx +\frac{2a}{x}\sin\left[\frac{2\pi f}{c}\frac{hy}{x}\right]\sin\left[2\pi f\left(t - \frac{x}{c}\right)\right]\end{aligned} \tag{3-31}$$

using the approximations from (3-30) that

$$\frac{r + r'}{2} \approx x$$

and

$$\frac{r' - r}{2} \approx \frac{yh}{x}$$

Under these conditions the combined effect of the direct and surface reflected waves is a wave propagating in the x direction near the surface whose amplitude varies according to the function

$$\frac{2a}{x} \sin \left[2\pi \left(\frac{f}{c} \right) \frac{hy}{x} \right] = \frac{2a}{x} \sin \left(2\pi \frac{hy}{\lambda x} \right) \tag{3-32}$$

If hy is much smaller than λx, the sine function may be approximated by its argument giving for (3-32)

$$\frac{4\pi a h y}{\lambda x^2} \tag{3-33}$$

The amplitude decreases as x^{-2}; or since the intensity is proportional to the square of the amplitude of the velocity potential, we may say that the intensity decreases as x^{-4}. Such a range dependence is sometimes referred to as a *Lloyd mirror* dependence. It is greater than the more conventional spherical spreading for a point source in an unbounded medium and is applicable only to the region near a free surface. We also see from (3-33), as we should expect, that the amplitude approaches zero as either h or y approaches zero. This is sometimes referred to as the *pressure-release* effect and is dependent on the wavelength. For long wavelengths (low frequencies) the pressure-release effect will become apparent at greater distances away from the boundary than for short wavelengths (high frequencies). Referring back to (3-32) for further distances h and y away from the boundary or shorter horizontal ranges x than that given by (3-33), the function ψ of (3-31) will vanish for

$$2 \frac{hy}{\lambda x} = 0, 1, 2, 3, \ldots .$$

and will be a maximum for

$$4 \frac{hy}{\lambda x} = 1, 3, 5, \ldots .$$

The above discussion refers to the steady-state response of a simple harmonic source. If the source is given by a function which is commenced at time $t = 0$, there will be no effect at the receiver before time $t = r/c$ and there will be no interference effects before the arrival of the reflected wave at time $t = r'/c$. The solution of the problem for a pulse source has already been given in the original formulation [Eq. (3-27)] since the interference effect is stated in terms of a simple time-difference effect. For example, if as in the previous example we take an exponential pulse defined by

$$\begin{aligned} f(\tau) &= e^{-\lambda\tau} && \text{for } \tau \geqslant 0 \\ &= 0 && \tau < 0 \end{aligned}$$

the solution (3-27) will be

$$\begin{aligned} \psi &= \frac{1}{r} f\left(t - \frac{r}{c}\right) - \frac{1}{r'} f\left(t - \frac{r'}{c}\right) \\ &= 0 && \text{for } t < \frac{r}{c} \\ &= \frac{1}{r} e^{-\lambda[t-(r/c)]} && \frac{r}{c} \leqslant t < \frac{r'}{c} \\ &= \frac{1}{r} e^{-\lambda[t-(r/c)]} - \frac{1}{r'} e^{-\lambda[t-(r'/c)]} && t \geqslant \frac{r'}{c} \qquad (3\text{-}34) \end{aligned}$$

Figure 3-31 is a graph of ψ for various values of $r'/c - r/c$. The spectrum, $\gamma(\omega)$, of an interference effect which is such as (3-34) of a pulse $f(t)$

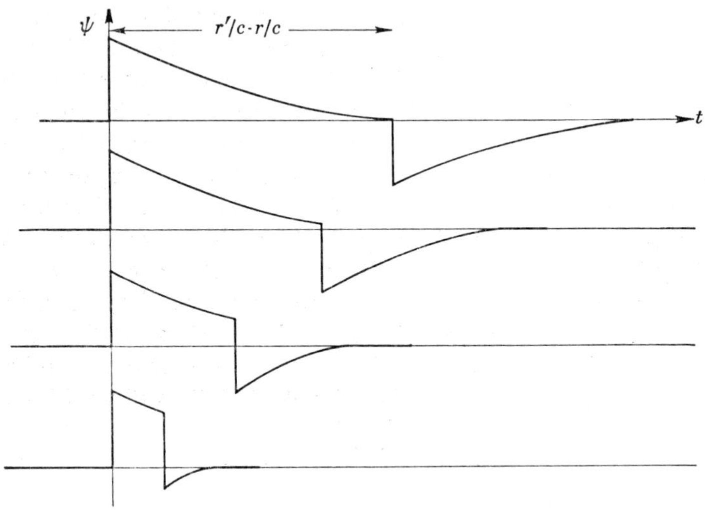

FIG. 3-31.

with a delayed pulse $-f(t + T)$ can be expressed simply by the Fourier transform (2-110)

$$\begin{aligned} \gamma(\omega) &= \int_0^\infty f(t)e^{-i\omega t}\,dt - \int_T^\infty f(t + T)e^{-i\omega t}\,dt \\ &= \int_0^\infty f(t)e^{-i\omega t}\,dt - e^{-i\omega T}\int_0^\infty f(t + T)e^{-i\omega(t+T)}\,d(t + T) \\ &= (1 - e^{-i\omega T})g(\omega) \qquad (3\text{-}35) \end{aligned}$$

where $g(\omega)$ is the Fourier transform, spectrum, of the original pulse $f(t)$. The amplitude of the interference spectrum is equal to the spectrum of the original pulse modulated by a term

$$[(1 - \cos(\omega T))^2 + \sin^2(\omega T)]^{1/2} = [2(1 - \cos(\omega T))]^{1/2} \quad (3\text{-}36)$$

The interference spectrum will be zero for values of the frequency f given by

$$f = 0, \frac{1}{T}, \frac{2}{T}, \frac{3}{T}, \ldots \quad (3\text{-}37)$$

and will be a maximum for values of f given by

$$f = \frac{1}{2T}, \frac{3}{2T}, \frac{5}{2T}, \ldots \quad (3\text{-}38)$$

If the amplitude of the original spectrum is given by Fig. 3-32*a*, the amplitude of the interference spectrum will resemble Fig. 3-32*b*.

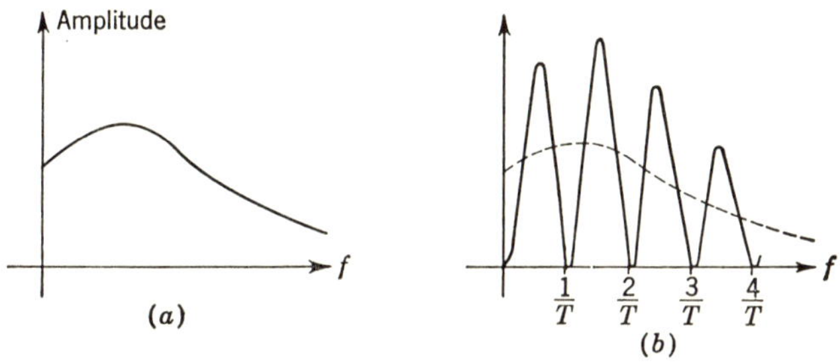

FIG. 3-32.

Normal incidence to one and two boundaries

This section deals with one of the simpler examples of plane waves incident to a boundary and a sequence of boundaries. It is concerned primarily with the interference effects that are developed between the incident and reflected waves and is included as an example of one of the more elementary normal-mode effects that may be encountered in sound transmission. It serves as an introduction to the general normal-mode solution for shallow-water propagation of the following section.

For a simple harmonic plane wave propagating in the y direction normal to an interface, Eq. (3-27) reduces to

$$\psi = A \cos\left[2\pi f\left(t - \frac{y}{c}\right)\right] - A \cos\left[2\pi f\left(t + \frac{y}{c}\right)\right] \quad (3\text{-}39)$$

and upon reduction, using the expression for the difference of two cosines.

$$\psi = -2A \sin\left(2\pi \frac{y}{\lambda}\right) \sin(2\pi f t) \quad (3\text{-}40)$$

Equation (3-40) is in a considerably different form from the more usual and general form of a propagating wave, such as the individual terms of (3-39). Equation (3-40) states that, at those values of time for which $\sin(2\pi ft)$ equals zero, the motion in the medium will be zero everywhere and similarly at those values of time for which $\sin(2\pi ft)$ equals unity the motion will be a maximum throughout all space. Correspondingly at those depths for which $\sin[2\pi(y/\lambda)]$ equals zero the motion will be zero for all time and at those depths for which $\sin[2\pi(y/\lambda)]$ equals unity the motion will go through a maximum at times for which $\sin(2\pi ft)$ is also unity. Such a vibrational system is referred to as a *standing wave*. All points in the medium are in the same phase of their motion at the same time. In this case we see that the standing wave is the combination of two simple harmonic waves of the same frequency traveling in opposite directions. The separation of the points of no motion, nodes, and those of maximum motion, antinodes, is dependent on the wavelength of the incident simple harmonic wave and the time for a complete cycle of oscillation of the system on the frequency. All frequencies are permitted.

With the introduction of a lower boundary certain restrictions are imposed upon allowable frequencies in the steady-state solution. The solution is the same as that for the similar problem of the vibrating string in Chap. 2 and carries through the same reasoning. In addition to the upper boundary condition, a lower boundary condition must be met as well. Let us presume that the lower boundary condition is also that of a free surface across which the pressure must vanish. It can be satisfied for all time only for those values of the wavelength λ for which the expression (3-40) is a node.

$$\sin\left(2\pi\frac{l}{\lambda}\right) = 0$$

$$f = \frac{c}{2l}, \frac{2c}{2l}, \frac{3c}{2l}, \ldots \qquad (3\text{-}41)$$

$$\lambda = 2l, \frac{2l}{2}, \frac{2l}{3}, \ldots$$

At no other frequencies is it possible to satisfy the boundary conditions in the steady-state solution. In a physical sense this means that for those frequencies (3-41) the waves generated from successive reflections will always meet in phase and add constructively. For all other frequencies this will necessarily not be the case, and in the limit of an infinite sum over all reflections the total effect for the latter frequencies will be zero.

With the aid of Fig. 3-33 we can see that the condition for in-phase reflections is simply Eq. (3-41). From Eq. (3-39) we have for the phase change due to the total path difference, $2l$, between one upgoing reflection and the next minus the two phase changes of π upon reflection from the

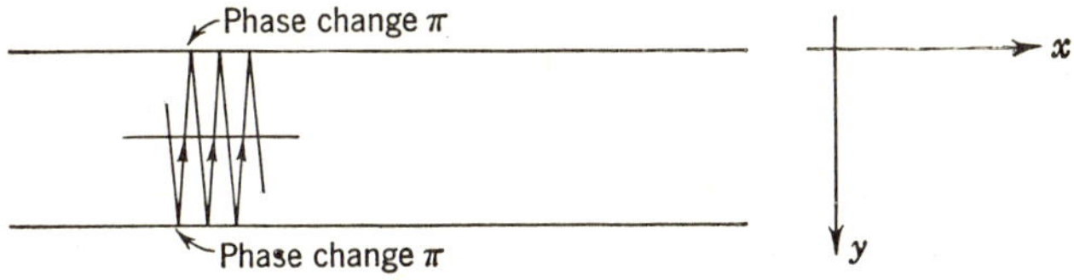

FIG. 3-33.

upper and lower free surfaces

$$2\pi f \frac{2l}{c} - \pi - \pi = 2n\pi$$

which gives

$$f = \frac{c}{2l}, \frac{2c}{2l}, \frac{3c}{2l}, \ldots \qquad (3\text{-}41)$$

Normal-mode solution

The normal-mode solution, as given here, is applicable to shallow-water transmission at the longer ranges and at moderate to low frequencies. At these frequencies the interference effects from the several multiple reflections must be obtained. The solution is not that of each ray path considered as a separate entity but the combined interference effects from all the ray paths incident to a particular receiver location. For the increasingly longer ranges the number of reflections at angles more grazing than the critical angle will be increased and the time intervals between the incidence of successive reflections will become shorter so that more and more ray paths must be considered to obtain an adequate description of the resultant effects. The normal-mode solution provides a convenient means for obtaining this description; in principle it is the same as that discussed in the previous section and is applied here to a somewhat more complex problem.

Before giving the formal solution it is instructive to take a simplified, somewhat more physical, look at the problem. At long ranges from a point source we may consider the incident wave fronts to be plane, and we may consider shallow-water propagation to consist of a large number of multiply reflected plane waves covering all angles of incidence. Let us consider one set of multiply reflected waves reflected from the bottom and surface at the angle θ (Fig. 3-34). The density and velocity of the ocean and subbottom material are designated by the constant values ρ_1, c_1, and ρ_2, c_2, respectively, and the depth of water by H. There will be a set of downgoing and upgoing multiply reflected waves whose direction is defined by the angle θ. The description of the interference effects for the set of multiply reflected rays will depend on the frequency. At those frequencies for which the phase difference between the downgoing wave associated with the point A and the downgoing wave which has undergone one additional bottom and surface reflection associated with the point B is an integral number of 360 deg, there will be constructive interference. Further, if this condition is met, there will be constructive interference for all the downgoing and upgoing waves associated with this set of rays for all time and for all r and z throughout the channel. For all other frequencies the waves will interfere out of phase and their net effect will be small compared with those frequencies for which there is always constructive interference.

These frequencies may be determined simply as follows. We desire the total phase change for the ray path from A to B minus the phase change

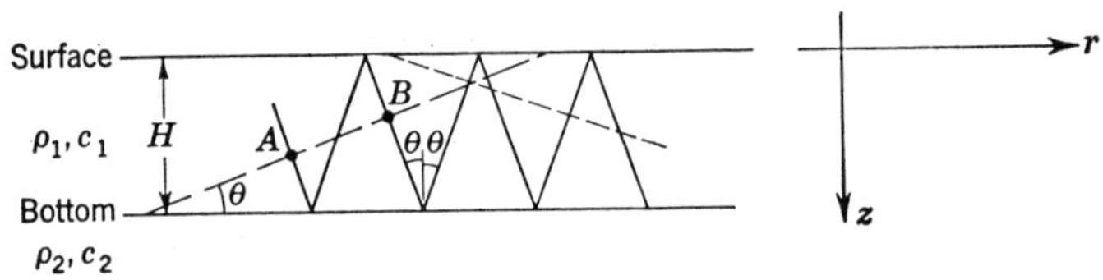

FIG. 3-34.

for reflection from the free surface π, minus the phase change for reflection from the bottom ϵ, to be equal to $2(n - 1)\pi$ where n is an integer and can take successively the values 1, 2, 3, ... This will be, from Fig. 3-34,

$$\frac{2\pi}{\lambda_n}\left[\frac{H}{\cos\theta} + \frac{H}{\cos\theta}\cos(2\theta)\right] - \epsilon - \pi = 2(n-1)\pi$$
$$\frac{2\pi f_n H}{c_1}\cos\theta - \frac{\epsilon}{2} = (2n-1)\frac{\pi}{2} \tag{3-42}$$

where λ_n and f_n, wavelength and frequency, respectively, are the solutions to (3-42) for the respective value of n. The phase change upon reflection from the bottom for angles more grazing than the critical angle is given from (2-162) as

$$\tan\frac{\epsilon}{2} = \frac{\rho_1\sqrt{\sin^2\theta - (c_1^2/c_2^2)}}{\rho_2\sqrt{1-\sin^2\theta}} \tag{3-43}$$

The spatial interference between the downgoing and upgoing waves, both at an angle θ with the horizontal, is such as to produce a standing-wave pattern in the z direction, similar to the results of Eqs. (3-39) and (3-40) of the previous section, and a net propagation in the r direction only. The velocity with which the wave front progresses in the r direction is dependent on the angle θ and will always be greater than the velocity c_1 of each upgoing or downgoing ray progressing along the ray direction. From Fig. 3-34 it is given simply as

$$V = \frac{c_1}{\sin\theta} \tag{3-44}$$

Substituting (3-44) and (3-43) into (3-42) gives

$$\frac{2\pi f_n H}{c_1}\sqrt{1 - \frac{c_1^2}{V^2}} = (2n - 1)\frac{\pi}{2} + \frac{\epsilon}{2}$$

$$\tan\left(\frac{2\pi f_n H}{c_1}\sqrt{1 - \frac{c_1^2}{V^2}}\right) = \tan\left[(2n - 1)\frac{\pi}{2} + \frac{\epsilon}{2}\right] = -\cot\left(\frac{\epsilon}{2}\right)$$

$$\tan\left(\frac{2\pi f_n H}{c_1}\sqrt{1 - \frac{c_1^2}{V^2}}\right) = -\frac{\rho_2\sqrt{1 - (c_1^2/V^2)}}{\rho_1\sqrt{(c_1^2/V^2) - (c_1^2/c_2^2)}}$$

or

$$\tan\left(\frac{2\pi f_n H}{Vc_1}\sqrt{V^2 - c_1^2}\right) = -\frac{\rho_2 c_2\sqrt{V^2 - c_1^2}}{\rho_1 c_1\sqrt{c_2^2 - V^2}} \tag{3-45}$$

or

$$\tan\left(\frac{2\pi f_n H}{V}\sqrt{\frac{V^2}{c_1^2} - 1}\right) = -\frac{\rho_2\sqrt{(V^2/c_1^2) - 1}}{\rho_1\sqrt{1 - (V^2/c_2^2)}}$$

Equation (3-42) is a relation between the angle of incidence of a particular group of multiply reflected rays and the set of discrete frequencies that will be prominent. For each angle of incidence there is a particular set of frequencies. Equation (3-45) is a relation between the velocity of propagation along the channel and the frequency. It states that the wave-guide velocity is a function of the frequency. The interference effects from the multiple reflections have led to a dispersive wave propagation. The velocity V is the phase velocity of (2-108).

Equation (3-45) is multiple-valued in f_n. For each value of V, as for each value of θ, there will be an infinite set of values for f_n. The dispersion curve, phase velocity versus frequency, for $n = 1$ is referred to as the first mode; the dispersion curve for $n = 2$ is referred to as the second mode, and so on. Equations (3-42) and (3-45) are valid only for angles of incidence more grazing than critical. For angles of incidence steeper than critical there will be no phase change upon reflection from the bottom. In contrast, however, there will be energy refracted into the bottom upon each bottom reflection so that at long ranges the net contributions from incidence steeper than critical will be small compared with incidence more grazing than critical and may be neglected. This result is similar to that obtained for high-frequency transmission at long ranges in the ray-solution section. If the problem is solved for incidence steeper than critical, a loss term as a function of range must be included. For unattenuated transmission, then, there will be a low-frequency cutoff for each mode corresponding to critical angle reflection. From (3-44) and (3-45) this will be

$$V = \frac{c_1}{\sin\theta_c} = c_2 \tag{3-46}$$

$$\frac{2\pi f_{cn}H}{c_1}\sqrt{1-\frac{c_1^2}{c_2^2}} = (2n-1)\frac{\pi}{2}$$
$$f_{cn} = \frac{(2n-1)c_1}{4H\sqrt{1-(c_1^2/c_2^2)}} \tag{3-47}$$

As has been discussed in Chap. 2, a complete description of dispersive wave propagation from an arbitrary initial disturbance requires a second evaluation to describe the change in shape of the dispersive wave train as it progresses outward in space and time. This evaluation led to the concept of the group velocity U, which is the velocity of propagation for a particular frequency in the train as contrasted with the phase velocity V of a particular point or phase of the train. From (2-121) we have for U,

$$\frac{1}{U} = \frac{1}{V} - \frac{f_n}{V^2}\frac{dV}{df_n} \tag{3-48}$$

From this physical formulation we might expect to be able also to obtain the amplitude dependence as a function of depth z between the interfering downgoing and upgoing waves. From Fig. 3-35a we shall

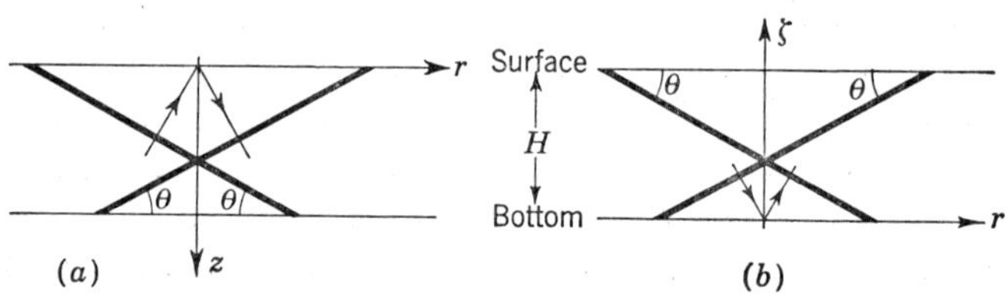

FIG. 3-35.

have for the interference between the upgoing wave and the downgoing wave reflected from the surface,

$$a\cos\left[2\pi f\left(t - \frac{r\sin\theta}{c_1} + \frac{z\cos\theta}{c_1}\right)\right] + a\cos\left[2\pi f\left(t - \frac{r\sin\theta}{c_1} - \frac{z\cos\theta}{c_1}\right) + \pi\right]$$
$$= 2a\cos\left[2\pi f\left(\frac{z\cos\theta}{c_1}\right) - \frac{\pi}{2}\right]\cos\left[2\pi f\left(t - \frac{r\sin\theta}{c_1}\right) + \frac{\pi}{2}\right]$$

where the second cosine factor is the propagation term with a velocity of propagation in the r direction of that given by (3-44) and the first cosine factor is the amplitude dependence in the z direction. For the first cosine factor we shall have then substituting from (3-44)

$$\cos\left[2\pi f\left(\frac{z\cos\theta}{c_1}\right) - \frac{\pi}{2}\right] = -\sin\left[\frac{2\pi fz}{V}\sqrt{\frac{V^2}{c_1^2} - 1}\right] \tag{3-49}$$

If instead of taking the surface reflection we had used the bottom reflection, we should of course have obtained the same amplitude dependence. From Fig. 3-35*b*, where the coordinate ζ is along the negative z direction such that $\zeta = H - z$ we shall have

$$a \cos\left[2\pi f\left(t - \frac{r \sin\theta}{c_1} + \frac{\zeta \cos\theta}{c_1}\right)\right] + a \cos\left[2\pi f\left(t - \frac{r \sin\theta}{c_1} - \frac{\zeta\cos\theta}{c_1}\right) + \epsilon\right]$$

$$2a \cos\left[2\pi f\left(\frac{\zeta\cos\theta}{c_1}\right) - \frac{\epsilon}{2}\right] \cos\left[2\pi f\left(t - \frac{r\sin\theta}{c_1}\right) + \frac{\epsilon}{2}\right]$$

so that the amplitude dependence is, using (3-42),

$$\cos\left[2\pi f\left(\frac{\zeta\cos\theta}{c_1}\right) - \frac{\epsilon}{2}\right] = \cos\left[2\pi f\frac{z\cos\theta}{c_1} - (2n+1)\frac{\pi}{2}\right] = \pm \sin\left[\frac{2\pi f z}{H}\sqrt{\frac{V^2}{c_1^2} - 1}\right]$$

As has been stated previously, incidence steeper than the critical angle results in attenuated propagation. It is of interest to deduce some of the mode characteristics for these angles. Depending on the incident angle and the density and velocity contrast there will be a phase change of 0 or 180 deg for each bottom reflection. In addition there will be a loss of energy out of the ocean due to the partial refraction of energy into the bottom. For each bottom reflection the amplitude will be reduced to R times its incident value where R is the Rayleigh reflection coefficient from (2-154),

$$R = \frac{\dfrac{\rho_2}{\rho_1} - \dfrac{\sqrt{(c_1^2/c_2^2) - \sin^2\theta}}{\sqrt{1 - \sin^2\theta}}}{\dfrac{\rho_2}{\rho_1} + \dfrac{\sqrt{(c_1^2/c_2^2) - \sin^2\theta}}{\sqrt{1 - \sin^2\theta}}} \tag{2-154}$$

From (3-44) R can be written as

$$R = \frac{\dfrac{\rho_2}{\rho_1} - \dfrac{\sqrt{(V^2/c_2^2) - 1}}{\sqrt{(V^2/c_1^2) - 1}}}{\dfrac{\rho_2}{\rho_1} + \dfrac{\sqrt{(V^2/c_2^2) - 1}}{\sqrt{(V^2/c_1^2) - 1}}} \tag{3-50}$$

For the normal condition of $\rho_2 > \rho_1$ and $c_2 > c_1$, the phase change for bottom reflection will be 0 deg. From Fig. 3-36 the equation for constructive interference, similar to (3-42), will be

$$\frac{2\pi}{\lambda_n}\left[\frac{H}{\cos\theta} + \frac{H}{\cos\theta}\cos(2\theta)\right] - \pi = 2(n-1)\pi$$
$$\frac{2\pi f_n H}{c_1}\cos\theta = (2n-1)\frac{\pi}{2} \tag{3-51}$$

or substituting from (3-44)

$$\frac{4f_nH}{V}\sqrt{\frac{V^2}{c_1^2} - 1} = (2n-1) \tag{3-52}$$

As can be seen from (3-44), the phase velocity continues to increase with decreasing θ and approaches infinity as θ approaches zero. From (3-52) the frequency decreases with increasing V, starting at the cutoff frequency (3-47) for unattenuated propagation. The frequency approaches

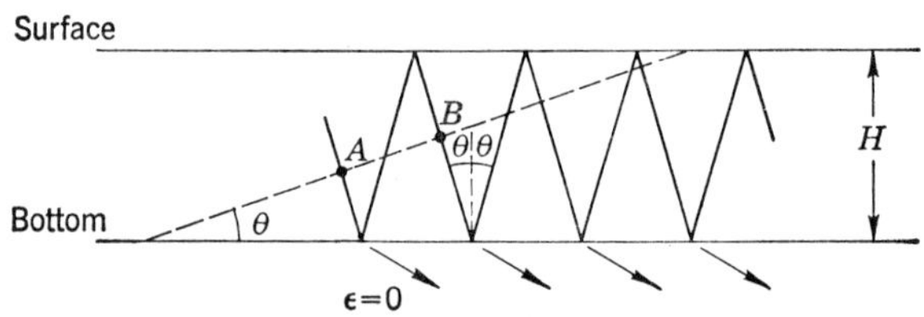

FIG. 3-36.

a lower limit as V approaches infinity. This is the cutoff frequency for attenuated propagation below which there is no mode propagation. From (3-52) the cutoff frequency for attenuated propagation is

$$f_{acn} = \frac{(2n-1)c_1}{4H} \tag{3-53}$$

The decrease in amplitude is proportional to the incident amplitude at any given time. The rate of decrease of amplitude will be proportional to the incident amplitude,

$$\frac{dA}{dt} = -pA$$
$$A = e^{-pt} \tag{3-54}$$

In the time Δt_{AB} for the ray to travel from A to B all rays will have undergone one bottom reflection. The amplitude will be decreased to R of its initial value. We shall then have for p,

$$e^{-p\Delta t_{AB}} = R$$
$$p = -\frac{1}{\Delta t_{AB}}\ln R = \frac{V}{2H\sqrt{(V^2/c_1^2) - 1}}\ln R^{-1} \tag{3-55}$$

where from Fig. 3-36 Δt_{AB} is given by

$$\Delta t_{AB} = \frac{2H\cos\theta}{c_1} = \frac{2H}{V}\sqrt{\frac{V^2}{c_1^2} - 1}$$

Equation (3-54) becomes

$$A = \exp\left[-\frac{Vt}{2H\sqrt{(V^2/c_1^2) - 1}} \ln R^{-1}\right] \tag{3-56}$$

Since any given frequency travels with the group velocity U, we may express (3-56) with a range dependence as

$$A = \exp\left[-\frac{Vr}{2HU\sqrt{(V^2/c_1^2) - 1}} \ln R^{-1}\right] \tag{3-57}$$

A sketch of the form of the phase and group velocities for the first three modes is given in Fig. 3-37. The attenuated portion below the cutoff

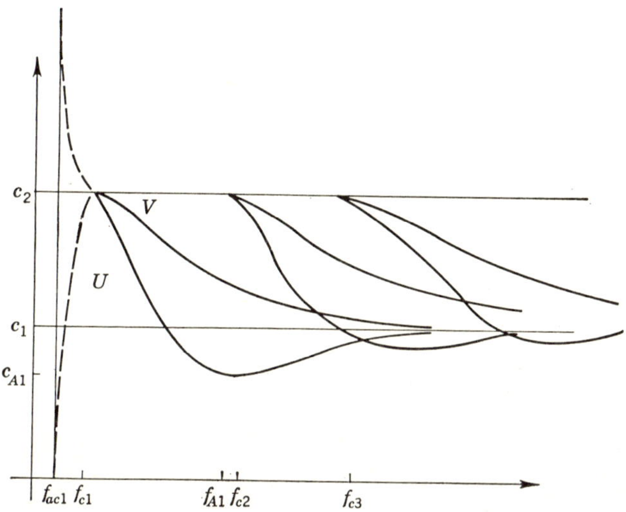

FIG. 3-37.

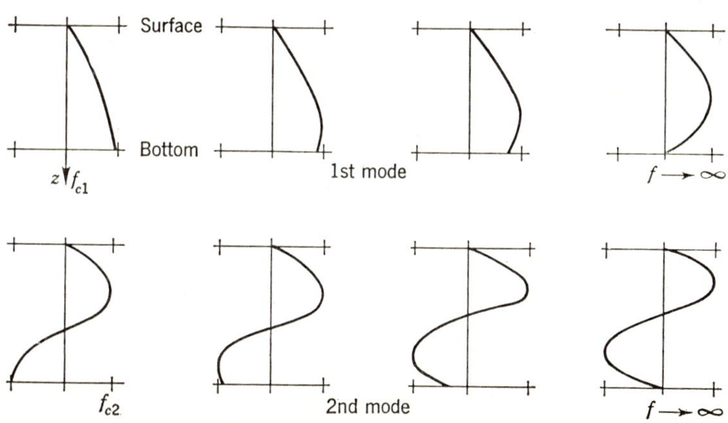

FIG. 3-38.

frequency (3-47) for the first mode has also been included. In Fig. 3-38 the depth dependence has been sketched for a number of frequencies for the first two modes from Eq. (3-49).

From Fig. 3-37 it is possible to comprehend qualitatively the form of the dispersive wave train. Considering the first mode only the train will start with the frequency f_{c1} having traveled with the velocity c_2. With increasing time the wave train will increase in frequency down to the part which has traveled with the velocity c_1. From there on there will be two superimposed wave trains to the termination traveling with the velocity c_{A1}. The first wave train is a continuation of the original with increasing frequencies down to f_{A1}; the second starts at very high frequencies with the velocity c_1 and decreases in frequency to f_{A1}. The resultant wave train would resemble that shown in Fig. 3-39. In addition there will be a third wave train, much attenuated in comparison with

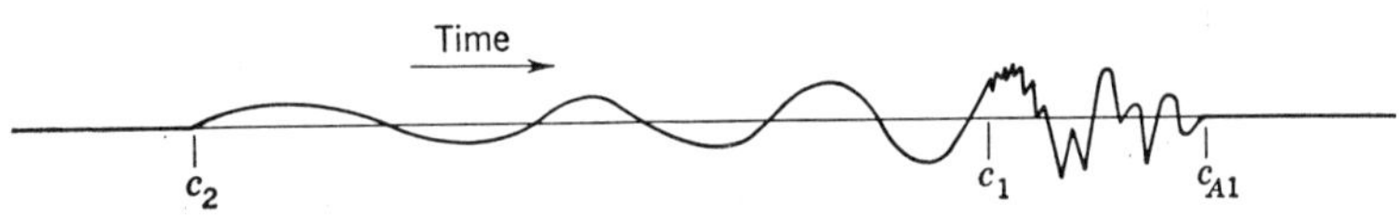

FIG. 3-39.

the first two, which starts with a frequency f_{c1} having traveled with a velocity c_2 and continues indefinitely with decreasing frequency toward f_{ac1}.

All the features for unattenuated propagation that we have discussed here in this simplified physical evaluation will appear in the formal solution to follow. They permit a physical understanding of this type of propagation which the formal solution by itself does not provide.

For the formal solution we desire a solution of the wave equation satisfying the boundary conditions at the surface and the bottom and satisfying the condition of a point source. Since the boundaries are plane parallel and the interest is primarily in radial propagation outward from a point source, it is convenient to solve the problem in cylindrical coordinates with symmetry about the polar axis. The wave equation in symmetrical cylindrical coordinates r and z is

$$\nabla^2\varphi = \frac{1}{r}\frac{\partial}{\partial r}\left(r\frac{\partial\varphi}{\partial r}\right) + \frac{\partial^2\varphi}{\partial z^2} = \frac{1}{c^2}\frac{\partial^2\varphi}{\partial t^2} \tag{3-58}$$

We shall take a product solution of (3-58) for a simple harmonic source of frequency ω,

$$\varphi = e^{i\omega t}\,M(r)F(z) \tag{3-59}$$

Substituting (3-59) into (3-58) gives, after a small amount of reduction,

$$\frac{(d^2F/dz^2) + (\omega^2/c^2)F}{F} = \frac{-(d^2M/dr^2) - (1/r)(dM/dr)}{M} \tag{3-60}$$

The right-hand side of (3-60) is independent of z and the left-hand side independent of r. Therefore, if the equality (3-60) is to be true for all values of r and for all values of z, the expressions on each side of the equality must be equal to a constant k^2 independent of r and z. We have then

$$\frac{d^2M}{dr^2} + \frac{1}{r}\frac{dM}{dr} + k^2M = 0 \tag{3-61}$$

and

$$\frac{d^2F}{dz^2} + \left(\frac{\omega^2}{c^2} - k^2\right)F = 0 \tag{3-62}$$

The solution to (3-61) is the zero-order Bessel function

$$M = J_0(kr) \tag{3-63}$$

The second solution to (3-61), the zero-order Neumann function, $Y_0(kr)$, becomes infinite at the origin and is not applicable to the present problems. The form of the solution to (3-62) depends on the sign of $[(\omega^2/c^2) - k^2]$. In complex notation a general form may be written as

$$F = e^{\pm i\beta z} \tag{3-64}$$

where

$$\beta = \left(\frac{\omega^2}{c^2} - k^2\right)^{1/2}$$

In terms of real variables F will be a sine and cosine term for β^2 positive and an exponential for β^2 negative.

The manner in which the condition of a point source can be expressed needs a little investigation. In the vicinity of our point source, taken here to be at the origin, $r = 0$, $z = 0$, emitting simple harmonic waves of frequency ω, we desire the solution to reduce to the form (1-39),

$$\Phi = \frac{\exp i\omega\left(t - \dfrac{\sqrt{r^2 + z^2}}{c_1}\right)}{\sqrt{r^2 + z^2}} \tag{3-65}$$

Our product form of the solution to the wave equation is from (3-63) and (3-64)

$$\varphi = e^{i\omega t}J_0(kr)e^{\pm i\beta_1 z} \tag{3-66}$$

where k is a variable of integration independent of r and z and unrestricted in value. Any function of k multiplied by the solution (3-66) is also a solution to the wave equation (3-58) and any summation of such solutions will also be a solution. A general form is then

$$\Phi = e^{i\omega t}\int_0^\infty A(k)J_0(kr)e^{\pm i\beta_1 z}\,dk \tag{3-67}$$

In an unbounded medium solution (3-65) is the complete solution for a source of unit amplitude and constant frequency ω. We wish to find the value of $A(k)$ which will reduce the integral (3-67) to (3-65)

$$e^{i\omega t} \frac{e^{-i(\omega/c_1)\sqrt{r^2+z^2}}}{\sqrt{r^2 + z^2}} = e^{i\omega t} \int_0^\infty A(k) J_0(kr) e^{\pm i\beta_1 z}\, dk \tag{3-68}$$

where the negative sign of the exponential is taken for z positive and the positive sign for z negative to ensure outgoing waves from the source. The amplitude factor $A(k)$ is independent of r and z, and its value may be determined for any values of r and z. It is convenient to evaluate Eq. (3-68) at $z = 0$.

$$\frac{e^{-ik_1 r}}{r} = \int_0^\infty g(k) J_0(kr) k\, dk \tag{3-69}$$

where $k_1 \equiv \omega/c_1$ and $g(k) \equiv A(k)/k$. The form of Eq. (3-69) is that of the Fourier-Bessel transform, similar to the Fourier transforms of Chap. 2. The Fourier-Bessel transforms are simply the pair of equations

$$f(r) = \int_0^\infty g(k) J_0(kr) k\, dk$$

and (3-70)

$$g(k) = \int_0^\infty f(\rho) J_0(k\rho) \rho\, d\rho$$

From Eqs. (3-70) the solution of the integral equation (3-69) is then

$$g(k) = \int_0^\infty e^{-ik_1\rho} J_0(k\rho)\, d\rho \tag{3-71}$$

This integral may be evaluated.[1] Expressing $J_0(kr)$ in its integral form gives

$$\begin{aligned} g(k) &= \frac{1}{2\pi} \int_0^\infty e^{-ik_1\rho} \int_{-\pi}^{\pi} e^{ik\rho\cos\theta}\, d\theta \\ &= \frac{1}{2\pi} \int_{-\pi}^{\pi} d\theta \int_0^\infty e^{-i(k_1 - k\cos\theta)\rho}\, d\rho \\ &= \frac{1}{2\pi i} \int_{-\pi}^{\pi} \frac{d\theta}{k_1 - k\cos\theta} \\ &= \frac{1}{2\pi i} \left(\frac{2\pi}{\sqrt{k_1^2 - k^2}} \right) = \frac{1}{i\beta_1} \end{aligned} \tag{3-72}$$

The proper value for the term arising from the upper limit in the ρ integration is zero, as can be seen by making k complex, i.e., including a small absorption term. The value of $A(k)$ is then

$$A(k) = \frac{k}{i\beta_1} \tag{3-73}$$

[1] B. O. Pierce, "A Short Table of Integrals," p. 41, Sec. 300, Ginn & Company, Boston, 1929.

and the form of the integral (3-67) for a point source is

$$\Phi = e^{i\omega t} \int_0^\infty J_0(kr) \frac{k\,dk}{i\beta_1} e^{\pm i\beta_1 z} \tag{3-74}$$

It is interesting to mention that we can see that (3-73) would be a correct form for $A(k)$ by carrying out the evaluation of (3-68) at $r = 0$.

$$\frac{e^{-ik_1 z}}{z} = \int_0^\infty A(k) e^{-i\sqrt{k_1^2 - k^2}\, z}\, dz \tag{3-75}$$

The form $A(k) = k/i\sqrt{k_1^2 - k^2}$ makes the integrand of the integral on the right side of Eq. (3-75) a perfect differential which when evaluated at the limits 0 and ∞ gives simply the left side of (3-75).

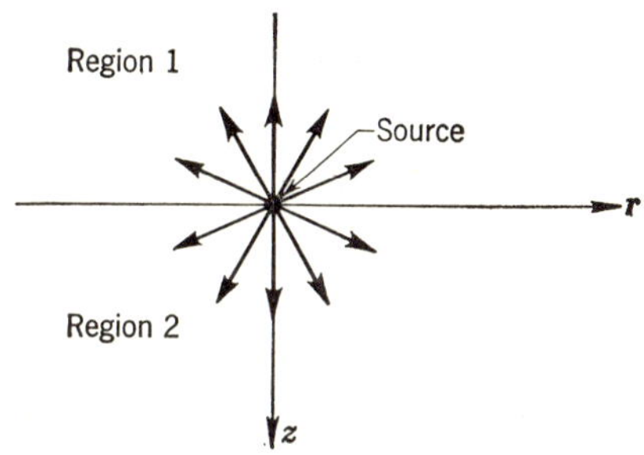

FIG. 3-40.

In a problem involving a number of boundary conditions we can be assured that the source condition is met and that the solution reduces to the form (3-65) in the vicinity of the source through the following subterfuge. Let us impose the conditions of continuity of pressure and a discontinuity of vertical particle velocity equal to $2k$ across the plane $z = 0$ at the depth of the source. From (3-59) these will be

$$F_1 = F_2$$
$$\frac{dF_1}{dz} - \frac{dF_2}{dz} = 2k \tag{3-76}$$

where φ is taken as the velocity potential. For outgoing waves from the source the form of F will be from Fig. 3-40

$$F_1 = ae^{i\beta_1 z} \qquad \text{and} \qquad F_2 = be^{-i\beta_1 z} \tag{3-77}$$

Substituting (3-77) into (3-76) gives

$$a - b = 0$$
$$i\beta_1 a + i\beta_1 b = 2k$$

so that

$$a = b = \frac{k}{i\beta_1} \tag{3-78}$$

Substituting (3-78) into (3-59) and integrating with respect to k from 0 to ∞ gives

$$\begin{aligned} \Phi &= e^{i\omega t} \int_0^\infty J_0(kr) \frac{k\,dk}{i\beta_1} e^{i\beta_1 z} \qquad \text{for } z < 0 \\ &= e^{i\omega t} \int_0^\infty J_0(kr) \frac{k\,dk}{i\beta_1} e^{-i\beta_1 z} \qquad z > 0 \end{aligned} \tag{3-79}$$

the desired source condition. We see that the source condition can be met by imposing an additional boundary condition (3-76) along the plane at the depth of the source and integrating the resultant with respect to k from 0 to ∞. This makes the final formulation considerably easier, for it is necessary to treat the formulation solely as a boundary-condition problem with the extra boundary condition (3-76) and integrate the final form from 0 to ∞ with respect to k.

The physical properties of the media for which the wave propagation is to be solved are shown in Fig. 3-41. The upper medium is defined by the

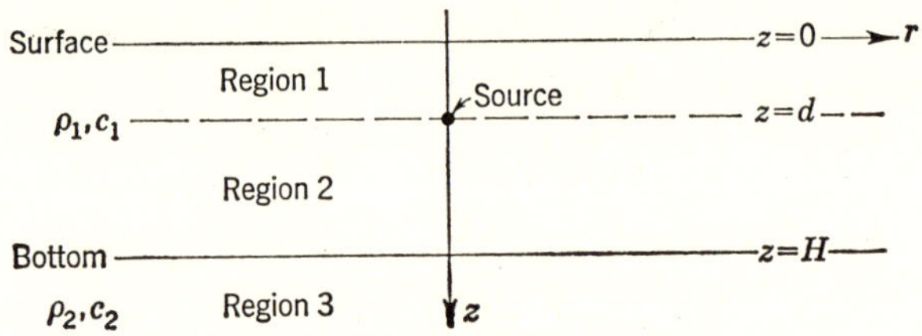

FIG. 3-41.

constant density ρ_1 and velocity c_1 and is bounded at $z = 0$ by a free surface and at the bottom, $z = H$, by a medium of constant density ρ_2 and velocity c_2 where c_2 is greater than c_1. The point source is located at $r = 0$, $z = d$. The propagation is defined by the wave equations

$$\nabla^2\varphi = \frac{1}{c_1^2}\frac{\partial^2\varphi}{\partial t^2}$$

and (3-80)

$$\Delta^2\varphi = \frac{1}{c_2^2}\frac{\partial^2\varphi}{\partial t^2}$$

in the upper and lower medium, respectively. We shall take solutions of the wave equations (3-80) of the form (3-59) for a simple harmonic source.

$$\varphi = e^{i\omega t}J_0(kr)F(z) \tag{3-81}$$

From the solution for F, β will be given by the following for the upper and lower medium, respectively,

$$\beta_1 = \sqrt{\frac{\omega^2}{c_1^2} - k^2} \qquad k < \frac{\omega}{c_1}$$

$$= -i\sqrt{k^2 - \frac{\omega^2}{c_1^2}} \qquad k > \frac{\omega}{c_1}$$

and (3-82)

$$\beta_2 = \sqrt{\frac{\omega^2}{c_2^2} - k^2} \qquad k < \frac{\omega}{c_2}$$

$$= -i\sqrt{k^2 - \frac{\omega^2}{c_2^2}} \qquad k > \frac{\omega}{c_2}$$

For F in the regions 1, 2, and 3 we shall take solutions (3-64) of the form

$$\begin{aligned} F_1 &= A \sin(\beta_1 z) + E \cos(\beta_1 z) \\ F_2 &= B \sin(\beta_1 z) + C \cos(\beta_1 z) \\ F_3 &= De^{-i\beta_2 z} + Ge^{i\beta_2 z} \end{aligned} \tag{3-83}$$

In Eqs. (3-83) G must be taken equal to zero in order to avoid the physically inconsistent picture of an exponential increase of F_3 with depth into the lower medium, and the negative imaginary is chosen in (3-82) so that for β_2 real the solution will represent propagation outward from the source into the lower medium. In the previous discussion we ascertained that the dominant contributions at long ranges came from multiple reflections within the water channel with no refraction of energy into the lower media. We should expect then that the major contributions at long ranges would be for β_1 real and β_2 imaginary.

We shall solve the problem in terms of the velocity potential φ where φ is related to the pressure and the horizontal and vertical velocity components from (1-83) and (1-85) by

$$p = \rho \frac{\partial \varphi}{\partial t}, \dot{q} = -\frac{\partial \varphi}{\partial r}, \dot{w} = -\frac{\partial \varphi}{\partial z} \tag{3-84}$$

The boundary conditions for pressure and vertical component of velocity are from (2-141) and (2-142)

$$F_1 = 0 \qquad \text{at } z = 0 \tag{3-85}$$

$$\left.\begin{aligned} F_1 &= F_2 \\ \frac{dF_1}{dz} - \frac{dF_2}{dz} &= 2k \end{aligned}\right\} \qquad \text{at } z = d \tag{3-86}$$

$$\left.\begin{aligned} \rho_1 F_2 &= \rho_2 F_3 \\ \frac{dF_2}{dz} &= \frac{dF_3}{dz} \end{aligned}\right\} \qquad \text{at } z = H \tag{3-87}$$

From (3-85) we see that E must be equal to zero. Substituting (3-83) into (3-85), (3-86), and (3-87) gives

$$\begin{aligned} A \sin(\beta_1 d) - B \sin(\beta_1 d) - C \cos(\beta_1 d) &= 0 \\ A\beta_1 \cos(\beta_1 d) - B\beta_1 \cos(\beta_1 d) + C\beta_1 \sin(\beta_1 d) &= 2k \\ B\rho_1 \sin(\beta_1 H) + C\rho_1 \cos(\beta_1 H) - D\rho_2 e^{-i\beta_2 H} &= 0 \\ B\beta_1 \cos(\beta_1 H) - C\beta_1 \sin(\beta_1 H) + iD\beta_2 e^{-i\beta_2 H} &= 0 \end{aligned} \tag{3-88}$$

Solving for A, B, C, and D we obtain

$$\begin{aligned}
A &= \frac{2k}{\beta_1}\left[\frac{\beta_1 \cos[\beta_1(H-d)] + ib\beta_2 \sin[\beta_1(H-d)]}{\beta_1 \cos(\beta_1 H) + ib\beta_2 \sin(\beta_1 H)}\right] \\
B &= \frac{2k \sin(\beta_1 d)}{\beta_1}\left[\frac{\beta_1 \sin(\beta_1 H) - ib\beta_2 \cos(\beta_1 H)}{\beta_1 \cos(\beta_1 H) + ib\beta_2 \sin(\beta_1 H)}\right] \\
C &= \frac{2k \sin(\beta_1 d)}{\beta_1} \\
D &= \frac{2bke^{-ik\beta_2 H} \sin(\beta_1 d)}{\beta_1 \cos(\beta_1 H) + ib\beta_2 \sin(\beta_1 H)}
\end{aligned} \tag{3-89}$$

where b has been defined as

$$b \equiv \frac{\rho_1}{\rho_2} \tag{3-90}$$

Substituting (3-89) into (3-83) and (3-81) and integrating from 0 to ∞ with respect to k gives for the formal solution for the velocity potential from a simple harmonic point source

$$\begin{aligned}
\Phi_1 &= e^{i\omega t}\Psi_1(r, z, \omega) \\
&= 2e^{i\omega t} \int_0^\infty J_0(kr)k\,dk \frac{\sin(\beta_1 z)}{\beta_1}\left[\frac{\beta_1 \cos[\beta_1(H-d)] + ib\beta_2 \sin[\beta_1(H-d)]}{\beta_1 \cos(\beta_1 H) + ib\beta_2 \sin(\beta_1 H)}\right]
\end{aligned} \tag{3-91}$$

$$\begin{aligned}
\Phi_2 &= e^{i\omega t}\Psi_2(r, z, \omega) \\
&= 2e^{i\omega t} \int_0^\infty J_0(kr)k\,dk \frac{\sin(\beta_1 d)}{\beta_1}\left[\frac{\beta_1 \cos[\beta_1(H-z)] + ib\beta_2 \sin[\beta_1(H-z)]}{\beta_1 \cos(\beta_1 H) + ib\beta_2 \sin(\beta_1 H)}\right]
\end{aligned} \tag{3-92}$$

$$\begin{aligned}
\Phi_3 &= e^{i\omega t}\Psi_3(r, z, \omega) \\
&= 2e^{i\omega t} \int_0^\infty J_0(kr)k\,dk \frac{\sin(\beta_1 d)e^{-i\beta_2(z-H)}}{\beta_1 \cos(\beta_1 H) + ib\beta_2 \sin(\beta_1 H)}
\end{aligned} \tag{3-93}$$

The solution for a point-source pulse $f(\tau)$ is given by (2-111) through the Fourier transforms (2-110)

$$\begin{aligned}
f(t) &= \frac{1}{2\pi}\int_{-\infty}^{\infty} g(\omega)e^{i\omega t}\,d\omega \\
g(\omega) &= \int_{-\infty}^{\infty} f(\tau)e^{-i\omega\tau}\,d\tau
\end{aligned} \tag{2-110}$$

as

$$P(r, z, t) = \frac{1}{2\pi}\int_{-\infty}^{\infty} g(\omega)\Psi(r, z, \omega)e^{i\omega t}\,d\omega \tag{3-94}$$

The solution is now reduced to an evaluation of the integrals (3-92) and (3-94). In general in such problems as this an exact solution to the resultant integrals is very difficult if not impossible to obtain without resort to the methods of numerical integration. Valid evaluations can be obtained, however, under the restriction of certain conditions. One such evaluation is to expand the solutions (3-92) into an infinite series of integrals, the integrands of which can be associated with the particular *rays*, including the multiple reflections, from a given source location to a given receiver location. The solution can be given by a finite number of these integrals at short to moderate ranges and at the higher frequencies. Another evaluation is given by the contributions from the infinite number of poles of the integrand (3-92) evaluated in the complex plane. This evaluation is valid at moderate to long ranges, and the solution can be given by a finite number of terms at the lower frequencies. This solution can be associated with the propagating *normal modes* discussed in the earlier part of this section.

Before discussing the normal-mode solution, which is the main part of this section, it is interesting to follow the reduction to a series of ray integrals. Let us define the quantity κ.

$$\kappa = \frac{\beta_1 - b\beta_2}{\beta_1 + b\beta_2} \tag{3-95}$$

Then, the expression below will be equal to the expression in the brackets of (3-92).

$$e^{-i\beta_1 z}\left[\frac{1 + \kappa e^{-i2\beta_1(H-z)}}{1 + \kappa e^{-i2\beta_1 H}}\right]$$

$$= e^{-i\beta_1 z}\left[\frac{(\beta_1 + b\beta_2) + (\beta_1 - b\beta_2)\{\cos^2[\beta_1(H-z)] - \sin^2[\beta_1(H-z)] - i2\sin[\beta_1(H-z)]\cos[\beta_1(H-z)]\}}{(\beta_1 + b\beta_2) + (\beta_1 - b\beta_2)\{\cos^2(\beta_1 H) - \sin^2(\beta_1 H) - i2\sin(\beta_1 H)\cos(\beta_1 H)\}}\right]$$

$$= \frac{\beta_1 \cos[\beta_1(H-z)] + ib\beta_2 \sin[\beta_1(H-z)]}{\beta_1\cos(\beta_1 H) + ib\beta_2\sin(\beta_1 H)}$$

Expanding the denominator of this expression, we have

$$e^{-i\beta_1 z}[1 + \kappa e^{-i2\beta_1(H-z)}][1 - \kappa e^{-i2\beta_1 H} + \kappa^2 e^{-i4\beta_1 H} - \cdot\cdot\cdot]$$

which when substituted into (3-92) gives

$$\begin{aligned}\Phi_2 = e^{i\omega t}\int_0^\infty J_0(kr)\frac{k\,dk}{i\beta_1}[e^{-i\beta_1(z-d)} - e^{-i\beta_1(z+d)}][1 + \kappa e^{-i2\beta_1(H-z)}][1 - \kappa e^{-i2\beta_1 H} \\ + \kappa^2 e^{-i4\beta_1 H} - \cdot\cdot\cdot] = e^{i\omega t}\int_0^\infty J_0(kr)\frac{k\,dk}{i\beta_1}\{[e^{-i\beta_1(z-d)} - e^{-i\beta_1(z+d)}] \\ + \kappa[e^{-i\beta_1(-z-d+2H)} - e^{-i\beta_1(-z+d+2H)} - e^{-i\beta_1(z-d+2H)} + e^{-i\beta_1(z+d+2H)}] \\ - \kappa^2[e^{-i\beta_1(-z-d+4H)} - e^{-i\beta_1(-z+d+4H)} - e^{-i\beta_1(z-d+4H)} + e^{-i\beta_1(z+d+4H)}] \\ + \kappa^3[\cdot\cdot\cdot] - \cdot\cdot\cdot\}\end{aligned} \tag{3-96}$$

The first two integrals are the same as the source integral (3-65); the first integral is the direct ray, and the second integral is the surface reflected ray. The next four terms, multiplied by the factor κ, are the four rays which have undergone one bottom reflection; the next four terms, multiplied by the factor κ^2, are the four rays which have undergone two bottom reflections, and so on. The change in sign of the integrands corresponds to an even (plus) or odd (minus) number of surface reflections, each surface reflection having a phase change of 180 deg. We should expect that the factor κ is related to the plane wave (Rayleigh) reflection coefficient. For values of k less than ω/c_1, k has the physical significance of being the wave number measured in the r direction (β being the wave number measured in the z direction). For these values of k, k can be written as

$$k = \frac{\omega}{c_1} \sin \theta \tag{3-97}$$

where θ is the ray angle. Substitution of (3-97) into (3-95) gives

$$\kappa = \frac{\dfrac{\rho_2}{\rho_1} - \dfrac{\beta_2}{\beta_1}}{\dfrac{\rho_2}{\rho_1} + \dfrac{\beta_2}{\beta_1}} = \frac{\dfrac{\rho_2}{\rho_1} - \dfrac{\sqrt{(c_1^2/c_2^2) - \sin^2 \theta}}{\sqrt{1 - \sin^2 \theta}}}{\dfrac{\rho_2}{\rho_1} + \dfrac{\sqrt{(c_1^2/c_2^2) - \sin^2 \theta}}{\sqrt{1 - \sin^2 \theta}}} \tag{3-98}$$

the plane wave reflection coefficient. Each of the integrals (3-96) expresses the reflection of a spherical wave from a plane boundary. We shall consider the evaluation of integrals of the form

$$\int_0^\infty J_0(kr) \frac{k\,dk}{i\beta_1} \left(\frac{\beta_1 - b\beta_2}{\beta_1 + b\beta_2} \right) e^{-i\beta_1(-z-d+2H)}$$

in a later chapter. The various ray paths from source to receiver are identified in Fig. 3-42.

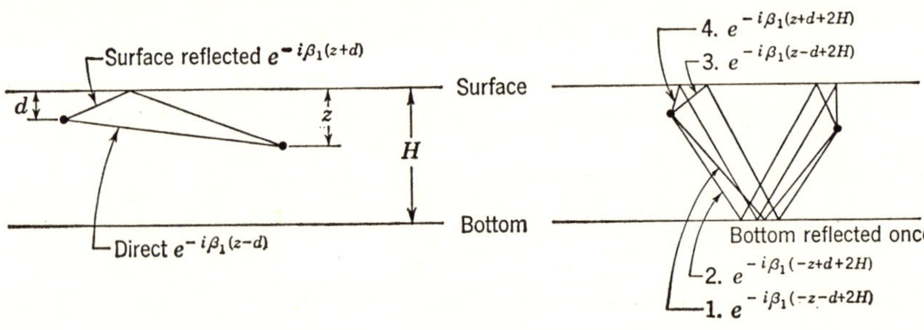

FIG. 3-42.

The normal-mode solution is obtained by a transformation and evaluation of the integral (3-92) in the complex k plane. As has been stated before, the normal-mode evaluation is useful at moderate to long ranges and at the lower frequencies. In carrying out this evaluation it must be remembered that the quantities β_1 and β_2 are multiple-valued in k with branch points at

$$k_1 = \frac{\omega}{c_1} \qquad \text{and} \qquad k_2 = \frac{\omega}{c_2} \tag{3-99}$$

respectively. We must therefore cut the k plane along branch lines from the branch points (3-99). Since we are interested in the evaluations satisfying the conditions (3-82), we shall take the branch lines as shown in Fig. 3-43. To the left of the k_1 branch line, β_1 will have the value

$$\beta_1 = \sqrt{k_1^2 - (k_1 - i\alpha)^2} = \sqrt{\alpha^2 - i2\alpha k_1} \qquad Re(k) < \frac{\omega}{c_1}$$

and to the right of the branch line β_1 will have the value

$$\beta_1 = -i\sqrt{(k_1 - i\alpha)^2 - k_1^2} = -i\sqrt{-(\alpha^2 - i2\alpha k_1)}$$
$$= -\sqrt{\alpha^2 - i2\alpha k_1} \qquad Re(k) > \frac{\omega}{c_1}$$

β_2 will remain unchanged. The sign of β_1 changes across the k_1 branch line. Similarly the sign of β_2 changes across the k_2 branch line. The signs of β_1 and β_2 across the k_1 and k_2 branch lines, respectively, are indicated in Fig. 3-43. The poles of the integrand (3-92) are located along the real

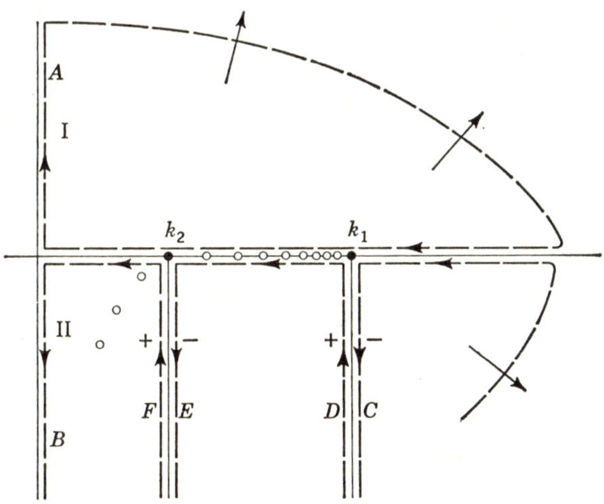

FIG. 3-43.

axis (unattenuated) between k_1 and k_2 and in the fourth quadrant (attenuated). For convenience in the formulation we shall include the real axis poles in the integration over the fourth quadrant. Substituting the relation

$$2J_0(kr) = H_0^{(1)}(kr) + H_0^{(2)}(kr) \tag{3-100}$$

into (3-92) where $H_0^{(1)}$ and $H_0^{(2)}$ are the first and second Hankel functions and integrating $H_0^{(1)}$ over path I in Fig. 3-43 and $H_0^{(2)}$ over path II, we shall have, from Cauchy's integral theorem

$$\int_0^\infty H_0^{(1)}(kr)F(\beta_1, \beta_2)k\,dk = \Big(\underset{A}{\int_0^{i\infty}} + \lim_{k\to\infty}\int_{i\infty}^{0}\Big) H_0^{(1)}(kr)F(\beta_1, \beta_2)k\,dk \tag{3-101}$$

and

$$\int_0^\infty H_0^{(2)}(kr)F(\beta_1, \beta_2)k\,dk$$

$$= \Big(\overset{B}{\int_0^{-i\infty}} + \lim_{k\to\infty}\overset{C}{\int_{-i\infty}^{0}} + \overset{D}{\int_{k_1}^{-i\infty}} + \overset{E}{\int_{-i\infty}^{k_1}} + \overset{F}{\int_{k_2}^{-i\infty}} + \int_{-i\infty}^{k_2}\Big) H_0^{(2)}(kr)F(\beta_1, \beta_2)k\,dk + \sum \text{residues} \tag{3-102}$$

where $F(\beta_1, \beta_2)$ is defined as

$$F(\beta_1, \beta_2) = \frac{\sin(\beta_1 d)}{\beta_1}\left[\frac{\beta_1 \cos[\beta_1(H-z)] + ib\beta_2 \sin[\beta_1(H-z)]}{\beta_1 \cos(\beta_1 H) + ib\beta_2 \sin(\beta_1 H)}\right] \tag{3-103}$$

The integral (3-92) will then be simply the sum of (3-101) and (3-102). The second integral of (3-101) is zero since $H_0^{(1)}$ vanishes at ∞ over the first quadrant, and the second integral of (3-102) is zero since $H_0^{(2)}$ vanishes at ∞ over the second quadrant. Since $H_0^{(1)}(iq) = -H_0^{(2)}(-iq)$, the first integrals on the right-hand side of (3-101) and (3-102) will cancel. Also since $F(\beta_1, \beta_2)$ is an even function of β_1, the third and fourth integrals of (3-102) will cancel. We are then left, for the sum of (3-101) and (3-102), with

$$\Psi_2 = \int_{-i\infty}^{k_2} H_0^{(2)}(kr)k\,dk[F(\beta_1, \beta_2) - F(\beta_1, -\beta_2)] + \sum \text{residues} \tag{3-104}$$

a branch-line integral along the k_2 branch line and the sum of the residues. The solution for the branch-line integral will be discussed in Chap. 5.

The residues are the normal modes. They are located at the poles of the integrand (3-103), the poles being those values of $k = k_n$ for which the denominator of (3-103) is equal to zero.

$$\beta_1 \cos(\beta_1 H) + ib\beta_2 \sin(\beta_1 H) = 0$$

or

$$\beta_2 = \frac{i\beta_1}{b \tan(\beta_1 H)} \tag{3-105}$$

From Cauchy's integral formula, the residues then are equal to

$$\begin{aligned}\Psi_2' &= \sum_n \text{residues} \int H_0^{(2)}(kr)F(\beta_1, \beta_2)k\,dk \\
&= -2\pi i \sum_n H_0^{(2)}(k_n r)k_n \frac{\sin(\beta_1 d)\{\beta_1 \cos[\beta_1(H-z)] + ib\beta_2 \sin[\beta_1(H-z)]\}}{\beta_1(\partial/\partial k)[\beta_1 \cos(\beta_1 H) + ib\beta_2 \sin(\beta_1 H)]} \\
&= -2\pi i \sum_n H_0^{(2)}(k_n r)\frac{k_n}{\beta_1}\sin(\beta_1 d) \frac{\cos(\beta_1 z)[\beta_1 \cos(\beta_1 H) + ib\beta_2 \sin(\beta_1 H)] + \sin(\beta_1 z)[\beta_1 \sin(\beta_1 H) - ib\beta_2 \cos(\beta_1 H)]}{-(k_n/\beta_1)\cos(\beta_1 H) + k_n H \sin(\beta_1 H) - i(bk_n/\beta_2)\sin(\beta_1 H) - i(b\beta_2 k_n H/\beta_1)\cos(\beta_1 H)} \\
&= -2\pi i \sum_n H_0^{(2)}(k_n r) \frac{\sin(\beta_1 d)\sin(\beta_1 z)[\beta_1/\sin(\beta_1 H)]}{H[\beta_1 \sin(\beta_1 H) - ib\beta_2 \cos(\beta_1 H)] - \cos(\beta_1 H) - ib\dfrac{\beta_1}{\beta_2}\sin(\beta_1 H)} \\
&= -\frac{2\pi i}{H}\sum_n H_0^{(2)}(k_n r)\frac{\beta_1 H \sin(\beta_1 d)\sin(\beta_1 z)}{\beta_1 H - \sin(\beta_1 H)\cos(\beta_1 H) - b^2 \sin^2(\beta_1 H)\tan(\beta_1 H)}\end{aligned} \tag{3-106}$$

using the relation (3-105) and remembering that Ψ_2' is evaluated at $k = k_n$. If instead of (3-92) we had used (3-91), which is the same as (3-92) by an interchange of z and d, we would have obtained the same result; Eq. (3-106) holds for all of medium 1. For large values of r the Hankel function may be approximated by its asympototic form

$$H_0^{(2)}(k_n r) \to \sqrt{\frac{2}{\pi k_n r}}\, e^{i[(\pi/4) - k_n r]} = i\sqrt{\frac{2}{\pi k_n r}}\, e^{-i[k_n r + (\pi/4)]} \tag{3-107}$$

Substituting (3-107) into (3-106) and (3-92) gives

$$\Phi_2' = \frac{2\pi}{H}\sqrt{\frac{2}{\pi r}}\sum_{n=1}^{\infty} e^{i[\omega t - k_n r - (\pi/4)]}\, G(k_n)\sin(\beta_1 d)\sin(\beta_1 z) \tag{3-108}$$

where $G(k_n)$ is defined by

$$G(k_n) \equiv \frac{1}{\sqrt{k_n}}\,\frac{\beta_1 H}{\beta_1 H - \sin(\beta_1 H)\cos(\beta_1 H) - b^2\sin^2(\beta_1 H)\tan(\beta_1 H)} \tag{3-109}$$

The meaning of the residues now becomes clear. They define a propagation in the r direction with a phase velocity V for each mode given by

$$V_n = \frac{\omega}{k_n} \tag{3-110}$$

The amplitude variation with respect to the depth of the source d and receiver z is given by $\sin(\beta_1 d)\sin(\beta_1 z)$. The function $G(k_n)$ defines the relative strength of each mode. Further the velocity potential decreases as $r^{-\frac{1}{2}}$ with range, cylindrical (two-dimensional) spreading.

Substituting (3-110), (3-90), and (3-82), into (3-105) we obtain for real values of k_n (unattenuated propagation)

$$\tan(\beta_1 H) = \frac{i\beta_1}{b\beta_2}$$

$$\tan\left(\omega H\sqrt{\frac{1}{c_1^2} - \frac{1}{V_n^2}}\right) = -\frac{\rho_2\sqrt{\frac{1}{c_1^2} - \frac{1}{V_n^2}}}{\rho_1\sqrt{\frac{1}{V_n^2} - \frac{1}{c_2^2}}}$$

or expressing f_n as a multiple-valued function of V rather than V_n as a multiple-valued function of f

$$\tan\left(\frac{2\pi f_n H}{V}\sqrt{\frac{V^2}{c_1^2} - 1}\right) = -\frac{\rho_2\sqrt{\frac{V^2}{c_1^2} - 1}}{\rho_1\sqrt{1 - \frac{V^2}{c_2^2}}} \tag{3-111}$$

which is the same as Eq. (3-45) obtained from the physical formulation. The depth dependence, z and d, of Eq. (3-108) is the same as (3-49) from the physical formulation upon substitution for β_1 from (3-82). In addition we might have expected from the physical formulation that, since the energy propagated was confined to the water by total reflections from the surface and bottom, the over-all spreading would be cylindrical as given in (3-108). Equation (3-111) is the characteristic equation; it is the dispersion equation relating the frequency f_n to the phase velocity V.

For the generalization to a point-source pulse, $f(\tau)$, we shall take as illustration an exponential pulse.

$$\begin{aligned} f(\tau) &= e^{-\lambda\tau} && \text{for } \tau \geqslant 0 \\ &= 0 && \tau < 0 \end{aligned} \tag{3-112}$$

Then, from (2-110) the Fourier transform $g(\omega)$ of $f(\tau)$ will be

$$g(\omega) = \int_0^\infty e^{-(\lambda+i\omega)\tau}\,d\tau = \frac{1}{\lambda + i\omega} \tag{3-113}$$

From (3-108) and (3-94) the normal-mode disturbance at long ranges will then be

$$P(r,z,t) = \sum_n P_n(r,z,t) = \sum_n \frac{1}{2\pi}\int_{-\infty}^{\infty} \frac{Q_n(r,z,\omega)}{\lambda + i\omega}\, e^{i[\omega t - k_n(\omega)r - (\pi/4)]}\,d\omega \tag{3-114}$$

where Q_n is defined as

$$Q_n(r, z, \omega) = \frac{2\pi}{H}\sqrt{\frac{2}{\pi k_n r}}\left[\frac{\beta_1 H \sin(\beta_1 d)\sin(\beta_1 z)}{\beta_1 H - \sin(\beta_1 H)\cos(\beta_1 H) - b^2\sin^2(\beta_1 H)\tan(\beta_1 H)}\right] \tag{3-115}$$

or the contribution to each mode may be written as

$$P_n(r, z, t) = \frac{1}{2\pi}\int_{-\infty}^{\infty}\frac{Q_n(r, z, \omega)}{\lambda + i\omega}e^{i[\omega t - k_n(\omega)r - (\pi/4)]}\,d\omega \tag{3-116}$$

Since the amplitude term, $Q_n(r, z, \omega)/(\lambda + i\omega)$, is a slowly varying function of ω with respect to the rapid oscillation of the exponential, the integral may be evaluated by the method of stationary phase, as discussed in Chap. 2. The principal contribution to the integral comes from the vicinity of those values of ω for which the modulus of the exponential is stationary. Only in this region will there be a nonoscillating contribution to the integral. We wish to evaluate the integral (3-116) at $\omega = \omega_0$ satisfying the condition

$$\frac{dp(\omega)}{d\omega} = \frac{d}{d\omega}\left[\omega t - k_n(\omega)r - \frac{\pi}{4}\right] = t - r\left(\frac{dk_n}{d\omega}\right)_{\omega=\omega_0} = 0 \tag{3-117}$$

In the vicinity of $\omega = \omega_0$ we may write $p(\omega)$ in a series expansion as

$$p(\omega) = p(\omega_0) - \frac{rk_n''(\omega_0)}{2}u^2 - \frac{rk_n'''(\omega_0)}{6}u^3 - \cdots \tag{3-118}$$

where

$$\omega = \omega_0 + u$$

Since the integral has an appreciable value only in this neighborhood, the amplitude term may be factored outside the integral and the limits of integration taken from $-\infty$ to ∞.

$$\begin{aligned}P_n(r, z, t) &= \frac{1}{2\pi}\frac{Q_n(\omega_0, r, z)e^{ip(\omega_0)}}{\lambda + i\omega_0}\int_{-\infty}^{\infty}e^{-i[rk''/2)u^2+\cdots]}\,du\\ &= \frac{1}{2\pi}\frac{Q_n(\omega_0, r, z)e^{ip(\omega_0)}}{\lambda + i\omega_0}\left(\sqrt{\frac{2\pi}{|rk''|}}\,e^{\pm i(\pi/4)} + \cdots\right)\\ &= \frac{1}{2\pi}\frac{Q_n(\omega_0, r, z)}{(\lambda^2 + \omega_0^2)^{1/2}}e^{i[\omega_0 t - k_n(\omega_0)r - (\pi/4) - \tan^{-1}(\omega_0/\lambda)]}\\ &\qquad\left(\sqrt{\frac{2\pi}{|rk''|}}\,e^{\pm i(\pi/4)} + \cdots\right)\end{aligned} \tag{3-119}$$

where the plus sign is taken for $k'' < 0$ and the negative sign for $k'' > 0$. This evaluation is valid only under the condition that the contribution from the third term of (3-118) is small compared with the second. Such will generally be the case except in the vicinity of values of k_n for which $k_n''(\omega_0)$ is also zero. The evaluation is not yet complete, for we have not considered the contributions for negative values of ω. For ω negative, the solution for the simple harmonic source problem would give for the Hankel function the negative of the phase in (3-107) in order to preserve the condition of outgoing waves from the source. The function will be odd in ω, and thus there will be another point of stationary phase at $\omega = -\omega_0$. Since $Q_n(\omega_0, r, z)$ is an even function in ω, this second contribution will be the complex conjugate of (3-119); and we shall have for the final solution

$$P_n(r, z, t) = \frac{Q_n(\omega_0, r, z)}{[2\pi r|k''(\omega_0)|(\lambda^2 + \omega_0^2)]^{1/2}} [e^{i[\omega_0 t - k_n(\omega_0)r - \tan^{-1}(\omega_0/\lambda)]} + e^{-i[\omega_0 t - k_n(\omega_0)r - \tan^{-1}(\omega_0/\lambda)]}]$$

$$= \frac{2Q_n(\omega_0, r, z) \cos [\omega_0 t - k_n(\omega_0)r - \tan^{-1}(\omega_0/\lambda)]}{[2\pi r|k''(\omega_0)|(\lambda^2 + \omega_0^2)]^{1/2}} \quad \text{for } k'' < 0 \qquad (3\text{-}120)$$

and

$$P_n(r, z, t) = \frac{2Q_n(\omega_0, r, z) \cos [\omega_0 t - k_n(\omega_0)r - \tan^{-1}(\omega_0/\lambda) - (\pi/2)]}{[2\pi r|k''(\omega_0)|(\lambda^2 + \omega_0^2)]^{1/2}} \quad \text{for } k'' > 0 \qquad (3\text{-}121)$$

Let us examine the solutions that have been obtained here for long-range propagation. For a simple harmonic point source of frequency ω the evaluation gives a series of normal modes (3-108) propagating in the r direction whose amplitude variation in the z direction is determined by the boundary conditions at $z = 0$ and $z = H$. The propagation is dispersive, the phase velocity V in the r direction being determined by Eq. (3-111) as a function of the frequency f. For an impulsive point source with a given spectrum of frequencies $g(\omega)$, it is necessary to consider the combined effect from all frequencies for each normal mode at any particular point of space-time. The result [Eqs. (3-120) and (3-121)] is that at any particular point of space-time, r_0, t_0, a dominant frequency, ω_0, determined by (3-117), will be observed. For any other point of space-time r_1, t_1, Eq. (3-117) will be in general satisfied for a different frequency, ω_1. If it is desired to follow a particular frequency ω_0, its velocity of propagation will be given from (3-117) and (3-110) as simply,

$$U = \frac{d\omega}{dk} = \frac{d(Vk)}{dk} = V + k\frac{dV}{dk} \qquad (3\text{-}122)$$

or in dimensionless form

$$\frac{U}{c_1} = \frac{V}{c_1} + (kH)\frac{d(V/c_1)}{d(kH)} \qquad (3\text{-}123)$$

The velocity U is the group velocity; its physical significance with respect to the phase velocity V has been discussed in Chap. 2.

Two of the important parameters defining the propagation are the phase velocity and the group velocity. From Eq. (3-111) we see that for unattenuated propagation the phase-velocity curve for a given normal mode begins at the low-frequency end with the cutoff frequency f_{cn}, given by Eq. (3-47), and the phase velocity $V = c_2$. It decreases with increasing frequency asymptotically toward $V = c_1$ as f_n approaches infinity. It is convenient to determine the group velocity graphically from the phase-velocity curve rather than arithmetically from an explicit expression (3-122). At the low-frequency end, $f_n = f_{cn}$, $dV/dk = 0$; and the group velocity is equal to the phase velocity, $U = c_2$. At the high-frequency end as $f_n \rightarrow \infty$, $dV/dk \rightarrow 0$; the group velocity approaches the phase velocity, $U \rightarrow c_1$. Since $dV/dk < 0$ for all values of f_n, the group velocity is always less than the phase velocity. At some intermediate value of frequency $|dV/dk|$ will be a maximum so that at some slightly larger value of frequency the group-velocity curve will go through a minimum. Figure 3-44 is a graph of the phase and group velocities versus a dimensionless parameter of the frequency for the first mode for various velocity contrasts c_2/c_1. Figure 3-45 is a graph of the phase and group velocities for the first three modes for a particular velocity contrast c_2/c_1.

From the expression for $P_n(r, z, t)$, we see that the disturbance for the frequency ω_0 observed at any particular point (r_0, z_0, t_0) is proportional to the spectral amplitude $(\lambda^2 + \omega_0^2)^{-1/2}$ of that frequency at the source. From (3-115) and (3-120) or (3-121) the range dependence is as r^{-1} as compared with $r^{-1/2}$ for the simple harmonic point-source solution. The intensity decreases as r^{-2} (spherical spreading) rather than r^{-1} (cylindrical spreading). At first glance this may seem contradictory to the condition of guided transmission with spreading in the r direction only. However, the dispersion phenomenon itself introduces an additional spreading. Any particular portion of the wave train defined between, say, the frequencies ω_1 and ω_2 will be spreading apart as a function of time as one follows the wave train; and that spreading will be proportional to $U\,dt$ as a function of time. The total spreading loss to a point (r, z, t) will then be $r^{-1} \times (Ut)^{-1}$ which from (2-118) is equal to

$$r^{-1} \times r^{-1} = r^{-2}$$

The total energy received at any particular point of space (r, z) integrated over all time will in general decrease as r^{-1}; the intensity at any particular value of time $t = t_0$ will, however, be proportional to r^{-2}. This result is similar to that obtained from the ray solution for the higher frequencies. The intensity for any given ray arrival is proportional to r^{-2}. The total

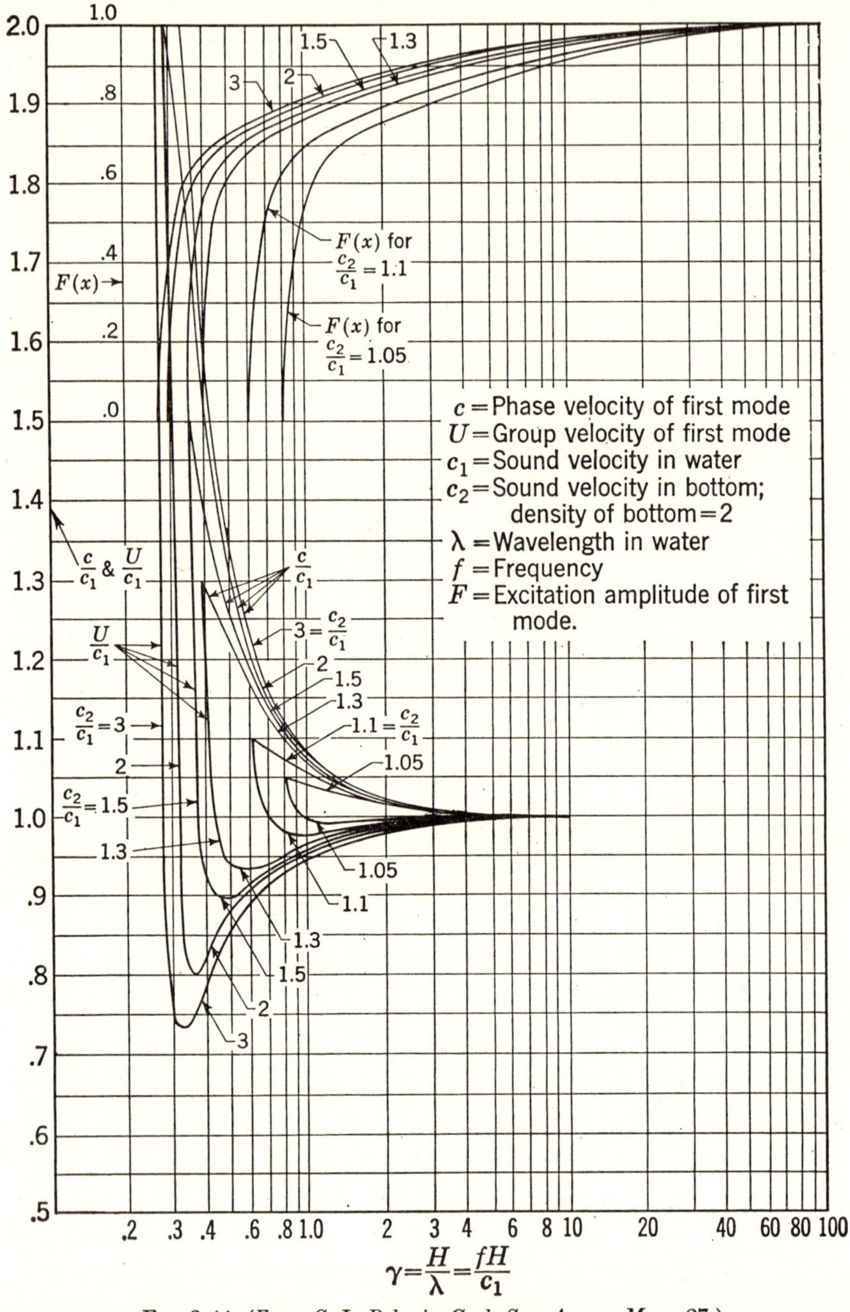

FIG. 3-44. (*From C. L. Pekeris, Geol. Soc. Amer., Mem. 27.*)

energy received over all ray paths is proportional to r^{-1}. Also from (3-120) or (3-121) we see that $P_n(r, z, t)$ is proportional to $|k''(\omega_0)|^{1/2}$. Such a dependence should be expected, for $k''(\omega_0)$ defines the relative rate of dispersion as a function of frequency. For $k''(\omega_0)$ large, the wave train is highly dispersive; adjacent frequencies ω_1 and ω_2 with group velocities U_1 and U_2 are spreading apart rapidly. For $k''(\omega_0)$ small, the dispersion spreading is smaller.

In the vicinity of the minimum group velocity, $U = U_m$, we see from (3-122) that $k'' \approx 0$. The condition that the second term of (3-118) is

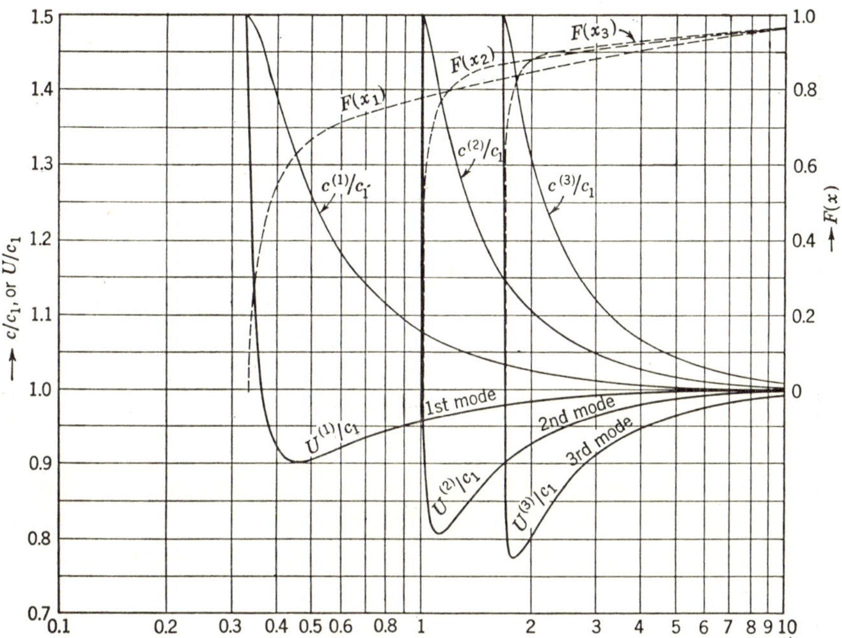

FIG. 3-45. (*From C. L. Pekeris, Geol. Soc. Amer., Mem.* 27.)

small with respect to the third term will no longer be satisfied, and the evaluation (3-119) will no longer be correct. A new evaluation of the integral (3-116) must be made for the region near the frequency $\omega = \omega_A$ of the minimum group velocity. This evaluation is moderately complex and will not be given here; the resultant contribution to the wave train is termed the *Airy phase*. The Airy phase contribution is

$$P_n(r, z, t) = \frac{2\pi Q_n(\omega_A, r, z)E(r, t, t_m)}{(\lambda^2 + \omega_A{}^2)^{1/2}[9rk'''(\omega_A)]^{1/3}}$$

$$\cos\left[\omega_A t - k_n(\omega_A)r - \tan^{-1}\left(\frac{\omega_A}{\lambda}\right) - \frac{\pi}{4}\right] \qquad \text{for } \omega = \omega_A \quad (3\text{-}124)$$

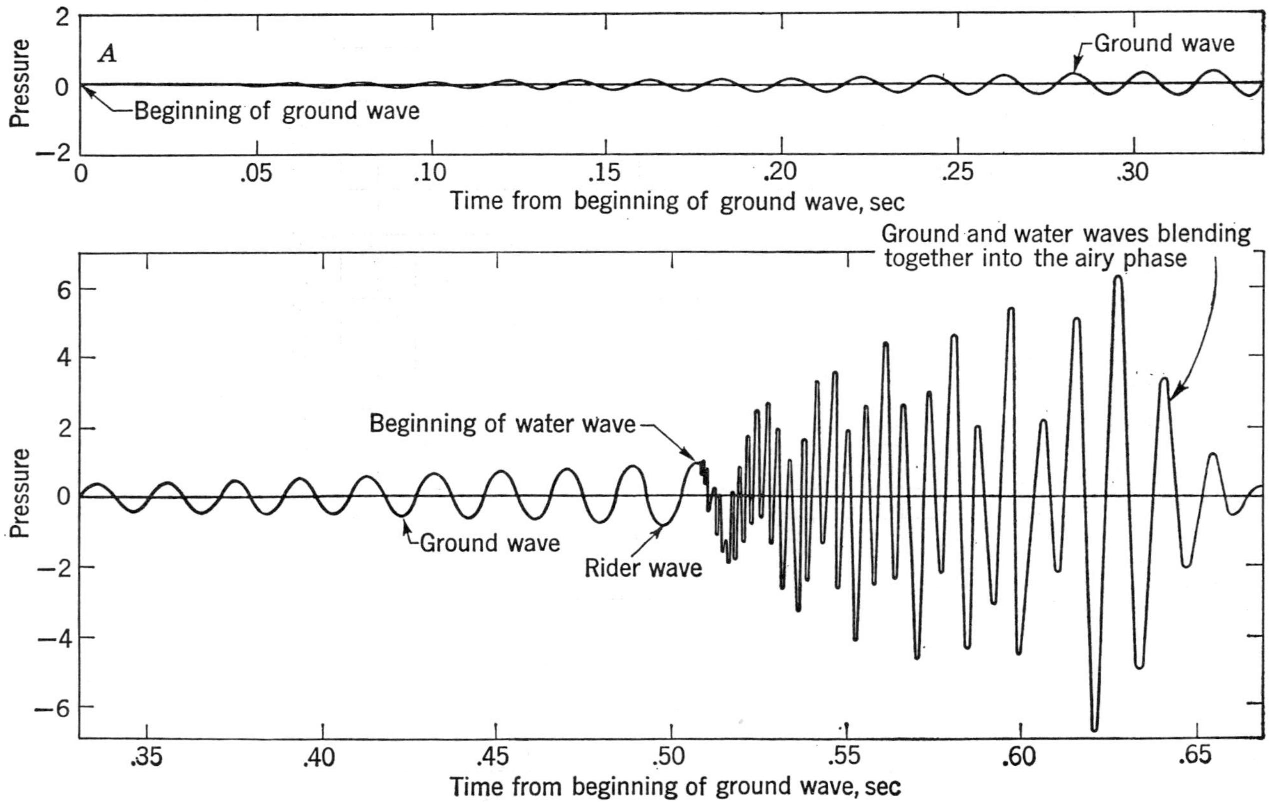

FIG. 3-46. (*From C. L. Pekeris, Geol. Soc. Amer., Mem. 27.*)

where $E(r, t, t_m)$ is a function of r, t, and the time $t_m = r/U_m$ about the minimum group velocity; the expression (3-124) is valid only for small values of $|t - t_m|$. The Airy phase contribution defines a small nondispersive wave packet of constant frequency, $\omega = \omega_A$. The wave train does not end abruptly at the minimum group velocity time, $t = t_m$, but continues with a small wave packet of frequency ω_A whose amplitude is modulated in time by the function $E(r, t, t_m)$. It is interesting to note from (3-115) and (3-124) that the range dependence is as $r^{-5/6}$ as compared with r^{-1} for the dispersive wave train. It might be expected that with increasing range the Airy phase would become the dominant part of the wave train; however, this is not necessarily the case, for the amplitude modulation $E(r, t, t_m)$ is also a function of r.

Figure 3-46 is a graph of the displacement as a function of time computed from Eqs. (3-120), (3-121), and (3-124) for a particular receiver location. The portion of the dispersive wave train beginning with the velocity c_2 and continuing with increasing frequency down to the minimum group velocity U_m is referred to as the *ground wave.* The portion beginning with velocity c_1 and in the limit with infinitely high frequency and continuing with decreasing frequency to the minimum group velocity is referred to as the *water wave.* For the water wave k'' is less than zero, and therefore Eq. (3-120) is applicable. For the ground wave k'' is greater than zero, and Eq. (3-121) is applicable. The last portion of the wave train is the Airy phase.

The formal solution for the attenuated portion of each mode is quite complicated. Solutions for complex values of k and β, rather than for real (unattenuated propagation) or pure imaginary (no propagation, exponential decrease) values, offer considerable difficulty. The previous conditions of positive real or negative imaginary values for k and β cannot be met simultaneously for complex values of k. The problem must be restated in a slightly different manner. Physically we can see that, for complex k, incidence at angles steeper than critical, there should be attenuated propagation in the r direction in the ocean channel. There should be a propagation coefficient in both the r and z directions in the lower medium, attenuated propagation in the r direction, and an increase in amplitude in the z direction to a limiting value.

The formal solution for complex values will not be given here. However, we can examine its form from solutions to Eq. (3-105). From Eq. (3-110) we see that since ω is real V must be complex for k to be complex. Solutions to Eq. (3-105) for complex V involve a large amount of algebra. We shall make the unjustified assumption for the formulation that has been given that V remains real and ω is complex. Equation (3-105) then becomes

$$\tan(\beta_1 H) = \frac{i\beta_1}{b\beta_2}$$

$$\tan\left[(\omega_r + i\omega_i)\frac{H}{V}\sqrt{\frac{V^2}{c_1^2} - 1}\right] = i\frac{\sqrt{(V^2/c_1^2) - 1}}{b\sqrt{(V^2/c_2^2) - 1}}$$

$$\frac{\tan[(\omega_r H/V)\sqrt{(V^2/c_1^2) - 1}] + \tan[i(\omega_i H/V)\sqrt{(V^2/c_1^2) - 1}]}{1 - \tan[(\omega_r H/V)\sqrt{(V^2/c_1^2) - 1}]\tan[i(\omega_i H/V)\sqrt{(V^2/c_1^2) - 1}]} = i\frac{\sqrt{(V^2/c_1^2) - 1}}{b\sqrt{(V^2/c_2^2) - 1}}$$

$$\frac{\tan[(\omega_r H/V)\sqrt{(V^2/c_1^2) - 1}] + i\tanh[(\omega_i H/V)\sqrt{(V^2/c_1^2) - 1}]}{1 - i\tan[(\omega_r H/V)\sqrt{(V^2/c_1^2) - 1}]\tanh[(\omega_i H/V)\sqrt{(V^2/c_1^2) - 1}]} = i\frac{\sqrt{(V^2/c_1^2) - 1}}{b\sqrt{(V^2/c_2^2) - 1}} \quad (3\text{-}125)$$

Solutions to (3-125) depend on whether the right-hand side is greater or less than 1. This condition exists because the hyperbolic tangent is restricted to values between -1 and 1. For $\sqrt{(V^2/c_1^2) - 1} > b\sqrt{(V^2/c_2^2) - 1}$ (no phase change upon reflection from the bottom) the solutions will be

$$\tan\left(\frac{\omega_r H}{V}\sqrt{\frac{V^2}{c_1^2} - 1}\right) = \pm\infty \quad (3\text{-}126)$$

and

$$\tanh\left(\frac{\omega_i H}{V}\sqrt{\frac{V^2}{c_1^2} - 1}\right) = \frac{b\sqrt{(V^2/c_2^2) - 1}}{\sqrt{(V^2/c_1^2) - 1}} \quad (3\text{-}127)$$

For $\sqrt{(V^2/c_1^2) - 1} < b\sqrt{(V^2/c_2^2) - 1}$ (phase change of 180 deg upon reflection from the bottom), the solutions will be

$$\tan\left[\frac{\omega_r H}{V}\sqrt{\frac{V^2}{c_1^2} - 1}\right] = 0 \quad (3\text{-}128)$$

and

$$\tanh\left(\frac{\omega_i H}{V}\sqrt{\frac{V^2}{c_1^2} - 1}\right) = \frac{\sqrt{(V^2/c_1^2) - 1}}{b\sqrt{(V^2/c_2^2) - 1}} \quad (3\text{-}129)$$

For the normal conditions of $\rho_2 > \rho_1$ and $c_2 > c_1$ we shall have from (3-126)

$$\frac{\omega_r H}{V}\sqrt{\frac{V^2}{c_1^2} - 1} = (2n - 1)\frac{\pi}{2} \quad (3\text{-}130)$$

the same as Eq. (3-51) from the physical derivation. From Eq. (3-127) we shall have

$$
\begin{aligned}
\frac{\omega_i H}{V}\sqrt{\frac{V^2}{c_1^2}-1} &= \tanh^{-1}\left[\frac{b\sqrt{(V^2/c_2^2)-1}}{\sqrt{(V^2/c_1^2)-1}}\right] \\
&= \frac{1}{2}\ln\left[1+\frac{b\sqrt{(V^2/c_2^2)-1}}{\sqrt{(V^2/c_1^2)-1}}\right] \\
&\qquad\qquad -\frac{1}{2}\ln\left[1-\frac{b\sqrt{(V^2/c_2^2)-1}}{\sqrt{(V^2/c_1^2)-1}}\right] \\
&= \frac{1}{2}\ln\left[\frac{\sqrt{(V^2/c_1^2)-1}+b\sqrt{(V^2/c_2^2)-1}}{\sqrt{(V^2/c_1^2)-1}-b\sqrt{(V^2/c_2^2)-1}}\right]
\end{aligned}
$$

and from Eq. (3-50)

$$\omega_i = \frac{V}{2H\sqrt{(V^2/c_1^2)-1}}\ln R^{-1} \tag{3-131}$$

The time exponential then becomes

$$e^{i(\omega_r+i\omega_i)t} = e^{i\omega_r t}\exp\left[-\frac{Vt}{2H\sqrt{(V^2/c_1^2)-1}}\ln R^{-1}\right] \tag{3-132}$$

the same as Eq. (3-56) from the physical derivation.

BIBLIOGRAPHY

Arons, A. B., and D. R. Yennie: Phase Distortion of Acoustic Pulses Obliquely Reflected from a Medium of Higher Sound Velocity, *J. Acoust. Soc. Amer.*, vol. 22, pp. 231–237, 1950.

Iselin, C. O'D.: The Application of Oceanography to Subsurface Warfare, Div. 6, vol. 6A, *Natl. Defense Research Council Summary Tech. Rept.*, Washington, 1946.

Kuenen, Ph. H.: "Marine Geology," John Wiley & Sons, Inc., New York, 1950.

Pekeris, C. L.: Theory of Propagation of Explosive Sound in Shallow Water *in* Propagation of Sound in the Ocean, *Geol. Soc. Amer., Mem.* 27, 1948.

Shepard, F. P.: "Submarine Geology," Harper & Brothers, New York, 1948.

Sommerfeld, A.: "Partial Differential Equations in Physics," Academic Press, Inc., New York, 1949.

Spitzer, L. (ed.): Physics of Sound in the Sea, Div. 6, vol. 8, *Natl. Defense Research Council Summary Tech. Rept.*, Washington, 1945.

Sverdrup, H. U., M. W. Johnson, and R. H. Fleming: "The Oceans," Prentice-Hall, Inc., Englewood Cliffs, N.J., 1942.

Tolstoy, I.: Submarine Topography in the North Atlantic, *Bull. Geol. Soc. Amer.*, vol. 62, pp. 441–450, 1951.

Tolstoy, I.: Resonant Frequencies and High Modes in Layered Wave Guides, *J. Acoust. Soc. Amer.*, vol. 28, pp. 1182–1192, 1956.

Woods Hole Oceanographic Institution: Sound Transmission in Sea Water, *Natl. Defense Research Council Rept.*, Washington, 1941.

4. *Transmission in Deep Water*

The description of sound transmission for deep water has, of course, many similarities to that for shallow water. The primary difference between the two is simply one of dimensions; the average depths of water for deep-water transmission are around 10,000 to 20,000 feet whereas those for shallow-water transmission are less than 300 feet. Nevertheless there are a number of features of both the ray and mode solutions which have not been discussed and which are more appropriately considered in connection with the environment of deep-water transmission.

Figure 3-1 is representative of deep-water bathymetry. Much of the area is covered by a broad basin at depths of between 2,500 and 3,000 fathoms. Isolated peaks are found, such as Caryn peak in the northwestern portion of the area; and ridges or lines of sea mounts, such as the group along the extension of a line southeast from Cape Cod, are not uncommon. Around the Bermuda islands there is a gentle rise from the basin up to the steeper submarine slope of the islands. Southeast from Bermuda the topography becomes fairly mountainous in a series of ridges. Bordering the continent of North America there is the gentle continental rise from the basin up to the steeper continental slope. In many areas the continental slope is incised by a number of submarine canyons, such as are shown by the indentations in the contours south of Cape Cod, which extend in some cases across a portion of the continental rise. Several of these canyons have an associated delta extending across the lower portions of the continental rise out to the basin. Bordering the continents or island arcs in some areas are the deep-sea trenches. They are features whose length is long compared with their width and whose depths are around 1,000 to 2,000 fathoms deeper than the adjacent basin. South of the area of this map and bordering the Antilles island arc is the Puerto Rico trench with depths of 4,400 fathoms. Another feature of importance is the mid-ocean ridge. To the east of the area of this map is the Mid-Atlantic ridge. It is the largest single topographic feature in the Atlantic and extends from Iceland through the entire length of the Atlantic to join with the Mid-Indian ridge. The ridge is fairly broad with dimensions of the order of 300 nautical miles, and the topography across the ridge is generally rugged, sometimes culminating in oceanic islands or island groups, such as the Azores. All these variations in submarine topography, sometimes rugged and sometimes gentle, are important to the description of deep-water sound transmission by bottom-reflected paths.

Over much of the deep-ocean areas the properties of the bottom materials are fairly similar. The sediments are predominantly of clay-sized particles with occasional interspersed sand and silt lenses. They contain a large fraction of calcareous material, principally the tests of such micro-organisms as globigerina and radiolaria. They are generally highly porous with porosities, i.e., the ratio of the volume of contained fluid to total fluid, in the neighborhood of 80 per cent. One principal distinction in the distribution of deep-sea sediments is that between the continental rise and the basin. The continental rise is the environment of most rapid deposition; over the basin the rate of deposition is considerably less. Figure 4-1 is a geologic section across the continental rise and basin from the east coast of North America to Bermuda.

The distribution of sound velocity in deep-ocean areas is not so variable as it is in the shallow-water areas that were discussed previously. In the majority of cases, the velocity as a function of depth has a more characteristic form. For the determination of the variations of velocity with depth, area, and time, salinity variations are generally unimportant; and the pressure function is simply a constant gradient proportional to depth. Temperature is the primary variable. Over much of the deep-ocean area the temperature structure can be divided into three parts, (1) a relatively warm *surface layer*, often showing isothermal structure, (2) the *main thermocline* in which the temperature decreases moderately rapidly with depth, and (3) the colder *deep-water layers* in which the temperature decreases slowly with depth. The surface layer is affected by meteorological changes and sea-roughness changes. Through wind action the surface layer often shows complete mixing; the sound velocity structure is then an isothermal channel with the ocean surface. The isothermal structure when developed generally extends down to a hundred feet or so. The surface layer is subject to daily, seasonal, and areal changes. The main thermocline is affected primarily only by areal changes. The deep-water layers also show variation only with respect to areal changes but of a lesser degree. The division between the main thermocline and the deep-water layers is generally considered to be around 600 fathoms.

Figure 4-2*a* is a graph of temperature versus depth for a mid-latitude area of the North Atlantic during the winter. The surface isothermal channel has been developed to an exceedingly deep depth. Figures 4-2*b*

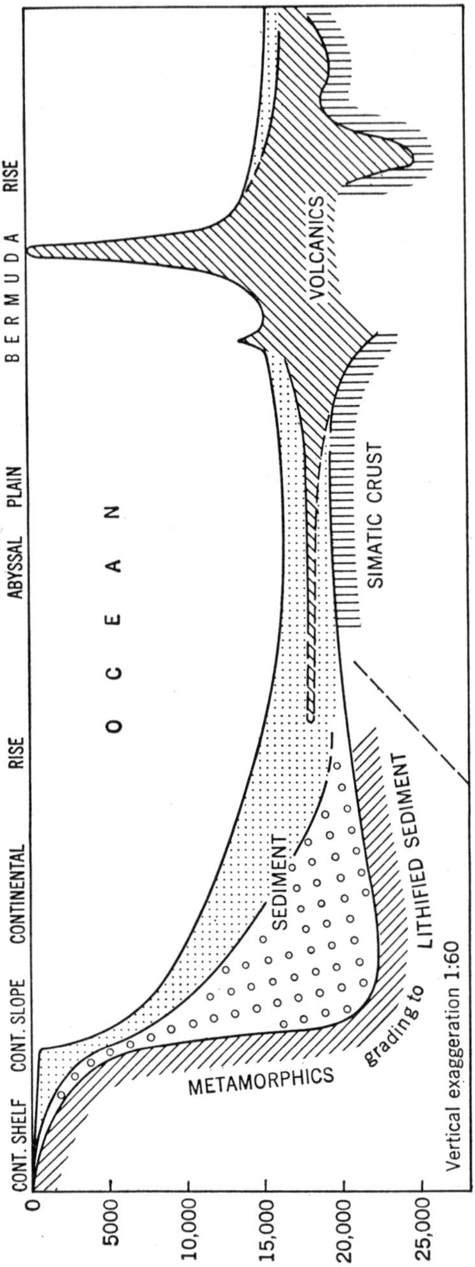

FIG. 4-1. (*From C. B. Officer, Deep Sea Research,* 2, 1955.)

and 4-2*c* are representative of summer conditions in the equatorial and high latitudes of the Atlantic. The surface channel is considerably smaller, and the main thermocline in equatorial waters is quite steep. Both the seasonal and daily temperature changes are marked, particularly in the mid-latitude regions. During the winter the surface water is colder and stirring greater so that the isothermal channel develops to its greatest depth. During the summer the surface water is warmer, and a *seasonal thermocline* often develops above the main thermocline. Figure

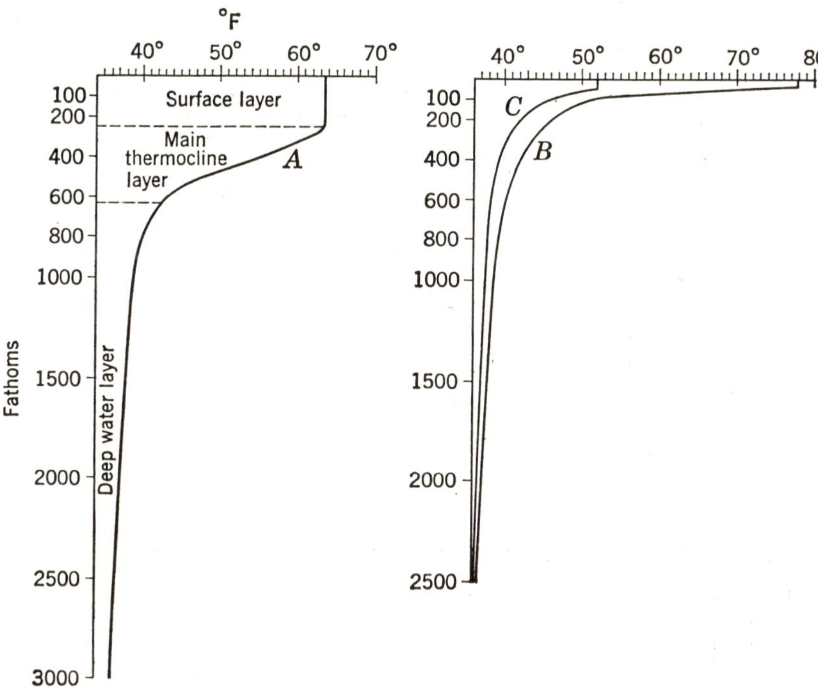

FIG. 4-2. (*From Woods Hole Oceanographic Institution, Natl. Defense Research Council,* 1941.)

4-3*a* is a temperature-depth plot for the mid-latitude area of Fig. 4-2*a* in mid-summer. Figure 4-3*b* shows the variation in the surface layer with season for the same area. Figures 4-4 and 4-5 are maps of isothermal and isovelocity contours versus depth and month of the year; they are included to illustrate the development of the seasonal thermocline. The diurnal effect is similar to the seasonal effect though of course on a smaller scale. The upper few fathoms of water warm up during the daytime and develop a slight thermocline of the order of a fraction of a degree. The thermocline disappears at night. Figure 4-6*a* is a plot of temperature versus depth for three times of day illustrating the diurnal effect. Figure 4-6*b* is an isothermal map versus depth and time of day of

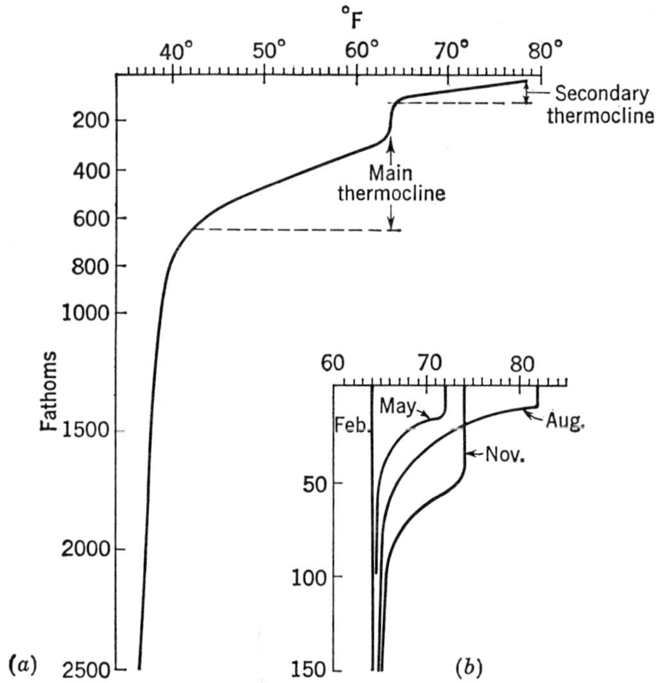

FIG. 4-3. (*From Woods Hole Oceanographic Institution, Natl. Defense Research Council,* 1941.)

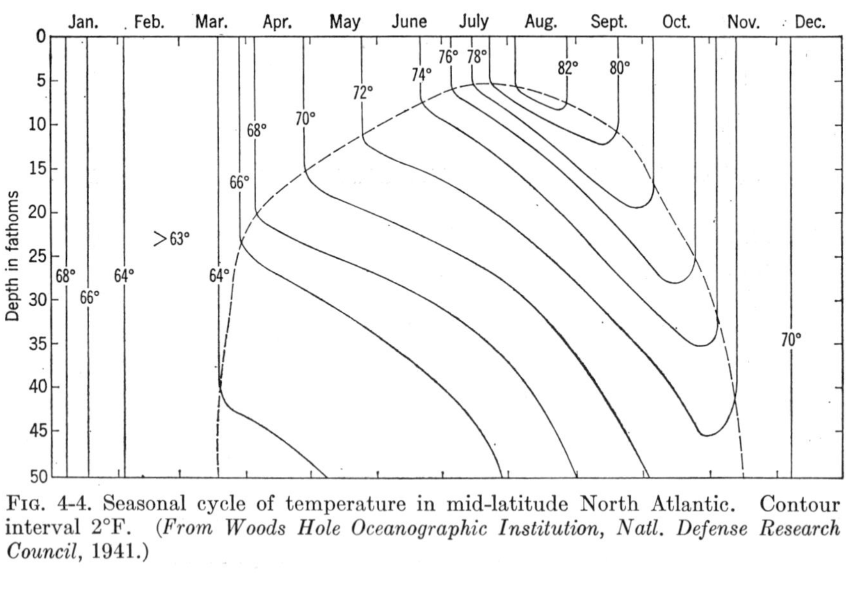

FIG. 4-4. Seasonal cycle of temperature in mid-latitude North Atlantic. Contour interval 2°F. (*From Woods Hole Oceanographic Institution, Natl. Defense Research Council,* 1941.)

the diurnal effect. A fair amount of the time no isothermal channel is present, and a negative temperature gradient exists from the surface downward. This gives a negative velocity gradient from the surface downward. There is no positive gradient (isothermal channel) with the ocean surface for long-distance horizontal propagation of sound.

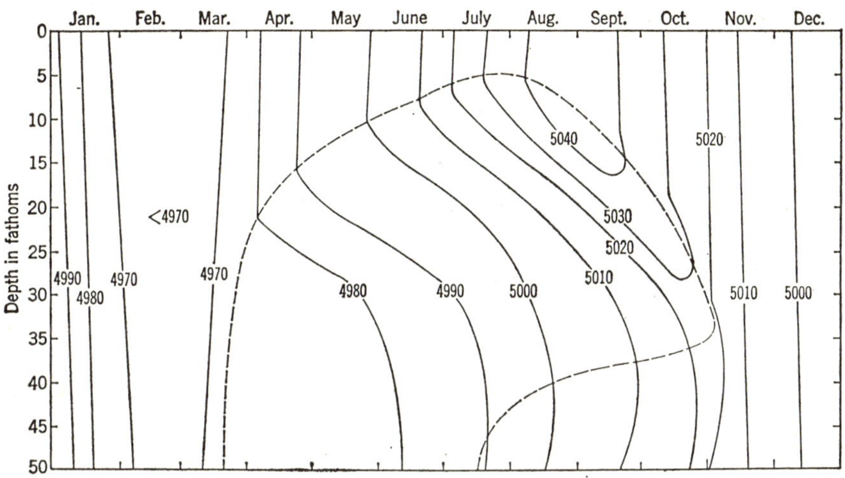

FIG. 4-5. Seasonal cycle of speed of sound in mid-latitude North Atlantic. Contour interval 10 ft/sec. (*From Woods Hole Oceanographic Institution, Natl. Defense Research Council,* 1941.)

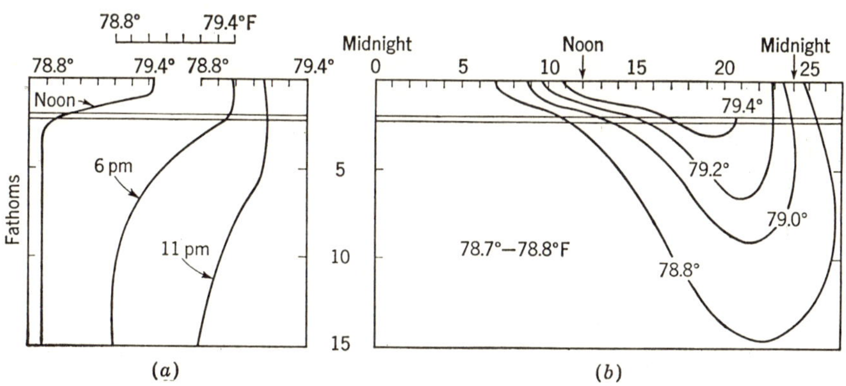

FIG. 4-6. (*From Woods Hole Oceanographic Institution, Natl. Defense Research Council,* 1941.)

The moderately rapid decrease of temperature with depth in the main thermocline produces a decrease of velocity with depth. In the deep-water layers the decrease in temperature is sufficiently slight that the increase of pressure becomes the dominant factor and produces an increase in velocity with increasing depth. The sound velocity profile for the deep ocean has a characteristic shape: first a decrease in velocity, followed by a more gradual increase in velocity. The depth of minimum velocity occurs near 600 fathoms in the mid-latitudes of the North Atlantic. The surface layer is variable, as discussed, resulting in general in either a sound channel with the ocean surface (positive velocity gradient) or no sound channel (negative velocity gradient) for horizontal transmission. Figure 4-7 is a characteristic sound velocity profile for the Mid-Atlantic.

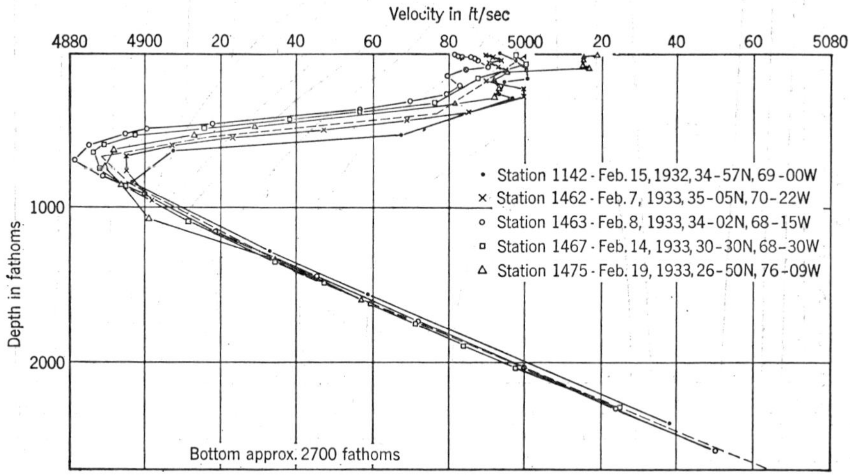

FIG. 4-7. (*From M. Ewing and J. L. Worzel, Geol. Soc. Amer., Mem.* 27.)

The range of validity of the ray and mode solutions with respect to frequency will be much different for deep-water transmission as compared with shallow-water transmission. Using the same approximate criteria as before the ray solution will be valid for frequencies above $f = 5{,}000 \times 10/15{,}000 = 3\frac{1}{3}$ cycles per second for an average deep-water depth of 15,000 feet. This covers the range of sonic and higher frequencies. The mode solution can be evaluated easily for frequencies below $f = 5{,}000 \times 2/15{,}000 = \frac{2}{3}$ cycles per second. These frequencies are very low and are in the range of earthquake frequencies. This comparison is valid only for an over-all consideration of deep-water transmission. There may be velocity variations within the ocean which restrict the ray solution to higher frequencies over the interval of these variations. In particular the surface layer shows such variations, and a description of long-range transmission in a surface sound channel, whose depths are generally of the order of 100 feet, must be given by mode theory for intermediate to higher sonic frequencies, i.e., frequencies from about a hundred to ten thousand cycles per second. For ultrasonic frequencies the effects of the underlying thermocline can be neglected in a description of channel propagation, and the considerations from ray theory will be applicable. For the lower sonic to subsonic frequencies the channel itself will have no effect, and the considerations from ray theory will again be applicable for a description of over-all transmission neglecting the velocity variations in the surface layer.

The order of procedure in this chapter is to give first a discussion of deep-water transmission from ray theory. The discussion includes the transmission effects from changes in the bottom depth and configuration, source depth, and receiver depth for the characteristic sound velocity profile of Fig. 4-7. Second, the normal-mode solution for a constant velocity ocean is given. Here, in contrast to the mode solution of Chap. 3, the underlying medium is considered to be elastic, propagating both dilatational (P) and irrotational (S) waves. This is taken to correspond with the crystalline rock basement underlying the deep-ocean basin areas; for the low frequencies of earthquake waves the effect of the relatively thin sediment, thin as compared with the depth of the ocean, can be neglected and the sediment thickness included in the ocean layer. Third, transmission near the ocean surface under the conditions of a negative velocity gradient from the surface downward is considered. Fourth, transmission near the ocean surface under the conditions of a positive followed by a negative velocity gradient is considered.

Ray solution

The description of sound transmission by ray theory in deep water given here is an extension of the last case considered in the first section of Chap. 3. For simplicity we shall begin with source and receiver at the ocean surface. In addition to the near-surface transmission, considered in more detail in the latter sections of this chapter, there are two possible sets of ray paths from source to receiver, the bottom-reflected paths and the RSR paths.

The bottom-reflected paths will always exist. Each order of bottom reflection will be received out to a limited range. The limiting ray defining the maximum range of reception of each order will depend on the bottom slope and depth and the sound velocity profile. Let us assume first that the bottom is horizontal. Two cases can then be distinguished for the definition of the limiting ray. First, the maximum velocity near the surface is greater than the maximum velocity at depth. The limiting ray is then that ray which is horizontal along its path at the depth of maximum velocity near the surface. If the surface layer has a negative velocity gradient from the surface downward, the ray will be horizontal at the surface. If the surface layer has an isothermal sound channel, the ray will be horizontal at the bottom of the sound channel. Second, the maximum velocity near the surface is less than the maximum velocity at depth. The limiting ray is then generally that ray which is horizontal (tangent) to the bottom. The exception is for those values of p in Fig. 4-8c slightly greater than zero where the ray tangent to the bottom is on the reversed sequent of the travel-time curve. The maximum ranges of reception for the higher orders of bottom reflection are simply multiples of that for the first order. These conditions all follow directly from

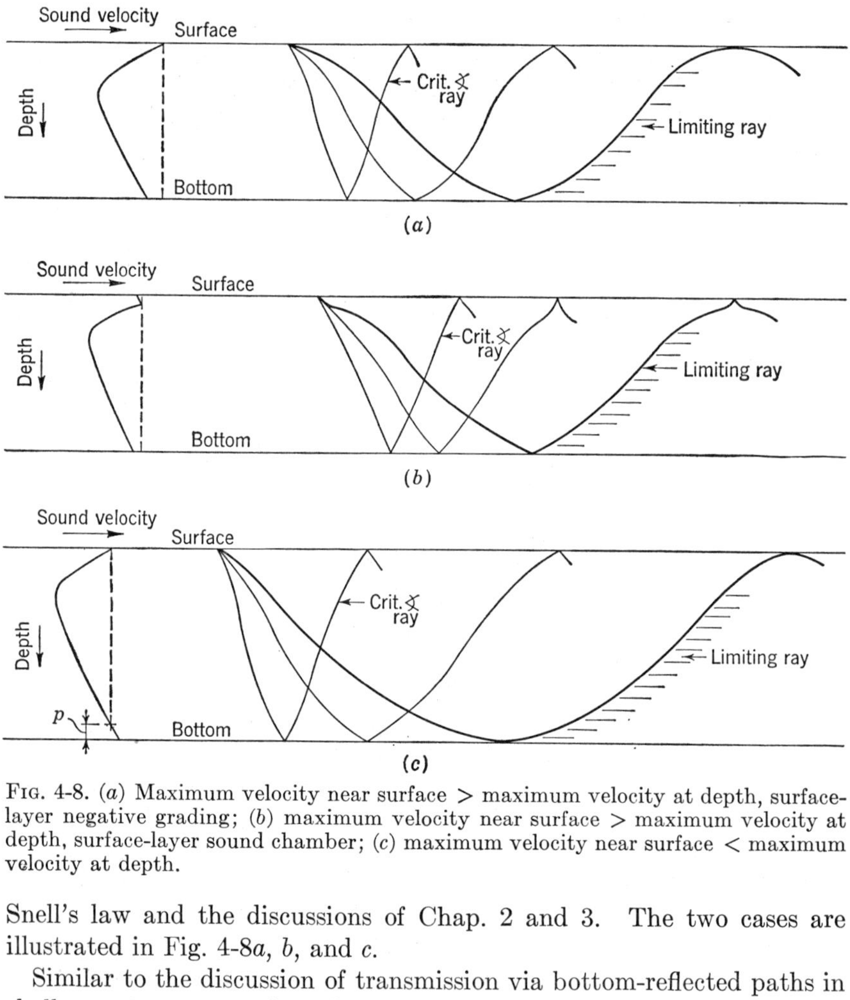

FIG. 4-8. (a) Maximum velocity near surface > maximum velocity at depth, surface-layer negative grading; (b) maximum velocity near surface > maximum velocity at depth, surface-layer sound chamber; (c) maximum velocity near surface < maximum velocity at depth.

Snell's law and the discussions of Chap. 2 and 3. The two cases are illustrated in Fig. 4-8a, b, and c.

Similar to the discussion of transmission via bottom-reflected paths in shallow water we must include the effects of the critical angle here as well. For incidence at angles steeper than the critical angle, there will be a partial refraction of energy into the bottom with a consequent loss of energy to the reflected ray. For angles more grazing than the critical angle there is total reflection. This has the effect of defining a range interval of best reception via bottom-reflected paths between the critical angle ray and the limiting ray. The definition of this interval at the lower limit becomes more prominent with increasing order of bottom reflection; the losses for angles steeper than critical are cumulative as a function of the order and are equal simply to the order n times the loss for the first order at the same incident angle.

The existence of the RSR rays depends on the sound velocity profile and the depth of water. The condition for the existence of RSR rays is that the maximum velocity at depth be greater than the maximum velocity near the surface. For a ray to be returned from depth without a bottom reflection it must pass through a horizontal angle ($\theta = 90$ deg) which from Snell's law gives

$$\sin \theta_0 = \frac{c_0}{c_d} \tag{4-1}$$

where c_d is the velocity at the depth at which the ray becomes horizontal, c_0 is, here, the maximum velocity near the surface, and θ_0 is the angle of the ray at the depth of the velocity c_0. For there to be values of θ_0

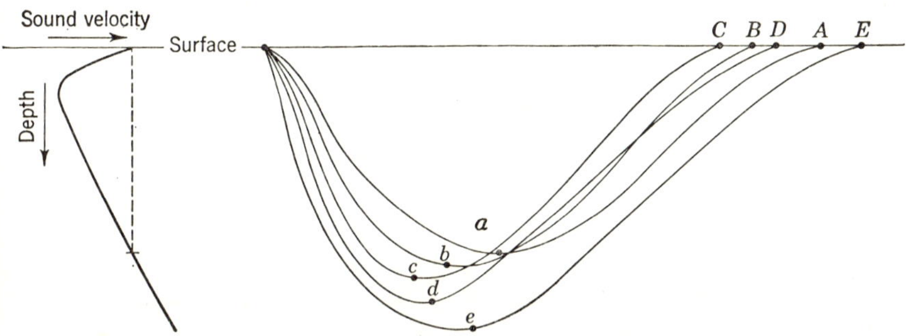

Fig. 4-9.

satisfying this relation there must be a range of c_d greater than c_0. For the sound velocity profiles of Fig. 4-8*a* and *b* there will be no RSR rays. For the sound velocity profile of Fig. 4-8*c* there will be RSR rays; the thickness of water p between the depth at which the maximum velocity near the surface is reached and the bottom is an index of the amount of RSR rays permitted. From Chap. 2 we know that the first RSR ray becomes horizontal at the depth for which the velocity is equal to the maximum velocity near the surface and that with succeedingly greater depths at which an RSR ray becomes horizontal the range at which the ray reaches the surface first decreases and then increases. The intensity of the RSR signals will be higher than that for the corresponding bottom reflections. The point at which the range of the RSR paths stops decreasing and begins to increase will be a focus point of intensity. Figure 4-9 is a ray diagram of the RSR paths for a bottomless ocean. For depths of water less than a there will be no RSR paths. For a depth of b there will be RSR paths to a surface receiver but no focus point; RSR reception will be over the range interval from B to A. For a depth of d there will be in addition a focus point at range C; RSR reception will be double over the range interval from C to D and single from D to A. For a depth of e there will be a focus point at range C, double RSR reception from C to A, and single reception from A to E. Figure 4-10 is illustrative of bottom-reflected and RSR coverage for a moderately great depth of ocean.

There are a number of plots which can be made for any particular problem which are useful in the understanding and definition of deep-water transmission. Among these are the plots of travel time from source to receiver versus range, emergent angle from the source versus range, intensity versus range, and the transmission diagram. All these can be computed with varying degrees of difficulty by hand computations,

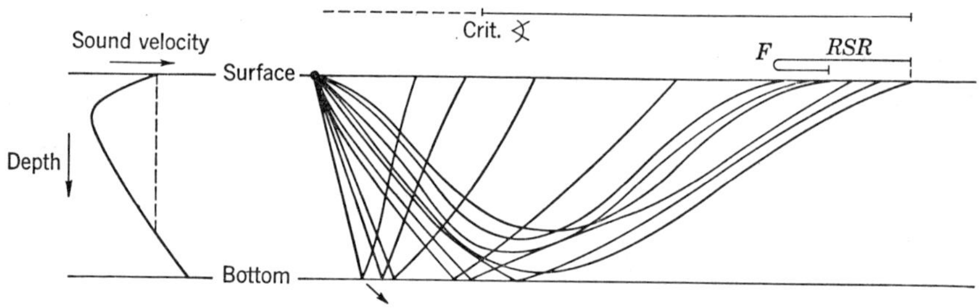

FIG. 4-10.

digital computers, or analogue computers for any given source and receiver depths, sound velocity profile, and bottom configuration using the formulas of Chap. 2. The usual procedure is to divide the sound velocity profile into a number of segments over each of which a constant velocity gradient can be reasonably assumed. The computations are then made stepwise with the formulas for a constant gradient. The following are representative of such plots. Figure 4-11 is a plot of travel time versus range for the bottom-reflected and RSR paths for the first order. A number of bottom-reflection curves corresponding to different depths have been drawn. The portion of the RSR curve is from the point W along to the point corresponding to the depth of water in the particular case; for depths less than w there will be no RSR portion. Figure 4-12 is a travel-time plot for the first few orders for a depth y. Figure 4-13 is a plot of the emergent angle from the source versus range for the first order and for various depths. Such a plot is a measure of the distribution of transmitted energy with range, for equal increments of angle from the source can be considered to define the initial ray bundles. A vertical or near vertical tangent to the curve of Fig. 4-13 defines a focusing region, and a near horizontal tangent defines a defocusing region. As in Fig. 4-11 the portion of the RSR curve to be included is that from the point W to the point corresponding to the actual depth. Figure 4-14 is the transmission diagram of the range interval of best reception for each order for a depth y plotted for the first few orders. It is interesting to note that with increasing range the RSR intervals will eventually

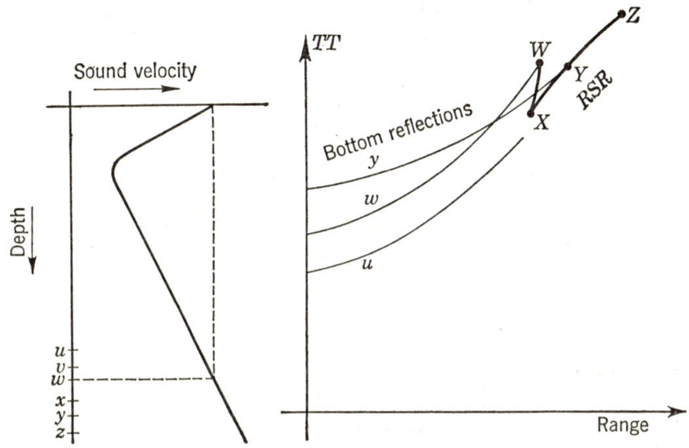

FIG. 4-11.

overlap giving continuous coverage by at least one RSR ray for that and greater ranges. The transmission diagram can be used as an estimate of the energy received at any range by simply summing up along a vertical line all the contributions to that range including the spherical ray-spreading factor and any individual focusing or defocusing effects.

If the bottom is not horizontal, the discussion of deep-water transmission becomes slightly more complex. For one cycle of bottom and surface reflection from a sloping bottom, the ray angle is increased or decreased by twice the inclination of the bottom, as is shown in Fig. 4-15. For transmission across a bottom which dips up from source to receiver the effect is to steepen the rays of each succeeding order by twice the inclination of the bottom. A portion of the energy from the region of total reflection is passed to the region of reflection at angles steeper than critical with each succeeding reflection until eventually, if the ranges are sufficiently long or the slope sufficiently steep, there will be no transmission which has not taken at least one reflection at angles steeper than

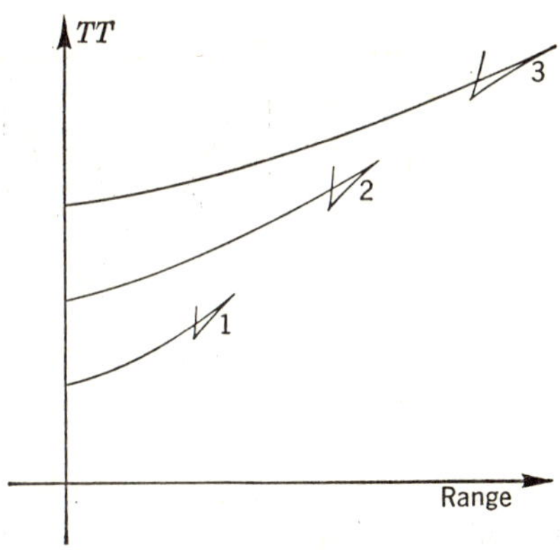

FIG. 4-12.

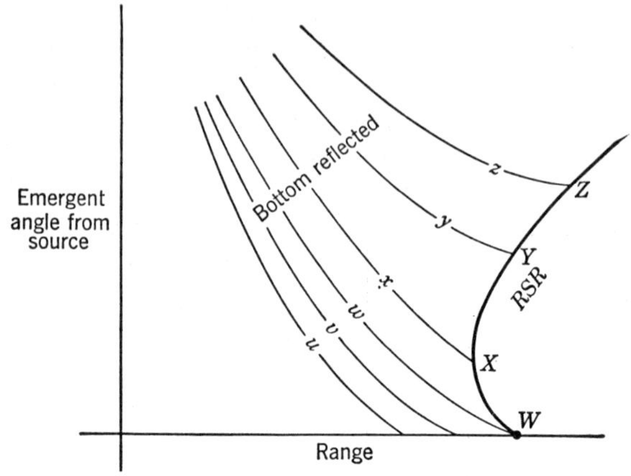

FIG. 4-13.

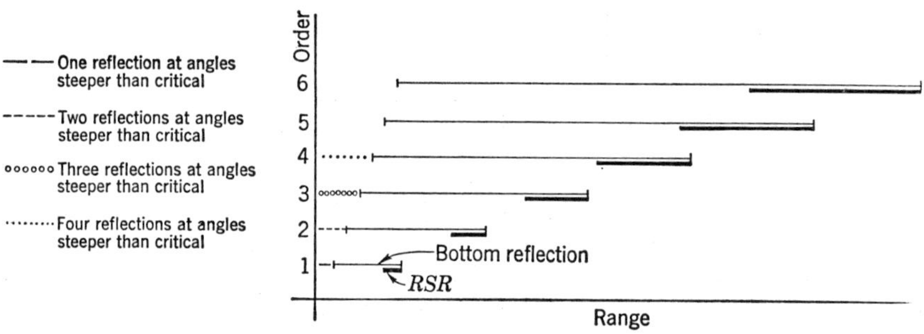

FIG. 4-14.

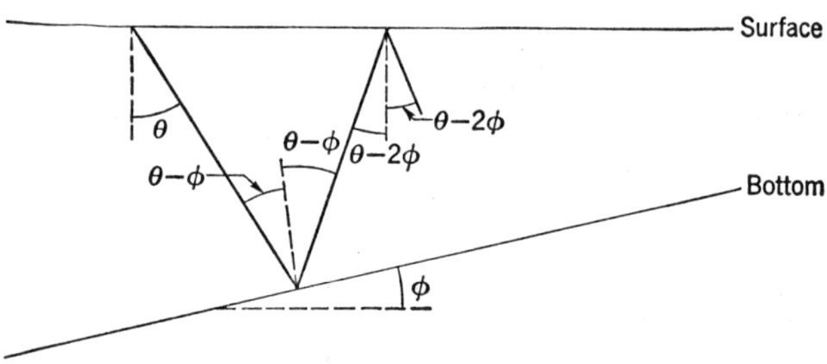

FIG. 4-15.

critical. For the RSR transmission the ray paths will become bottom reflections with decreasing depth away from the source. Figure 4-16 is a ray diagram of an original limiting ray from a source transmitting upslope. Figure 4-17 shows the transmission diagram for such a case. For transmission downslope the effect is simply the reverse. Energy, which has taken one or more reflections at angles steeper than critical, is successively passed over into the region of total reflection. Bottom-reflection energy is successively added to RSR transmission when the water becomes deep enough to allow RSR paths; and if the slope is sufficiently steep, some bottom-reflected energy may be converted to SOFAR paths.

Changes in the source and receiver depths introduce another set of effects. Lowering the source and receiver below the surface introduces four rays for each of the ones discussed previously. These correspond to the four combinations of upgoing and downgoing rays from the source and receiver, as shown in Fig. 4-18. For each order n of bottom reflection or RSR two of the rays will have taken n surface reflections, one $n - 1$ surface reflections, and one $n + 1$ surface reflections. The description

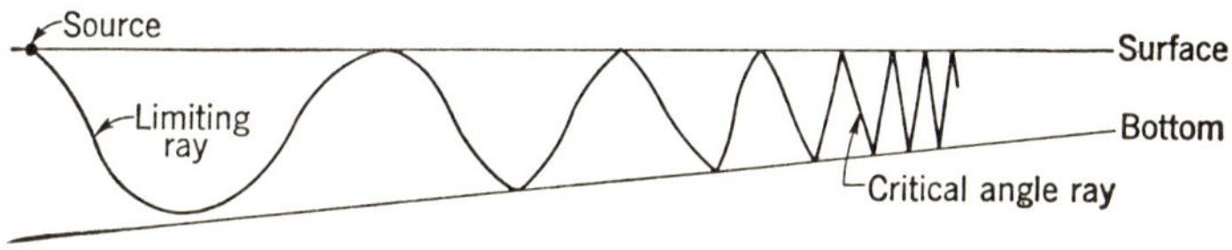

FIG. 4-16.

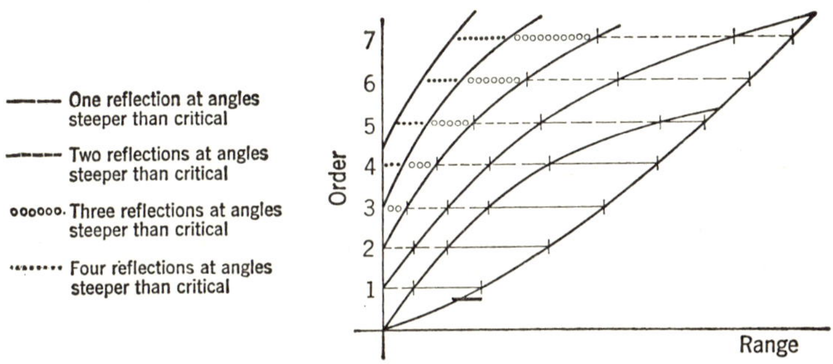

FIG. 4-17.

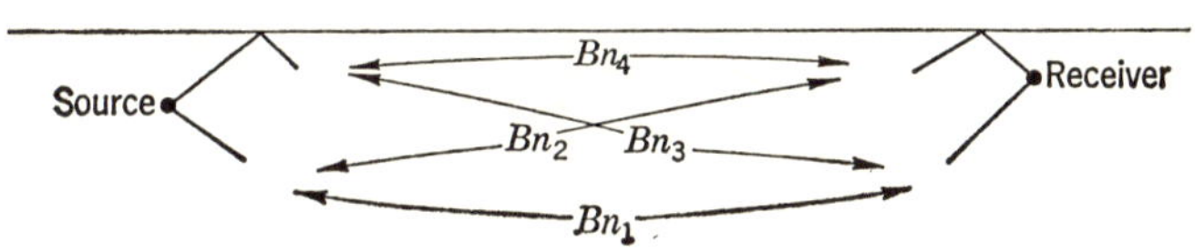

FIG. 4-18.

for each is as given above, the ranges for each being slightly different depending on the source and receiver depths.

A second effect is the introduction of SOFAR transmission. If the source or receiver is in a surface sound channel, no SOFAR transmission can exist between them. If the source and receiver are below the sound channel or if a negative gradient exists from the surface downward and if the velocity at both the source and receiver is exceeded at depth, SOFAR transmission can exist. These conditions follow directly from Snell's law and the previous discussions. Figure 4-19 illustrates the conditions under which SOFAR transmission will or will not exist.

The description of SOFAR transmission is much the same as that given for the bottom-reflected and RSR transmission. The essential difference is that the SOFAR rays are refracted away from each boundary without undergoing bottom or surface reflections. The sequence of arrivals is

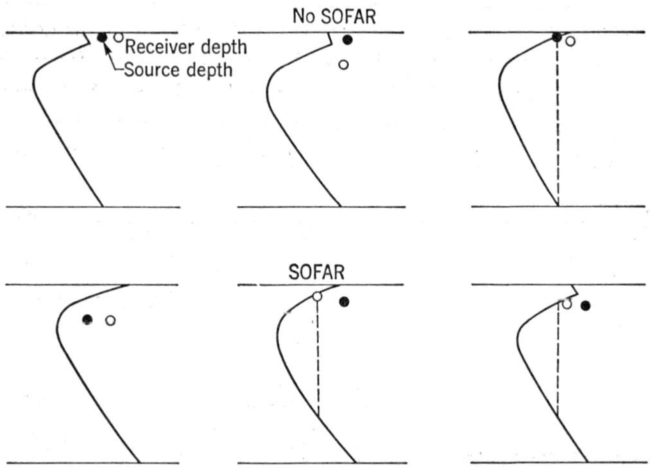

FIG. 4-19.

divided into orders of refractions from below. For each order n of refraction from below there will be four rays, one with $n - 1$ refractions from above, two with n refractions from above, and one with $n + 1$ refractions from above, as shown in Fig. 4-20. When SOFAR rays exist, they are in addition to the bottom-reflected and RSR rays.

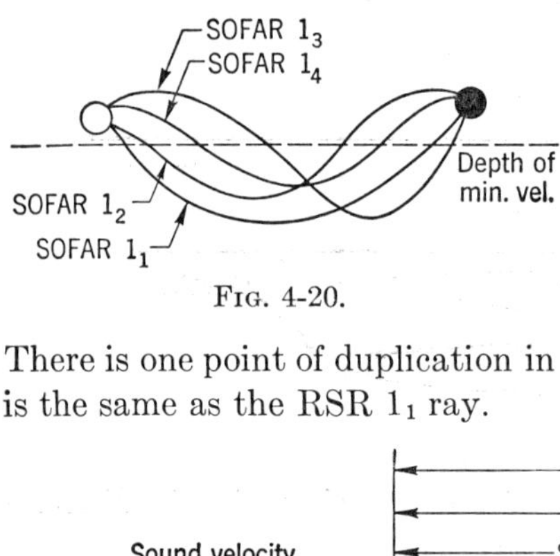

FIG. 4-20.

There is one point of duplication in the nomenclature; the SOFAR 1_1 ray is the same as the RSR 1_1 ray.

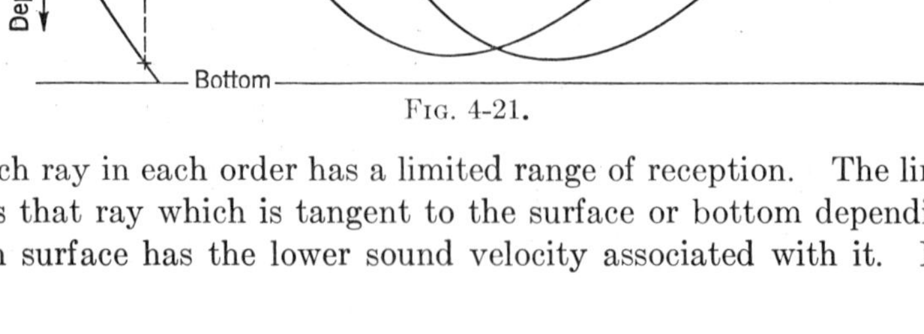

FIG. 4-21.

Each ray in each order has a limited range of reception. The limiting ray is that ray which is tangent to the surface or bottom depending on which surface has the lower sound velocity associated with it. Figure

4-21 is illustrative of the limiting ranges for the first order. The ray paths are the same for all four rays, the total ranges being simply the addition or subtraction of one refraction from above.

The sequence of progression from one order to the next is different for the SOFAR rays as compared with the bottom-reflected or RSR rays. This difference is illustrated in Fig. 4-22. The SOFAR rays converge toward the horizontal with increasing order. The time difference between succeeding orders will decrease with increasing order and approach zero. The SOFAR transmission ends abruptly. The bottom-reflected rays converge toward the vertical. The time difference between succeeding orders increases with increasing order and approaches the

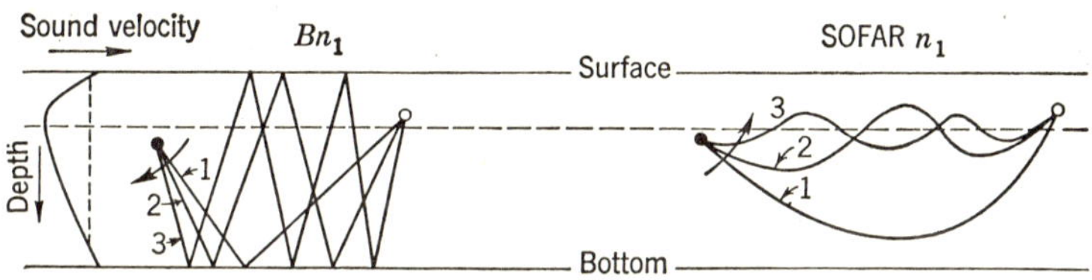

FIG. 4-22.

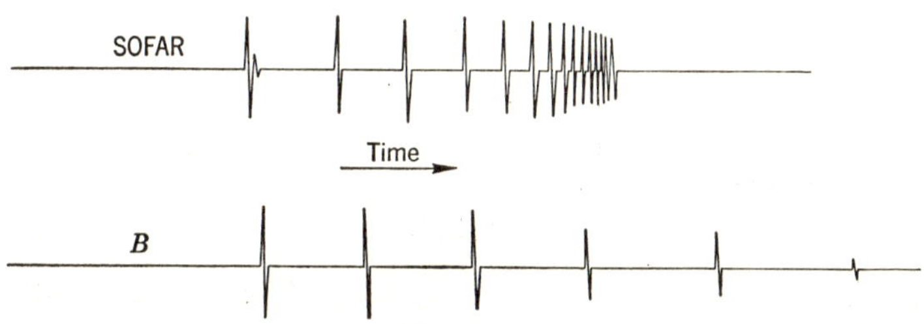

FIG. 4-23.

value of twice the vertical travel time from surface to bottom. The sequence of arrivals for the rays of Fig. 4-22 is shown on the same relative time scale in Fig. 4-23.

Normal-mode solution

We shall derive the normal-mode solution for deep-water transmission where the ocean is considered to be underlain by an elastic medium. The solution follows directly from the previous normal-mode solution in Chap. 3 applied to shallow-water transmission where the ocean was considered to be underlain by a fluid (sedimentary) medium. With reference to the procedures of that section the solution here will be stated in brief. The formulation is slightly different, but the evaluation of the resultant mode integrals is the same, and the Fourier synthesis and evaluation are the same. In this problem it is convenient to use the displacement potentials of Chap. 1 and the equivalent boundary condition of equality of vertical component of particle displacement. With these two changes the formulation (3-76) for the source condition remains the same. In addition to the normal boundary conditions of equality of pressure and vertical component of particle displacement (velocity) it is necessary to impose an additional boundary condition that the tangential component of stress in the elastic medium vanish at the bottom in order to preserve the boundary and avoid an unbalanced horizontal acceleration in the overlying ocean.

The physical properties of the media are shown in Fig. 4-24. The upper medium is defined by a constant density ρ_1 and velocity c_1 and is bounded at $z = 0$ by a free surface and at the bottom, $z = H$, by an elastic medium of constant density ρ_2, irrotational velocity α_2, and

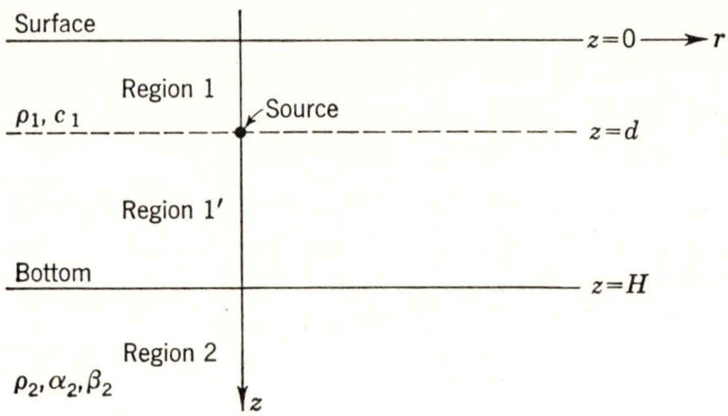

FIG. 4-24.

equivoluminal velocity β_2 where $\alpha_2 > \beta_2 > c_1$. The displacement potentials φ and ψ must satisfy the wave equations

$$\begin{aligned} \nabla^2\varphi_1 &= \frac{1}{c_1^2}\frac{\partial^2\varphi_1}{\partial t^2} \\ \nabla^2\varphi_2 &= \frac{1}{\alpha_2^2}\frac{\partial^2\varphi_2}{\partial t^2} \\ \nabla^2\psi_2 &= \frac{1}{\beta_2^2}\frac{\partial^2\psi_2}{\partial t^2} \end{aligned} \tag{4-2}$$

in the upper and lower media, respectively. In symmetrical cylindrical coordinates the solutions will be of the form

$$\varphi, \psi = e^{i\omega t}\bar{\varphi}, \bar{\psi} = e^{i\omega t}J_0(kr)F(z) \tag{4-3}$$

for a simple harmonic source. The functions $\bar{\varphi}$ and $\bar{\psi}$ satisfy the time-independent equations

$$\left(\nabla^2 + \frac{\omega^2}{c_1^2}\right)\bar{\varphi}_1 = 0, \left(\nabla^2 + \frac{\omega^2}{\alpha_2^2}\right)\bar{\varphi}_2 = 0, \left(\nabla^2 + \frac{\omega^2}{\beta_2^2}\right)\bar{\psi}_2 = 0 \tag{4-4}$$

The forms of $F(z)$ will be

$$\bar{\varphi}_1 = A \sin(\xi z) J_0(kr) \tag{4-5}$$
$$\bar{\varphi}_1' = [B \sin(\xi z) + C \cos(\xi z)] J_0(kr) \tag{4-6}$$
$$\bar{\varphi}_2 = D e^{-\eta z} J_0(kr) \tag{4-7}$$
$$\bar{\psi}_2 = E e^{-\zeta z} J_0(kr) \tag{4-8}$$

where

$$\begin{aligned} \xi &= \sqrt{\frac{\omega^2}{c_1{}^2} - k^2} \\ \eta &= \sqrt{k^2 - \frac{\omega^2}{\alpha_2{}^2}} \\ \zeta &= \sqrt{k^2 - \frac{\omega^2}{\beta_2{}^2}} \end{aligned} \tag{4-9}$$

The second solution for (4-5) has been omitted to satisfy the condition of zero pressure at the surface; the second solutions to (4-7) and (4-8) have been omitted to satisfy the boundary condition toward infinity of outgoing waves and decreasing amplitude. As before the signs of Eqs. (4-9) are either positive real or negative imaginary.

In terms of the displacement potentials the radial and vertical components of particle displacement in the two media are from Eqs. (1-86) and (1-97), (4-3), and (4-4)

$$\begin{aligned} \bar{q}_1 &= \frac{\partial \bar{\varphi}_1}{\partial r} \\ \bar{w}_1 &= \frac{\partial \bar{\varphi}_1}{\partial z} \\ \bar{q}_2 &= \frac{\partial \bar{\varphi}_2}{\partial r} + \frac{\partial^2 \bar{\psi}_2}{\partial r\, \partial z} \\ \bar{w}_2 &= \frac{\partial \bar{\varphi}_2}{\partial z} + \frac{\partial^2 \bar{\psi}_2}{\partial z^2} + \frac{\omega^2}{\beta_2{}^2} \bar{\psi}_2 \end{aligned} \tag{4-10}$$

From Eqs. (1-12), and (1-97), and (4-10) the radial and vertical components of stress on a plane normal to the z axis become

$$\begin{aligned} (\sigma_{zz})_2 &= \lambda_2 \Delta_2 + 2\mu_2 \frac{\partial w_2}{\partial z} \\ &= \lambda_2 \nabla^2 \bar{\varphi}_2 + 2\mu_2 \left(\frac{\partial^2 \bar{\varphi}_2}{\partial z^2} + \frac{\partial^3 \bar{\psi}_2}{\partial z^3} + \frac{\omega^2}{\beta_2{}^2} \frac{\partial \bar{\psi}_2}{\partial z} \right) \\ &= \rho_2 \left[-\omega^2 \bar{\varphi}_2 + 2\beta_2{}^2 \left(\frac{\partial^2 \bar{\varphi}_2}{\partial z^2} + \frac{\omega^2}{\alpha_2{}^2} \bar{\varphi}_2 + \frac{\partial^3 \bar{\psi}_2}{\partial z^3} + \frac{\omega^2 \partial \bar{\psi}_2}{\beta_2{}^2 \partial z} \right) \right] \\ (\sigma_{zz})_1 &= -\rho_1 \omega^2 \bar{\varphi}_1 \\ (\sigma_{zr})_2 &= \mu_2 \left(\frac{\partial q_2}{\partial z} + \frac{\partial w_2}{\partial r} \right) \\ &= \rho_2 \beta_2{}^2 \frac{\partial}{\partial r} \left(2 \frac{\partial \bar{\varphi}_2}{\partial z} + 2 \frac{\partial^2 \bar{\psi}_2}{\partial z^2} + \frac{\omega^2}{\beta_2{}^2} \bar{\psi}_2 \right) \end{aligned} \tag{4-11}$$

The boundary conditions that remain to be satisfied are

$$\begin{aligned} (\sigma_{zz})_1 &= (\sigma_{zz})_1' \\ w_1 - w_1' &= 2k \end{aligned} \qquad \text{at } z = d \tag{4-12}$$

and

$$\begin{aligned} (\sigma_{zz})_1' &= (\sigma_{zz})_2 \\ (\sigma_{zr})_2 &= 0 \\ w_1' &= w_2 \end{aligned} \qquad \text{at } z = H \tag{4-13}$$

Substituting (4-10) and (4-11) into (4-12) and (4-13) in order to determine the coefficients of (4-5), (4-6), (4-7), and (4-8) gives

$$\begin{aligned}
&-A\sin(\xi d) + B\sin(\xi d) + C\cos(\xi d) = 0 \\
&A\cos(\xi d) - B\cos(\xi d) + C\sin(\xi d) = \frac{2k}{\xi} \\
&B\frac{\rho_1\omega^2}{\rho_2\beta_2{}^2}\sin(\xi H) + C\frac{\rho_1\omega^2}{\rho_2\beta_2{}^2}\cos(\xi H) \\
&\quad + D\left(2k^2 - \frac{\omega^2}{\beta_2{}^2}\right)e^{-\eta H} - E2k^2\zeta e^{-\zeta H} = 0 \\
&-D2\eta e^{-\eta H} + E\left(2k^2 - \frac{\omega^2}{\beta_2{}^2}\right)e^{-\zeta H} = 0 \\
&B\xi\cos(\xi H) - C\xi\sin(\xi H) \\
&\quad + D\eta e^{-\eta H} - Ek^2 e^{-\zeta H} = 0
\end{aligned} \tag{4-14}$$

Solving Eqs. (4-14) for A, B, C, D, and E, substituting into Eqs. (4-5), (4-6), (4-7), and (4-8), and integrating with respect to k from 0 to ∞ in order to satisfy the source condition, we obtain for the formal solution of the displacement potential due to a periodic point source,

$$\Phi_1 = 2e^{i\omega t}\int_0^\infty J_0(kr)k\,dk\,\frac{\sin(\xi z)}{\xi}
\left\{\frac{\dfrac{\rho_1\omega^4\eta}{\rho_2\beta_2{}^4\xi}\sin[\xi(H-d)] - \left[4k^2\eta\zeta - \left(2k^2 - \dfrac{\omega^2}{\beta_2{}^2}\right)^2\right]\cos[\xi(H-d)]}{\dfrac{\rho_1\omega^4\eta}{\rho_2\beta_2{}^4\xi}\sin(\xi H) - \left[4k^2\eta\zeta - \left(2k^2 - \dfrac{\omega^2}{\beta_2{}^2}\right)^2\right]\cos(\xi H)}\right\} \tag{4-15}$$

$$\Phi_1' = 2e^{i\omega t}\int_0^\infty J_0(kr)k\,dk\,\frac{\sin(\xi d)}{\xi}
\left\{\frac{\dfrac{\rho_1\omega^4\eta}{\rho_2\beta_2{}^2\xi}\sin[\xi(H-z)] - \left[4k^2\eta\zeta - \left(2k^2 - \dfrac{\omega^2}{\beta_2{}^2}\right)^2\right]\cos[\xi(H-z)]}{\dfrac{\rho_1\omega^4\eta}{\rho_2\beta_2{}^2\xi}\sin(\xi H) - \left[4k^2\eta\zeta - \left(2k^2 - \dfrac{\omega^2}{\beta_2{}^2}\right)^2\right]\cos(\xi H)}\right\} \tag{4-16}$$

$$\Phi_2 = 2e^{i\omega t} \int_0^\infty J_0(kr)k\,dk \frac{\sin(\xi d)}{\xi}$$
$$\left\{ \frac{\dfrac{\rho_1}{\rho_2}\dfrac{\omega^2}{\beta_2{}^2}\left(2k^2 - \dfrac{\omega^2}{\beta_2{}^2}\right) e^{-\eta(z-H)}}{\dfrac{\rho_1}{\rho_2}\dfrac{\omega^4}{\beta_2{}^4}\dfrac{\eta}{\xi}\sin(\xi H) - \left[4k^2\eta\zeta - \left(2k^2 - \dfrac{\omega^2}{\beta_2{}^2}\right)^2\right]\cos(\xi H)} \right\} \tag{4-17}$$

$$\Psi_2 = 2e^{i\omega t} \int_0^\infty J_0(kr)k\,dk \frac{\sin(\xi d)}{\xi}$$
$$\left\{ \frac{\dfrac{\rho_1}{\rho_2}\dfrac{\omega^2}{\beta_2{}^2}\eta e^{-\zeta(z-H)}}{\dfrac{\rho_1}{\rho_2}\dfrac{\omega^4}{\beta_2{}^4}\dfrac{\eta}{\xi}\sin(\xi H) - \left[4k^2\eta\zeta - \left(2k^2 - \dfrac{\omega^2}{\beta_2{}^2}\right)^2\right]\cos(\xi H)} \right\} \tag{4-18}$$

The evaluation of these integrals is the same as before. The normal modes are the residues from an integration in the complex k plane. They occur at those values of $k = k_n$ for which the denominator of the integrals is equal to zero.

$$\frac{\rho_1}{\rho_2}\frac{\omega^4}{\beta_2{}^4}\frac{\eta}{\zeta}\sin(\xi H) - \left[4k^2\eta\zeta - \left(2k^2 - \frac{\omega^2}{\beta_2{}^2}\right)^2\right]\cos(\xi H) = 0 \tag{4-19}$$

The normal-mode contributions are, using the asymptotic expansion for the Hankel function,

$$\Phi_1 = \Phi_1' = \frac{2\pi}{H}\sqrt{\frac{2}{\pi r}}\sum_n e^{i[\omega t - k_n r - (\pi/4)]}\, G(k_n)\sin(\xi_n d)\sin(\xi_n z) \tag{4-20}$$

where

$$G(k_n) \equiv \frac{k_n H\left\{-\dfrac{\rho_1}{\rho_2}\dfrac{\omega^4}{\beta_2{}^4}\dfrac{\eta_n}{\xi_n}\cos(\xi_n H) - \left[4k_n{}^2\eta_n\zeta_n - \left(2k_n{}^2 - \dfrac{\omega^2}{\beta_2{}^2}\right)^2\right]\sin(\xi_n H)\right\}}{\sqrt{k_n}\dfrac{\partial}{\partial k}\left\{\dfrac{\rho_1}{\rho_2}\dfrac{\omega^4}{\beta_2{}^4}\dfrac{\eta_n}{\xi_n}\sin(\xi_n H) - \left[4k_n{}^2\eta_n\zeta_n - \left(2k_n{}^2 - \dfrac{\omega^2}{\beta_2{}^2}\right)^2\right]\cos(\xi_n H)\right\}}$$
$$= -\frac{k_n H \dfrac{\rho_1}{\rho_2}\dfrac{\omega^4}{\beta_2{}^4}\dfrac{\eta_n}{\xi_n}}{\sqrt{k_n}\cos(\xi_n H)\dfrac{\partial}{\partial k}\{\cdot\cdot\cdot\cdot\}} \tag{4-21}$$

Equation (4-19) is the characteristic equation, corresponding to Eq. (3-105) in the two-liquid case. It is the dispersion equation, defining the relation between the phase velocity V_n where

$$V_n = \frac{\omega}{k_n} \tag{4-22}$$

and the frequency ω. Substituting (4-22) into (4-19) we obtain

$$\tan\left(\frac{2\pi f_n H}{V}\sqrt{\frac{V^2}{c_1^2}-1}\right) = \frac{\rho_2\beta_2^4\sqrt{(V^2/c_1^2)-1}}{\rho_1 V^4\sqrt{1-(V^2/\alpha_2^2)}}\left[4\sqrt{1-\frac{V^2}{\alpha_2^2}}\sqrt{1-\frac{V^2}{\beta_2^2}}-\left(2-\frac{V^2}{\beta_2^2}\right)^2\right] \qquad (4\text{-}23)$$

This equation has the physical significance corresponding to Eq. (3-111) and may be derived from the same physical formulation of constructive interference of multiply reflected waves using the Rayleigh reflection coefficient from a liquid to a solid.

The Fourier synthesis of Eq. (4-20) from a simple harmonic to a pulse source is the same as before and will not be repeated. The resultant integral over ω is evaluated by the method of stationary phase and has contributions for those values of $\omega = \omega_0$ for which

$$t - r\left(\frac{dk_n}{d\omega}\right)_{\omega=\omega_0} = 0 \qquad (4\text{-}24)$$

which defines the group velocity U following the propagation of a particular frequency, where t and r are variable,

$$U = \frac{r}{t} = \frac{d\omega}{dk} = V + k\frac{dV}{dk} \qquad (4\text{-}25)$$

Characteristic phase velocity curves, determined numerically from Eq. (4-23), and group velocity curves, determined graphically from the phase velocity calculations and Eq. (4-25), are shown in Figs. 4-25 and 4-26 for the zero and first mode. These two graphs give only the real-value solutions for k, unattenuated propagation.

The mode patterns here are somewhat different from those for the liquid-liquid case of the last chapter. First, there is the introduction of a lower mode below the first mode. This mode does not have a lower cutoff frequency but extends down to zero frequency with a finite phase and group velocity. It does not extend out to the limit of infinite frequency as all the higher modes do but has a high-frequency cutoff. Second, the first and higher modes have a high velocity limit of the shear velocity in the lower medium for unattenuated propagation.

It is instructive to an understanding of these differences to give the physical derivation of Eq. (4-23) and to discuss some of the characteristics of the quantities used in the derivation. From Fig. 4-27 the condition for constructive interference is, as before,

$$\frac{2\pi}{\lambda_n}\left[\frac{H}{\cos\theta}+\frac{H}{\cos\theta}\cos(2\theta)\right]-\epsilon-\pi = 2(n-1)\pi$$

$$\frac{2\pi f_n H}{c_1}\cos\theta - \frac{\epsilon}{2} = (2n-1)\frac{\pi}{2} \qquad (4\text{-}26)$$

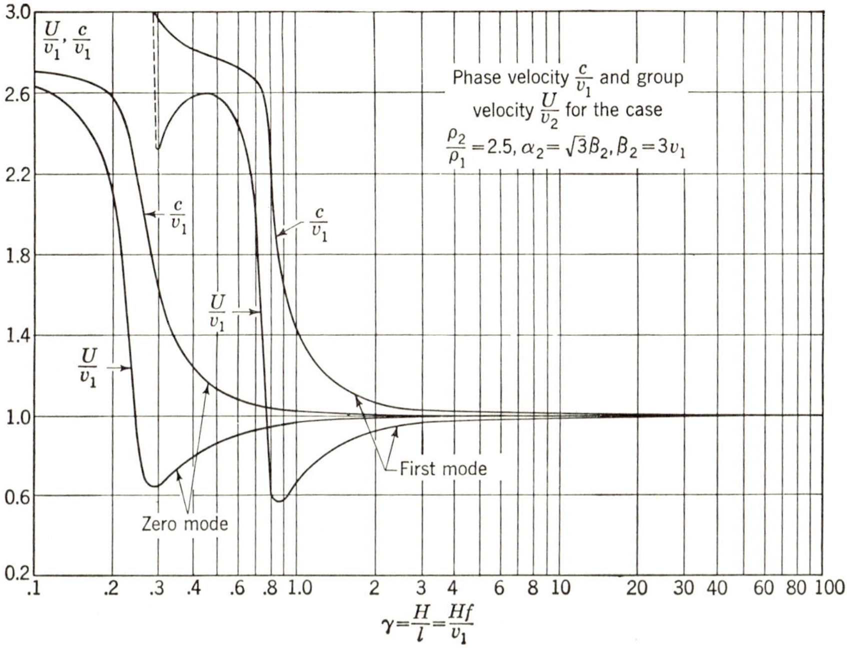

FIG. 4-25. (*From F. Press and M. Ewing, Geophysics, vol.* 15, 1950.)

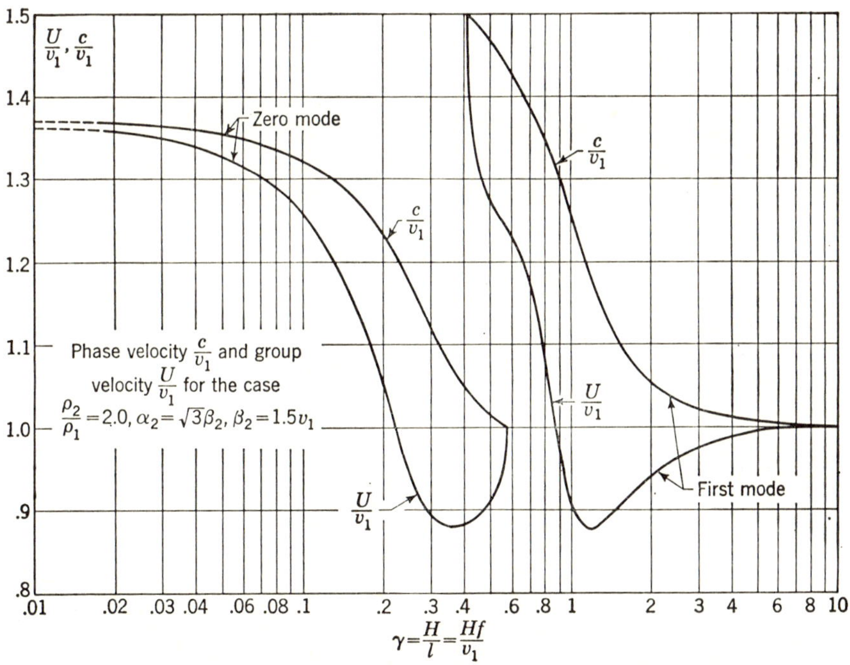

FIG. 4-26. (*From F. Press and M. Ewing, Geophysics, vol.* 15, 1950.)

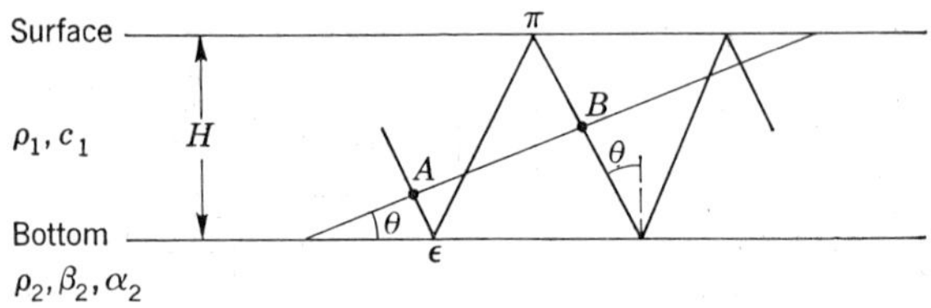

FIG. 4-27.

Using the familiar relation between the phase velocity in the r direction and the angle of incidence of the ray that

$$V = \frac{c_1}{\sin \theta} \tag{4-27}$$

we have

$$\frac{2\pi f_n H}{c_1} \sqrt{1 - \frac{c_1^2}{V^2}} = (2n - 1)\frac{\pi}{2} + \frac{\epsilon}{2}$$

$$\tan \left[\frac{2\pi f_n H}{V} \sqrt{\frac{V^2}{c_1^2} - 1} \right] = \tan \left[(2n - 1)\frac{\pi}{2} + \frac{\epsilon}{2} \right] = -\cot\left(\frac{\epsilon}{2}\right) \tag{4-28}$$

From Chap. 5 the phase change upon reflection from a liquid-solid interface at angles of incidence beyond the critical angle for total reflection is given by

$$\begin{aligned} \cot\left(\frac{\epsilon}{2}\right) &= \frac{\rho_2 \beta_2^4 \sqrt{1 - \sin^2 \theta}}{\rho_1 \sqrt{\sin^2 \theta - (c_1^2/c_2^2)}} \left[\left(\frac{2 \sin^2 \theta}{c_1^2} - \frac{1}{\beta_2^2} \right)^2 \right. \\ &\qquad \left. - \frac{4 \sin^2 \theta}{c_1^4} \sqrt{\sin^2 \theta - \frac{c_1^2}{\alpha_2^2}} \sqrt{\sin^2 \theta - \frac{c_1^2}{\beta_2^2}} \right] \\ &= \frac{\rho_2 \beta_2^4 \sqrt{(V^2/c_1^2) - 1}}{\rho_1 V^4 \sqrt{1 - (V^2/\alpha_2^2)}} \left[\left(2 - \frac{V^2}{\beta_2^2}\right)^2 - 4 \sqrt{1 - \frac{V^2}{\alpha_2^2}} \sqrt{1 - \frac{V^2}{\beta_2^2}} \right] \end{aligned} \tag{4-29}$$

Substituting (4-29) into (4-28) we obtain

$$\begin{aligned} \tan \left(\frac{2\pi f_n H}{V} \sqrt{\frac{V^2}{c_1^2} - 1} \right) = \frac{\rho_2 \beta_2^4 \sqrt{(V^2/c_1^2) - 1}}{\rho_1 V^4 \sqrt{1 - (V^2/\alpha_2^2)}} & \left[4 \sqrt{1 - \frac{V^2}{\alpha_2^2}} \sqrt{1 - \frac{V^2}{\beta_2^2}} \right. \\ & \left. - \left(2 - \frac{V^2}{\beta_2^2} \right)^2 \right] \end{aligned} \tag{4-30}$$

the same as Eq. (4-23).

From Eq. (4-26) we see that for $n = 1$ the right-hand side equals $\pi/2$. Both quantities on the left-hand side are positive so that there will always be solutions for f_n. We can also see that, as $\cos \theta \to 0$, f_n must approach infinity so that the equality will be preserved. There will be no high-frequency cutoff. From Eq. (4-23) we see that real-value solutions (unattenuated propagation) are restricted to values of V such that V is less than

$$V = \beta_2 \tag{4-31}$$

From Eq. (4-27) it may be alternatively stated that real-value solutions exist for values of θ such that θ is greater than

$$\sin \theta_\beta = \frac{c_1}{\beta_2} \tag{4-32}$$

the critical angle for shear refraction into the lower medium (total reflection in the upper medium). For higher values of n the same situation will be true with solutions for a given value of V or θ giving successively higher values of f_n.

Let us examine what conditions must prevail for a real-value solution to exist for $n = 0$. When $n = 0$, the right-hand side of (4-26) is equal to $-(\pi/2)$. Solutions can then exist only if $\epsilon/2$ is greater than $\pi/2$ over some range of values of θ. If this condition does exist, there will be solutions of (4-26) for $f_0 = 0$ at those values of θ for which $\epsilon/2$ equals $\pi/2$ unless this occurs at $\theta = \pi$. The values of f_0 will be restricted to a finite range from zero up to a high-frequency cutoff; for the left-hand side must be equal to $-(\pi/2)$, which restricts the first term to finite values of f_0. The frequency effects for the zeroth mode are just the opposite of those for the higher modes. Algebraically, this is due simply to the change in sign of the right-hand side of (4-26) from $n = 0$ to $n = 1, 2, \ldots$.

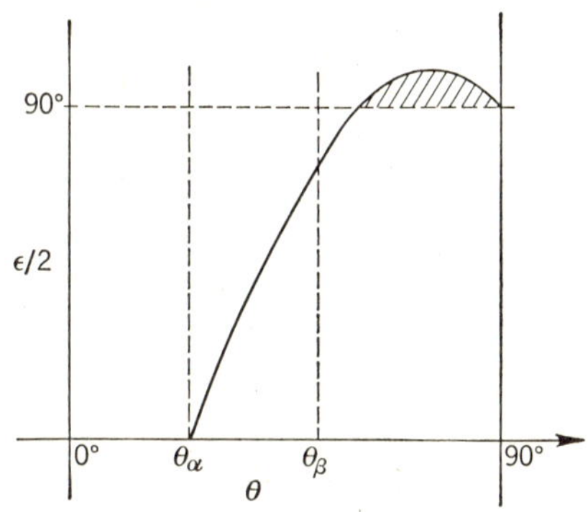

FIG. 4-28.

From Chap. 5 we have that the phase change is given by

$$\frac{\epsilon}{2} = \cot^{-1}\left\{\frac{\rho_2\beta_2{}^4\sqrt{(V^2/c_1{}^2) - 1}}{\rho_1 V^4 \sqrt{1 - (V^2/\alpha_2{}^2)}}\left[\left(2 - \frac{V^2}{\beta_2{}^2}\right)^2 - 4\sqrt{1 - \frac{V^2}{\alpha_2{}^2}}\sqrt{1 - \frac{V^2}{\beta_2{}^2}}\right]\right\} \tag{4-33}$$

Starting from the critical angle for shear refraction, $V = \beta_2$, the term in the brackets of (4-33) is positive and decreases with increasing V. If the second term in the brackets becomes greater than the first term, $\epsilon/2$ will become greater than $\pi/2$; and the condition for an $n = 0$ solution to exist will prevail. For normal values of β_2 and α_2 encountered in nature this condition will exist over a limited range of θ. Figure 4-28 is a sketch of $\epsilon/2$ versus θ.

Of further interest for the first and higher modes are the additional deductions that can be made from the discussions of Chap. 5 on the reflection for a liquid-solid interface. Below the angle θ_β there will be attenuated propagation. First the phase and group velocities approach the value $V = U = \alpha_2$ at θ_α. Then V approaches infinity toward the cutoff for attenuated propagation [Eq. (3-53)]. Figure 4-29 is a sketch of the first three modes including the attenuated portion for the first mode.

The existence of a normal-mode solution in the limit of zero frequency is an interesting conclusion from this investigation. The portion of the mode train for $n = 0$ extending out from zero frequency toward the minimum group velocity is often referred to as the *Rayleigh wave* train.

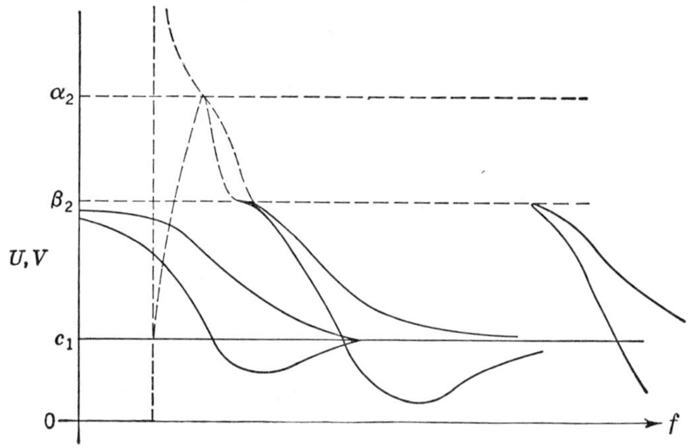

FIG. 4-29.

Since, in the limit of zero frequency and finite velocity, the wavelength approaches infinity, the effect of the ocean becomes negligible. The problem reduces to that of a free surface bounded by an elastic medium. We can then examine as a separate problem the conditions for the existence of a wave bound to the free surface and propagating outward in the radial direction only. (The normal-mode conditions reduce to this for a single boundary.)

We shall give this derivation for plane waves. For dependence on the x and z space coordinates only, the displacements u and w in terms of the displacement potentials are from (1-107)

$$u = \frac{\partial \varphi}{\partial x} + \frac{\partial \psi}{\partial z}, \quad w = \frac{\partial \varphi}{\partial z} - \frac{\partial \psi}{\partial x} \tag{4-34}$$

where φ and ψ satisfy the equations

$$\alpha^2 \nabla^2 \varphi = \frac{\partial^2 \varphi}{\partial t^2} \qquad \text{and} \qquad \beta^2 \nabla^2 \psi = \frac{\partial^2 \psi}{\partial t^2} \tag{4-35}$$

The normal and tangential components of stress are from (1-12)

$$\begin{aligned} \sigma_{zz} &= \lambda\Delta + 2\mu\frac{\partial w}{\partial z} = \rho\left[\alpha^2\nabla^2\varphi - 2\beta^2\left(\frac{\partial^2\varphi}{\partial x^2} + \frac{\partial^2\psi}{\partial x\,\partial z}\right)\right] \\ \sigma_{zr} &= \mu\left(\frac{\partial w}{\partial x} + \frac{\partial u}{\partial z}\right) = \mu\left[2\frac{\partial^2\varphi}{\partial x\,\partial z} - \frac{\partial^2\psi}{\partial x^2} + \frac{\partial^2\psi}{\partial z^2}\right] \end{aligned} \tag{4-36}$$

In order for the solution to be a wave bound to the free surface and propagating outward in the x direction only we want φ and ψ to be of the form

$$\begin{aligned} \varphi &= Ae^{-\eta z}e^{i(\omega t - kx)} \\ \psi &= Be^{-\zeta z}e^{i(\omega t - kx)} \end{aligned} \tag{4-37}$$

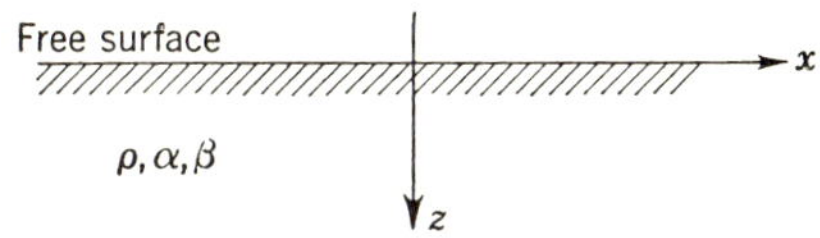

FIG. 4-30.

for a simple harmonic wave. The coefficients η and ζ are as before

$$\eta = \sqrt{k^2 - \frac{\omega^2}{\alpha^2}} \qquad \text{and} \qquad \zeta = \sqrt{k^2 - \frac{\omega^2}{\beta^2}} \tag{4-38}$$

For such a wave to exist all the coefficients k, η, and ζ must have real-value solutions from the conditional equations satisfying the boundary conditions. From Fig. 4-30 the boundary conditions are that σ_{zz} and σ_{zr} equal zero at $z = 0$. Substituting (4-37) into (4-36) gives

$$\begin{aligned} \alpha^2(-k^2 + \eta^2)A + 2\beta^2(k^2A - ik\zeta B) &= 0 \\ i2k\eta A + (k^2 + \zeta^2)B &= 0 \end{aligned} \tag{4-39}$$

Substituting from (4-38) and from the relation between the phase velocity V in the x direction and the wave number k measured in the x direction,

$$V = \frac{\omega}{k} \tag{4-40}$$

we obtain

$$\begin{aligned} \left(2 - \frac{V^2}{\beta^2}\right)A - i2\sqrt{1 - \frac{V^2}{\beta^2}}\,B &= 0 \\ i2\sqrt{1 - \frac{V^2}{\alpha^2}}\,A + \left(2 - \frac{V^2}{\beta^2}\right)B &= 0 \end{aligned} \tag{4-41}$$

For solutions other than $A = B = 0$ to exist for Eqs. (4-41) the determinant of the coefficients must vanish, or simply

$$\left(2 - \frac{V^2}{\beta^2}\right)^2 - 4\sqrt{1 - \frac{V^2}{\alpha^2}}\sqrt{1 - \frac{V^2}{\beta^2}} = 0 \tag{4-42}$$

This is the same condition that we had before for an $n = 0$ mode and consequently an $f = 0$ solution to exist. Generally a real-value solution for V will exist such that $V < \beta$, satisfying the conditions that k, η, and ζ

be real. When Poisson's ratio holds, we will have that $\lambda = \mu$ [Eq. (1-17)] so that

$$\alpha = \sqrt{3}\,\beta \tag{4-43}$$

and there will be one real-value solution to (4-42) satisfying the above conditions. For this case the solution is

$$V = 0.92\beta \tag{4-44}$$

Near-surface transmission, negative velocity gradient

Let us consider here some of the transmission effects which may occur in the vicinity of the ocean surface. We shall be interested in the direct and surface-reflected transmission from a near-surface source to a near-surface receiver. The total depth range of interest is of the order of a few hundred feet, and the contributions from reflections from the ocean bottom at a depth of 10,000 to 20,000 feet will be ignored. For this problem we have an upper boundary at the ocean surface and no lower boundary. There are two characteristic sound velocity profiles for this depth range in a deep-ocean environment. In one the main or seasonal thermocline extends to the ocean surface so that there is a negative velocity gradient from the surface downward. In the other a mixed water layer overlies the thermocline so that there is a slight positive velocity gradient in the isothermal layer underlain by a steeper negative gradient. Transmission for the first case will be considered in this section, and that for the second case in the next section.

For a negative velocity gradient there is downward refraction of the rays (Fig. 4-31). A shadow boundary beyond which there will be no ray transmission is formed with the ocean surface. The travel times and intensities for reception inside the shadow boundary can be determined by the formulas of Chap. 2. Beyond the shadow boundary ray theory states that there will be no reception. However, in the neighborhood of the predicted discontinuous decrease in sound intensity at the shadow boundary the condition (2-18) for the validity of the ray theory breaks down.

We must return to the original wave equation to determine the energy diffracted into the shadow zone. This problem is similar to the problem of diffraction around an edge in a constant velocity medium, which will be considered in Chap. 6. Here the diffraction is brought about because of the presence of the ocean surface and a negative velocity gradient. The solution to the wave equation for the velocity as a function of depth is quite difficult to obtain, even for the simplest velocity functions. For an approximate linear gradient the normal-mode solution is given in terms of a series of imaginary-order Bessel functions of complex argument. The solution of the characteristic equation for the complex values of the wave number k in terms of these functions is also quite difficult. An approximate solution to the wave equation, referred to as the WKBJ approximation, is useful in such problems as this. We shall not discuss this method here. Its derivation and description, especially for the determination of complex k, is difficult. Its application is usually straightforward. It is important to reiterate, however, that it is a useful method for the approximate solution of variable velocity sound problems. For this problem both the normal mode and WKBJ evaluations determine that the amplitude decreases exponentially into the shadow zone. The exponent of the leading term is proportional to the horizontal distance away from the shadow boundary, the two-thirds power of the sound velocity gradient at the surface, and the one-third power of the frequency and inversely proportional to the surface sound velocity.

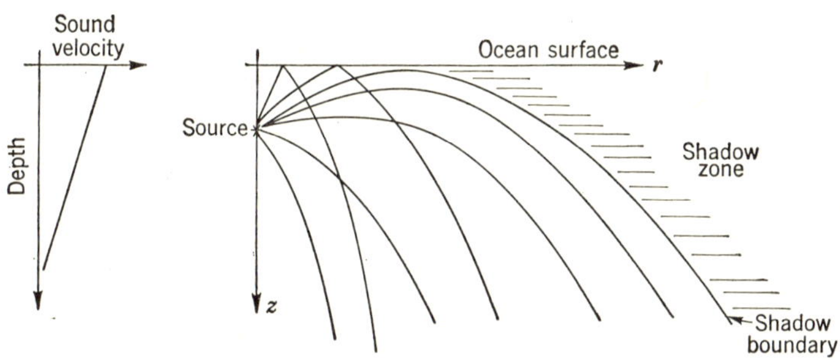

FIG. 4-31.

The intensity decreases rapidly away from the shadow boundary into the shadow zone. For such a low-intensity region the details of the sound velocity profile can become important. In particular if the sound velocity gradient decreases in the immediate vicinity of the free surface and reaches a zero gradient, there will be no shadow boundary. The entire region in what was formerly the shadow zone will be weakly insonified. There will be a ray path for any receiver location. This comparison is illustrated in Figs. 4-32 and 4-33. Only the direct rays have been drawn in these figures; there will be a similar set of rays for the surface reflections. Inside the ray A of Fig. 4-33 or the shadow boundary ray of Fig. 4-32 the ray intensities will be normal. The total area inside ray A is less than that inside the shadow boundary ray. Between rays A and B the intensity will be considerably less than that in the normal region but will not be zero as predicted by ray theory for the region beyond the shadow boundary of Fig. 4-32.

It is of interest, as an illustration of the methods of ray theory, to derive the equations for ray intensity for the conditions of Fig. 4-33. We shall take a linear velocity function below the depth $z = l$ and a parabolic velocity function with axis at the surface for $z \leqslant l$,

$$\begin{aligned} c &= c_1(1 - \alpha z^2) \qquad 0 \leqslant z \leqslant l \\ c &= c_2(1 - \beta z) \qquad l \leqslant z \end{aligned} \tag{4-45}$$

For α and β small, valid for the ocean, Eqs. (4-45) may be approximated by

$$\begin{aligned} c &\approx c_1(1 + 2\alpha z^2)^{-1/2} \qquad 0 \leqslant z \leqslant l \\ c &\approx c_2(1 + 2\beta z)^{-1/2} \qquad l \leqslant z \end{aligned} \tag{4-46}$$

We stipulate that the velocity functions shall join smoothly at $z = l$, i.e., that the functions and their derivatives be continuous. This gives two

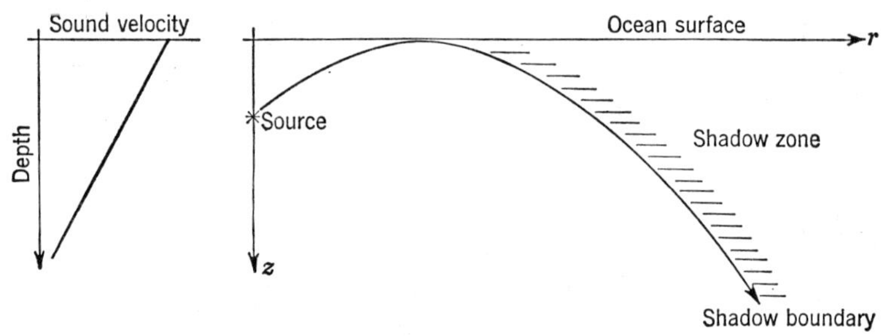

FIG. 4-32.

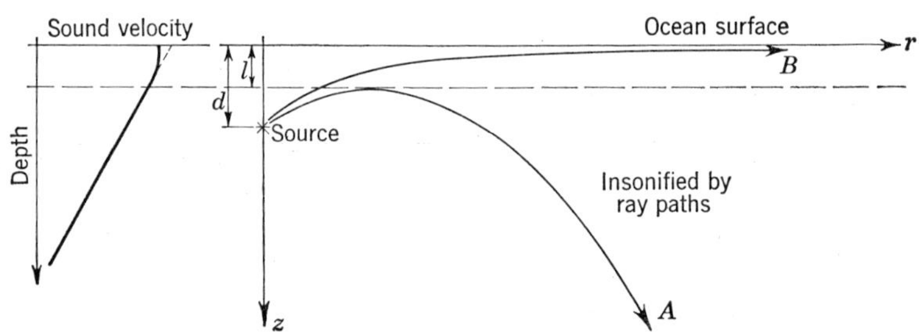

FIG. 4-33.

relations between the five quantities c_1, c_2, α, β, and l defining the velocity as a function of depth. Let us solve for c_2 and α in terms of the other three. From (4-46) we have for continuity of the velocity and its derivative at $z = l$,

$$\begin{aligned} \frac{c_1}{(1 + 2\alpha l^2)^{1/2}} &= \frac{c_2}{(1 + 2\beta l)^{1/2}} \\ c_1 \frac{4\alpha l}{(1 + 2\alpha l^2)^{3/2}} &= c_2 \frac{2\beta}{(1 + 2\beta l)^{3/2}} \end{aligned} \tag{4-47}$$

Solving for α and c_2 we obtain

$$\frac{4\alpha l}{1 + 2\alpha l^2} = \frac{2\beta}{1 + 2\beta l}$$

$$2\alpha = \frac{\beta}{l(1 + \beta l)} \tag{4-48}$$

$$c_2^2 = c_1^2 \frac{1 + 2\beta l}{1 + 2\alpha l^2}$$

$$= c_1^2 \frac{\beta}{2\alpha l} \tag{4-49}$$

Since most of the rays that we shall be concerned with are at grazing angles, it is convenient to use the complement φ of the Snell's law angle θ.

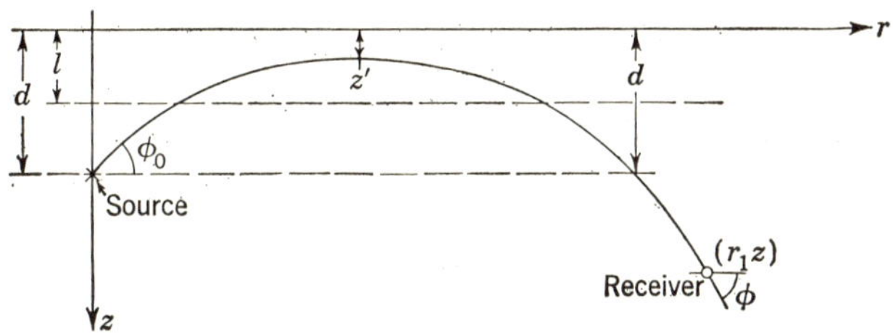

FIG. 4-34.

The angle φ is the angle to the ray path measured from the horizontal. For φ Snell's law (2-51) becomes

$$\frac{\cos \varphi_0}{c_0} = \frac{\cos \varphi}{c} \tag{4-50}$$

where φ_0 and c_0 are the angle and sound velocity at the source and φ and c the angle and velocity at any point along the ray path emergent from the source at φ_0. In terms of φ the ray intensity (2-54) becomes

$$I(r, \varphi) = -\frac{P \cos \varphi_0}{r \sin \varphi(\partial r/\partial \varphi_0)} \tag{4-51}$$

For a ray which does not penetrate above the depth $z = l$, the velocity will always be a linear function of depth. The intensity will be given by (2-89)

$$I = P \frac{\cos^2 \varphi_0}{r^2} \tag{4-52}$$

For a ray which penetrates above the depth $z = l$ we must first derive the relation for r in terms of the ray parameters. From Fig. 4-34 this will be

$$r = \int^{\text{ray path}} dr = 2 \int_{z'}^{d} \frac{dr}{dz} dz + \int_{d}^{z} \frac{dr}{dz} dz \tag{4-53}$$

From (4-50) and (4-46) dr/dz will be given by the following for $z \leqslant l$.

$$\begin{aligned}\frac{dr}{dz} &= \cot\varphi = \frac{\cos\varphi}{(1-\cos^2\varphi)^{1/2}} = \left(\frac{1}{\cos^2\varphi} - 1\right)^{-1/2} \\ &= \left[\frac{c_0{}^2}{c_1{}^2\cos^2\varphi_0}(1+2\alpha z^2) - 1\right]^{-1/2} \\ &= \frac{c_1\cos\varphi_0}{c_0}\left(1 + 2\alpha z^2 - \frac{c_1{}^2\cos^2\varphi_0}{c_0{}^2}\right)^{-1/2} \\ &= (2\alpha)^{-1/2}\frac{c_1\cos\varphi_0}{c_0}(z^2 - \lambda_1)^{-1/2} \qquad (4\text{-}54)\end{aligned}$$

where λ_1 is defined by

$$\lambda_1 = \frac{1}{2\alpha}\left(\frac{c_1{}^2\cos^2\varphi_0}{c_0{}^2} - 1\right) \qquad (4\text{-}55)$$

Similarly for $z \geqslant l$ we shall have

$$\frac{dr}{dz} = (2\beta)^{-1/2}\frac{c_2\cos\varphi_0}{c_0}(z - \lambda_2)^{-1/2} \qquad (4\text{-}56)$$

where λ_2 is defined by

$$\lambda_2 \equiv \frac{1}{2\beta}\left(\frac{c_2{}^2\cos^2\varphi_0}{c_0{}^2} - 1\right) \qquad (4\text{-}57)$$

Substituting (4-54) and (4-56) into (4-53) and integrating,[1] we obtain

$$\begin{aligned} r &= \frac{\cos\varphi_0}{c_0}\left[\frac{2c_1}{(2\alpha)^{1/2}}\int_{z'}^{l}(z^2 - \lambda_1{}^2)^{-1/2}\,dz + \frac{2c_2}{(2\beta)^{1/2}}\int_{l}^{d}(z-\lambda_2)^{-1/2}\,dz \right.\\ &\qquad\left. + \frac{c_2}{(2\beta)^{1/2}}\int_{d}^{z}(z-\lambda_2)^{-1/2}\,dz\right] \\ &= \frac{2\cos\varphi_0}{c_0}\left\{\frac{c_2}{(2\beta)^{1/2}}[(z-\lambda_2)^{1/2} + (d-\lambda_2)^{1/2} + 2(l-\lambda_2)^{1/2}]\right.\\ &\qquad\left. + \frac{c_1}{(2\alpha)^{1/2}}\ln\left[\frac{l + (l^2-\lambda_1)^{1/2}}{z' + (z'^2 - \lambda_1)^{1/2}}\right]\right\} \\ &= \frac{2c_2\cos\varphi_0}{c_0(2\beta)^{1/2}}\left\{(z-\lambda_2)^{1/2} + (d-\lambda_2)^{1/2} - 2(l-\lambda_2)^{1/2}\right.\\ &\qquad\left. + (2l)^{1/2}\ln\left[\frac{l + (l^2-\lambda_1)^{1/2}}{z' + (z'^2 - \lambda_1)^{1/2}}\right]\right\} \qquad (4\text{-}58)\end{aligned}$$

using the relation (4-49). It is convenient for the determination of I to express r in terms of a single ray parameter λ_2 and the depth of the

[1] B. O. Pierce, "A Short Table of Integrals," p. 16, Sec. 95, p. 20, Sec. 126, Ginn & Company, Boston, 1929.

receiver z. From (4-46), (4-50), and (4-55) we have for z' in terms of λ_1

$$\frac{1}{c_1(1 + 2\alpha z'^2)^{-1/2}} = \frac{\cos \varphi_0}{c_0}$$
$$2\alpha z'^2 = \frac{c_1^2 \cos^2 \varphi_0}{c_0^2} - 1$$
$$z' = (\lambda_1)^{1/2} \qquad (4\text{-}59)$$

From (4-57) we have for φ_0 in terms of λ_2

$$\frac{c_2^2 \cos^2 \varphi_0}{c_0^2} = 1 + 2\beta\lambda_2 \qquad (4\text{-}60)$$

From (4-55), (4-60), (4-49), and (4-48) we have for λ_1 in terms of λ_2

$$\begin{aligned} \lambda_1 &= \frac{1}{2\alpha}\left[\frac{c_1^2}{c_2^2}(1 + 2\beta\lambda_2) - 1\right] \\ &= \frac{l}{\beta}(1 + 2\beta\lambda_2) - \frac{1}{2\alpha} \\ &= \frac{l}{\beta}(1 + 2\beta\lambda_2 - 1 - \beta l) \\ &= l(2\lambda_2 - l) \end{aligned} \qquad (4\text{-}61)$$

Substituting (4-59), (4-60), and (4-61) into (4-58) gives

$$\begin{aligned} r &= 2\left(\lambda_2 + \frac{1}{2\beta}\right)^{1/2} \left\{(z - \lambda_2)^{1/2} + (d - \lambda_2)^{1/2} + 2(l - \lambda_2)^{1/2} \right. \\ &\qquad \left. + (2l)^{1/2} \ln\left[\frac{l + (l^2 - \lambda_1)^{1/2}}{\lambda_1^{1/2}}\right]\right\} \\ &= 2\left(\lambda_2 + \frac{1}{2\beta}\right)^{1/2} \left\{(z - \lambda_2)^{1/2} + (d - \lambda_2)^{1/2} + 2(l - \lambda_2)^{1/2} \right. \\ &\qquad \left. + (2l)^{1/2} \ln\left[\frac{l^{1/2} + (2l - 2\lambda_2)^{1/2}}{(2\lambda_2 - l)^{1/2}}\right]\right\} \end{aligned} \qquad (4\text{-}62)$$

For the ray B of Fig. 4-33 $z' = 0$; for the ray A, $z' = l$. From Eq. (4-59) λ_1 can have values between the limits 0 and l^2. From Eq. (3-61) the corresponding range of values for λ_2 is between l and $l/2$.

In order to determine the intensity I we must find the value of

$$\frac{\partial r}{\partial \varphi_0} = \frac{\partial r}{\partial \lambda_2}\frac{d\lambda_2}{d\varphi_0} \qquad (4\text{-}63)$$

From (4-62) we have

$$
\begin{aligned}
\frac{\partial r}{\partial \lambda_2} &= \left(\lambda_2 + \frac{1}{2\beta}\right)^{-1/2} \left(\cdot \cdot \cdot \cdot\right) + 2\left(\lambda_2 + \frac{1}{2\beta}\right)^{1/2} \left\{ -\frac{1}{2}(z - \lambda_2)^{-1/2} \right. \\
&\quad - \frac{1}{2}(d - \lambda_2)^{-1/2} + (l - \lambda_2)^{-1/2} - (2l)^{1/2} \left[\frac{l^{1/2} + (2l - 2\lambda_2)^{1/2}}{(2\lambda_2 - l)^{1/2}} \right]^{-1} \\
&\qquad \left. \left[\frac{(2\lambda_2 - l)^{1/2}(2l - 2\lambda_2)^{-1/2} + [l^{1/2} + (2l - 2\lambda_2)^{1/2}](2\lambda_2 - l)^{-1/2}}{(2\lambda_2 - l)} \right] \right\} \\
&= \frac{1}{2}\left(\lambda_2 + \frac{1}{2\beta}\right)^{-1} r - \left(\lambda_2 + \frac{1}{2\beta}\right)^{1/2} \left\{ (z - \lambda_2)^{-1/2} + (d - \lambda_2)^{-1/2} \right. \\
&\quad \left. + 2\, \frac{-2^{1/2}(2\lambda_2 - l)[l^{1/2} + (2l - 2\lambda_2)^{1/2}] + (2l)^{1/2}(2\lambda_2 - l) + (2l)^{1/2}(2l - 2\lambda_2)^{1/2}[l^{1/2} + (2l - 2\lambda_2)^{1/2}]}{(2l - 2\lambda_2)^{1/2}(2\lambda_2 - l)[l^{1/2} + (2l - 2\lambda_2)^{1/2}]} \right\} \\
&= \frac{1}{2}\left(\lambda_2 + \frac{1}{2\beta}\right)^{-1} r - \left(\lambda_2 + \frac{1}{2\beta}\right)^{1/2} \left((z - \lambda_2)^{1/2} + (d - \lambda_2)^{1/2} \right. \\
&\quad \left. + \frac{4(l - \lambda_2)^{1/2}\{-(2\lambda_2 - l) + l^{1/2}[l^{1/2} + (2l - 2\lambda_2)^{1/2}]\}}{(2l - 2\lambda_2)^{1/2}(2\lambda_2 - l)[l^{1/2} + (2l - 2\lambda_2)^{1/2}]} \right) \\
&= \frac{1}{2}\left(\lambda_2 + \frac{1}{2\beta}\right)^{-1} r - \left(\lambda_2 + \frac{1}{2\beta}\right)^{1/2} \left((z - \lambda_2)^{-1/2} + (d - \lambda_2)^{-1/2} \right. \\
&\quad \left. + \frac{4(l - \lambda_2)^{1/2}\{(2l - 2\lambda_2)^{1/2}[l^{1/2} + (2l - 2\lambda_2)^{1/2}]\}}{(2l - 2\lambda_2)^{1/2}(2\lambda_2 - l)[l^{1/2} + (2l - 2\lambda_2)^{1/2}]} \right) \\
&= \frac{1}{2}\left(\lambda_2 + \frac{1}{2\beta}\right)^{-1} r - \left(\lambda_2 + \frac{1}{2\beta}\right)^{1/2} \\
&\qquad [(z - \lambda_2)^{-1/2} + (d - \lambda_2)^{-1/2} + 4(l - \lambda_2)^{1/2}(2\lambda_2 - l)^{-1}] \qquad (4\text{-}64)
\end{aligned}
$$

From (4-57) and (4-60) we have

$$
\begin{aligned}
\frac{d\lambda_2}{d\varphi_0} &= -\frac{c_2^2 \cos \varphi_0 \sin \varphi_0}{\beta c_0^2} \\
&= -2\left(\lambda_2 + \frac{1}{2\beta}\right) \tan \varphi_0 \qquad (4\text{-}65)
\end{aligned}
$$

Substituting (4-63), (4-64), and (4-65) into (4-51) we obtain finally for the intensity

$$
\begin{aligned}
I = P \frac{\cos^2 \varphi_0}{-\sin \varphi_0 \sin \varphi} &\left\{ 2\left(\lambda_2 + \frac{1}{2\beta}\right)^{3/2} \right. \\
&\left. [(z - \lambda_2)^{-1/2} + (d - \lambda_2)^{-1/2} + 4(l - \lambda_2)^{1/2}(2\lambda_2 - l)^{-1}] - r \right\} \qquad (4\text{-}66)
\end{aligned}
$$

The derivation given here is for the direct ray. A similar derivation can be given for the surface-reflected ray.

Near-surface transmission, positive velocity gradient overlying a negative gradient

For a positive velocity gradient overlying a negative gradient there is upward refraction of the rays in the positive-gradient region and downward refraction beneath (Fig. 4-35). In Fig. 4-35 a discontinuous change in the derivative of the velocity has been assumed at the depth

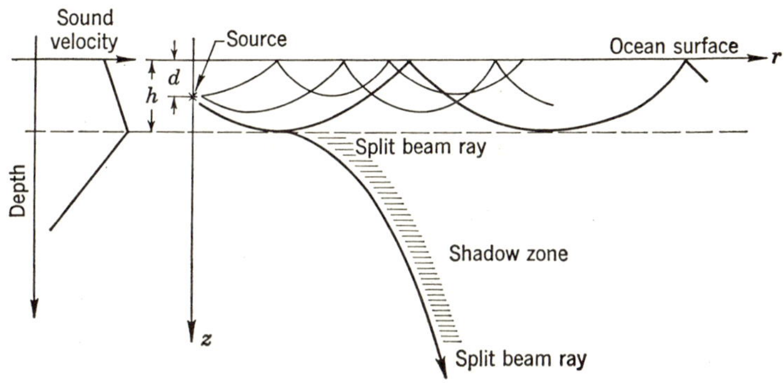

FIG. 4-35.

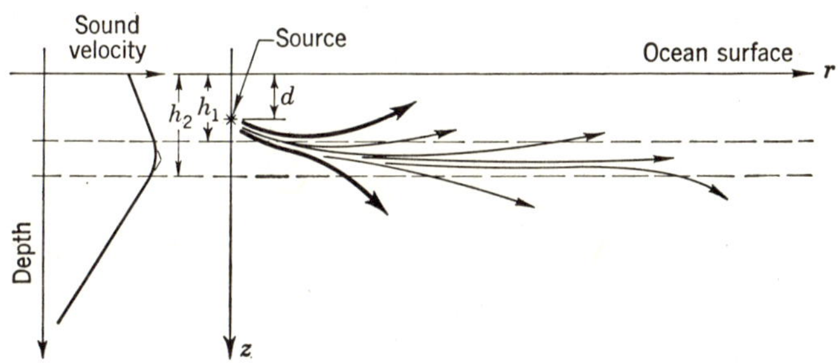

FIG. 4-36.

$z = h$. For this case the split-beam ray is of interest. The split-beam ray defines a shadow zone similar to the previous example. If the derivative of the velocity were continuous across the depth $z = h$, the region of the shadow zone would be weakly insonified by ray paths. These two cases are illustrated in Figs. 4-35 and 4-36. For clarity in illustration only the rays originally downgoing from the source have been shown. There will be a similar set of ray paths for the rays originally upgoing from the source and reflected at the ocean surface.

There are two points of particular interest here connected with transmission in the positive-gradient region. One is the definition of a sound channel for depths less than $z = h$; the other is the formation of caustics within the sound channel. The rays emergent from the source at angles more grazing than the split-beam ray are confined to propagate within the depth range from $z = 0$ to $z = h$. They are refracted upward at depth and reflected downward from the ocean surface. They form a set of RSR rays within the surface layer. The energy emitted from the source between the angles $-\varphi_{0s}$ and φ_{0s} where φ_{0s} is the angle at the source of the split-beam ray is confined to propagate within the depths $z = 0$ to $z = h$ for all r. Neglecting diffraction losses to the underlying negative-gradient region, there is a net spreading and a net propagation in in the r direction only. In principle the wave-guide propagation in this surface sound channel is similar to the wave-guide propagation in shallow water treated in the last chapter. It exhibits the same features but to a different degree. Instead of a discontinuous increase in velocity at the depth $z = h$ there is a gradual increase in velocity throughout the layer. At no frequencies will there be strictly no attenuation in the r direction, for there will always be some loss of energy by diffraction out of the sound channel to the underlying negative-gradient region. The split-beam ray corresponds to the critical-angle ray in the shallow-water case. It defines the frequency region of low attenuation. Since the split-beam ray is at a much more grazing angle than the critical-angle ray, the cutoff frequency for low-attenuation propagation in the surface sound channel occurs at a much higher frequency. The formal solution to wave-guide propagation in this example may be carried out in a manner similar to that for the normal-mode solution of Chap. 3. The wave-equation solutions in terms of a series of imaginary-order Bessel functions of complex argument are difficult to evaluate. A WKBJ solution offers a convenient method of approximation.

The other point of interest is the definition of a series of caustics in the sound channel. Each series of surface-reflected rays forms a caustic envelope. The direct rays, having undergone no surface reflections, do not form a caustic. For a surface-reflected ray we may think of the problem in terms of the image source. Then the velocity profile to be considered is first a decrease in velocity followed by an increase. As described in Chap. 2 for such a velocity function, a caustic is formed. Along the caustic the intensity of the received signal is high; from ray theory it is infinite. It is of interest to define the locations of the caustics.

Any ray path can be defined by a single ray parameter, such as the initial angle φ_0 at the source. A set of rays can then be defined by an equation

$$f(r, z, \alpha) = 0 \tag{4-67}$$

where α is the ray parameter. In three-dimensional r, z, α space equation (4-67) will define a curved surface. Such a curved surface may be that shown in Fig. 4-37. Its traces on a series of planes perpendicular to the α

axis are the ray paths (Fig. 4-38). The projection on the r, z plane of the curved line in this surface defined by the relation

$$\frac{df}{d\alpha} = 0 \tag{4-68}$$

is the envelope of the rays. The traces of this curved surface on a series of planes perpendicular to the r axis will form a set of curves as shown in

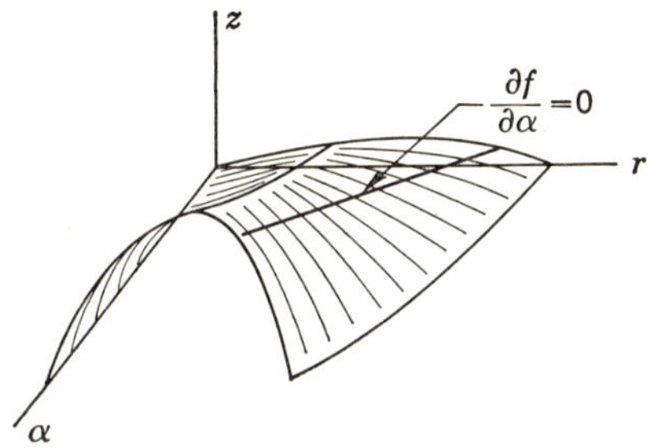

FIG. 4-37.

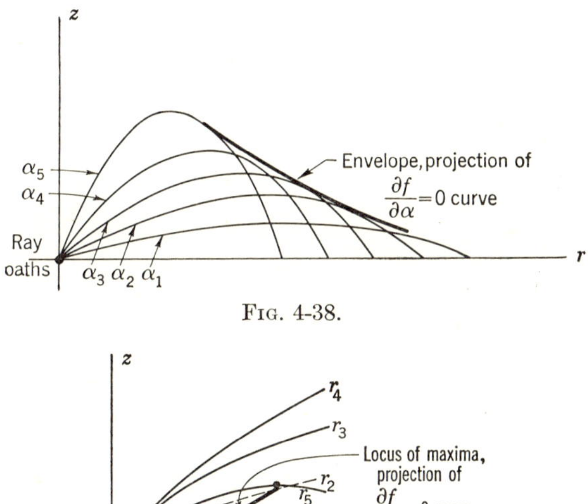

FIG. 4-38.

FIG. 4-39.

Fig. 4-39. The parameter α is a measure of the energy of any given ray bundle. At the maxima points of Fig. 4-39 there is a finite amount of energy arriving over an infinitesimal change in depth. The locus of the maxima is the projection of the $df/d\alpha = 0$ curve on the z, α plane. Similarly, the locus of the maxima for traces on the r, α plane is defined by this curve. The ray intensity is infinite along the caustic. The caustic is defined by the discrete values of α, r, and z satisfying both eqs. (4-67) and (4-68).

For our case of an isothermal sound channel, there is a linear increase in velocity with depth. From Chap. 2 we know that the ray paths will be arcs of circles. The centers of the circles will all lie along the line for which the extrapolated velocity is zero. From Fig. 4-40 the equation for the direct rays will be

$$\begin{aligned} (r \mp \beta)^2 + z^2 &= \alpha^2 \\ (r \mp \sqrt{\alpha^2 - z_0^2})^2 + z^2 &= \alpha^2 \end{aligned} \tag{4-69}$$

The minus sign applies to the rays originally downgoing from the source, the plus sign to the rays originally upgoing. The center of the circle

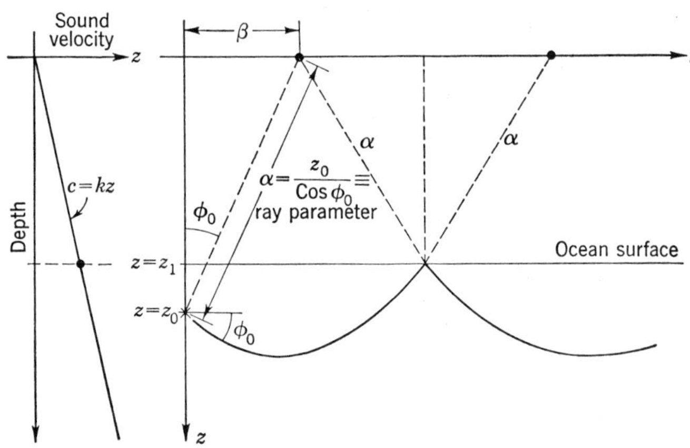

FIG. 4-40.

defining the same ray after it has been reflected once from the surface will be located at a horizontal distance

$$\beta_1 = \pm \sqrt{\alpha^2 - z_0^2} + 2\sqrt{\alpha^2 - z_1^2} \tag{4-70}$$

from Fig. 4-40. For n surface reflections the distance will be simply

$$\beta_n = \pm \sqrt{\alpha^2 - z_0^2} + 2n\sqrt{\alpha^2 - z_1^2} \tag{4-71}$$

The equation for the set of n surface-reflected rays will be

$$(r \mp \sqrt{\alpha^2 - z_0^2} - 2n\sqrt{\alpha^2 - z_1^2})^2 + z^2 = \alpha^2 \tag{4-72}$$

Taking a derivative of (4-72) with respect to α, we obtain

$$[r \mp \sqrt{\alpha^2 - z_0^2} - 2n\sqrt{\alpha^2 - z_1^2}][\mp(\alpha^2 - z_0^2)^{-\frac{1}{2}} - 2n(\alpha^2 - z_1^2)^{-\frac{1}{2}}] = 1 \tag{4-73}$$

The equation for the envelopes is obtained by eliminating α between Eqs. (4-72) and (4-73). For a given calculation it is generally more convenient to use the equations in the above parametric form. For a given order of surface reflection, n, a value of α is chosen. Then the value of r can be determined from Eq. (4-73) and z from Eq. (4-72). For a ray originally downgoing from the source the second bracket in Eq. (4-73) is always negative. Therefore, the first bracket must also be negative. The range r will be less than βn; the envelope will form on the downgoing side of the rays. For rays originally upgoing, the second bracket can be positive in the vicinity of $\alpha \approx z_0$, i.e., $\varphi_0 \approx 0$. For this condition r will be greater than β_n; an envelope will form on the upgoing side of the rays. For larger α the second bracket will be negative, and an envelope will form

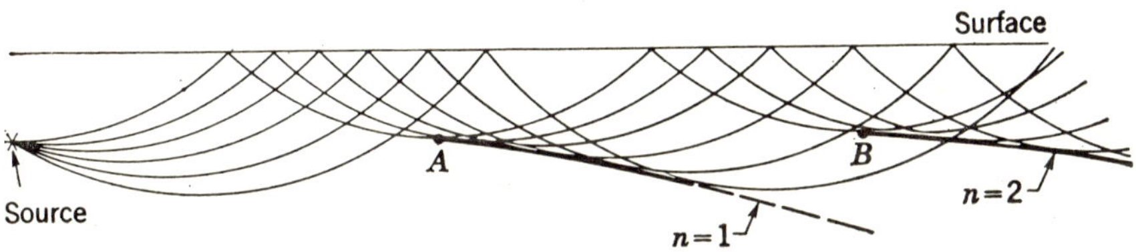

FIG. 4-41.

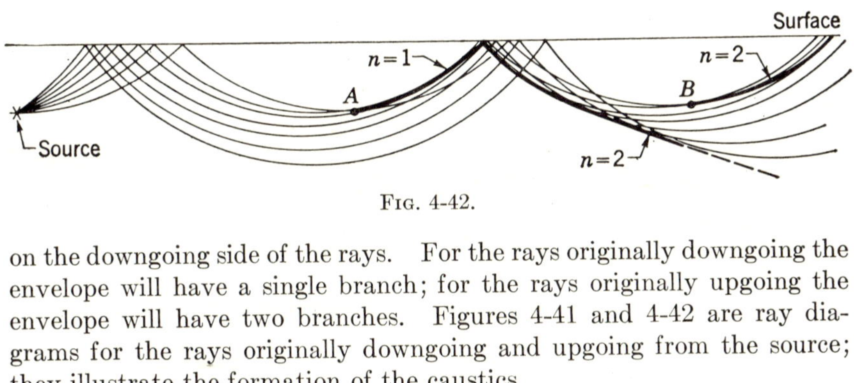

FIG. 4-42.

on the downgoing side of the rays. For the rays originally downgoing the envelope will have a single branch; for the rays originally upgoing the envelope will have two branches. Figures 4-41 and 4-42 are ray diagrams for the rays originally downgoing and upgoing from the source; they illustrate the formation of the caustics.

For a source at the surface, $z_0 = z_1$, Eqs. (4-72) and (4-73) reduce to

$$[r - (2n + 1)\sqrt{\alpha^2 - z_1^2}]^2 + z^2 = \alpha^2 \tag{4-74}$$

$$[r - (2n + 1)\sqrt{\alpha^2 - z_1^2}]\left[\frac{-(2n + 1)}{\sqrt{\alpha^2 - z_1^2}}\right] = 1 \tag{4-75}$$

Solving for α in terms of r in (4-75) we obtain

$$\sqrt{\alpha^2 - z_1^2} = \frac{(2n + 1)r}{(2n + 1)^2 - 1}$$

$$\alpha^2 = z_1^2 + \frac{(2n + 1)^2 r^2}{[(2n + 1)^2 - 1]^2}$$

Substituting into (4-74) gives

$$\frac{-r^2}{[(2n+1)^2-1]^2} + z^2 = z_1{}^2 + \frac{(2n+1)^2r^2}{[(2n+1)^2-1]^2}$$

$$\frac{-r^2}{(2n+1)^2-1} + z^2 = z_1{}^2 \tag{4-76}$$

Equation (4-76) is the equation for the envelope; it is the equation of a hyperbola.

BIBLIOGRAPHY

Bremmer, H.: "Terrestrial Radio Waves," Elsevier Publishing Company, Amsterdam, 1949.

Bullen, K. E.: "Introduction to the Theory of Seismology," Cambridge University Press, New York, 1947.

Burrows, C. R., and S. S. Attwood (eds.): "Radio Wave Propagation," Academic Press, Inc., New York, 1949.

Ewing, M., and J. L. Worzel: Long Range Sound Transmission, *in* Propagation of Sound in the Ocean, *Geol. Soc. Amer., Mem.* 27, 1948.

Furry, W. H.: Theory of Characteristic Functions in Problems of Anomalous Propagation, *Mass. Inst. Technol. Radiation Lab. Repts.* 680, 1945.

Hersey, J. B., and M. Ewing: Seismic Reflections from Beneath the Ocean Floor, *Trans. Am. Geophys. Union,* vol. 30, pp. 5–14, 1949.

Lamb, H.: On the Propagation of Tremors over the Surface of an Elastic Solid, *Phil. Trans. Roy. Soc. London,* ser. A, vol. 203, pp. 1–42, 1904.

Marsh, H. W.: Theory of the Anomalous Propagation of Acoustic Waves in the Ocean, *U.S. Navy Underwater Sound Lab. Rept.* 111, 1950.

Morse, R. W.: The Dependence of Shadow Zone Sound on the Surface Velocity Gradient, *J. Acoust. Soc. Amer.,* vol. 22, pp. 857–860, 1950.

Officer, C. B.: Geologic Interpretation of a Series of Seismic Reflection Profiles from Bermuda to the Continental Margins, *Deep Sea Research,* vol. 2, pp. 253–260, 1955.

Pekeris, C. L.: Theory of Propagation of Sound in a Half-space of Variable Sound Velocity under Conditions of Formation of a Shadow Zone, *J. Acoust. Soc. Amer.,* vol. 18, pp. 295–315, 1946.

Press, F., and M. Ewing: A Theory of Microseisms with Geologic Applications, *Trans. Am. Geophys. Union,* vol. 29, pp. 163–174, 1948.

Press, F., and M. Ewing: Propagation of Explosive Sound in a Liquid Layer Overlying a Semi-infinite Elastic Solid, *Geophysics,* vol. 15, pp. 426–446, 1950.

Roberg, J.: Sound Transmission in the Surface Isothermal Channel, Ray Theory Discussion, *Woods Hole Oceanographic Inst. Rept.* 52–96, 1952.

Woods Hole Oceanographic Institution: Sound Transmission in Sea Water, *Natl. Defense Research Council Rept.,* 1941.

5. *Reflectivity*

A great many of the observed sound-transmission effects occur because of the presence of a discontinuity between two media of differing acoustical properties or a region of rapid change in acoustical properties. A number of these have been discussed in the previous chapters. Here we shall continue this discussion. This chapter contains the derivation and discussion of a number of phenomena; they are loosely related by the fact that all have to do with the transmission effects that can occur in the vicinity of a change in acoustical properties.

The chapter is divided roughly into four parts. In the first part the derivation of the formal solution for a point source in the vicinity of a plane boundary is given. This is followed by an approximate evaluation of the resultant integral, valid in the limit of high frequencies and for angles of incidence less than the critical angle. Next a second evaluation is given, valid at long ranges and at angles of incidence more grazing than the critical angle. The evaluation introduces a second wave, the refraction arrival, in addition to the reflected wave. The final section of this part is an evaluation of the branch-line integral from the normal-mode solution of Chap. 3; this evaluation, valid at long ranges, leads to the refraction arrival for this case of two-layered media.

The second part discusses plane wave reflection from single boundaries which cannot be characterized simply by a density and velocity contrast across a plane interface. First, the solution for normal incidence to a rapid change in velocity is given. Second, reflection from a plane boundary underlain by an elastic rather than a fluid medium is given. Third, reflection from a corrugated boundary is discussed. Fourth, reflection from a boundary underlain by a porous medium is discussed. And, fifth, the solution for reflection from an interface for which the boundary condition is continuity of acoustic impedance independent of the angle of incidence is given.

The third part contains a discussion of some of the phenomena that can occur in the presence of two or more boundaries. First, the solution for plane wave reflection from a sequence of three layers is given. Second, the phenomena of and conditions for the coupling of energy from horizontal propagation in an overlying semi-infinite medium to an underlying wave guide are discussed.

The fourth part deals with the determination of geologic structure from echo sounding measurements and from geophysical reflection and refraction profiles. The formulas necessary for these determinations are derived, and the methods of interpretation are discussed.

Formal solution, point source in the vicinity of a plane boundary

From Fig. 5-1 we have a point source at $r = 0$, $z = d$. It is taken to be a simple harmonic point source of frequency ω and unit amplitude. At $z = 0$ we have a plane boundary between two media of densities ρ_1 and ρ_2 and velocities c_1 and c_2, respectively. Following the arguments in Chap. 3 we look for fundamental solutions to the cylindrically symmetrical wave equation of the form

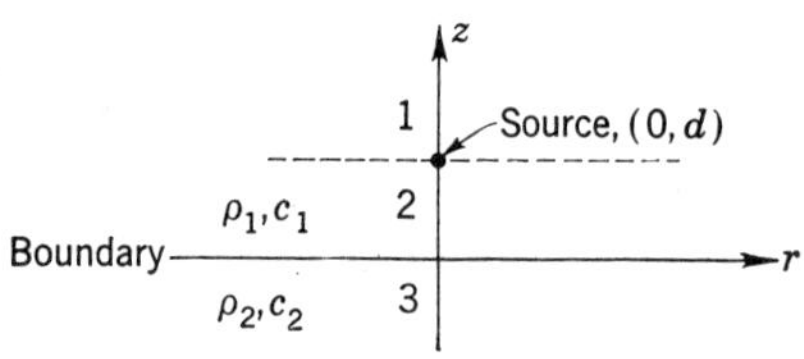

FIG. 5-1.

$$\varphi = e^{i\omega t} J_0(kr) F(z) \tag{5-1}$$

where φ is the velocity potential. The forms of $F(z)$ for the three regions of Fig. 5-1 will be

$$\begin{aligned} F_1 &= Ae^{-i\beta_1 z} \\ F_2 &= Be^{-i\beta_1 z} + Ce^{i\beta_1 z} \\ F_3 &= De^{i\beta_2 z} \end{aligned} \tag{5-2}$$

where β_1 and β_2 are given by

$$\begin{aligned} \beta_1 &= \sqrt{\frac{\omega^2}{c_1{}^2} - k^2} = \sqrt{k_1{}^2 - k^2} \\ \beta_2 &= \sqrt{\frac{\omega^2}{c_2{}^2} - k^2} = \sqrt{k_2{}^2 - k^2} \end{aligned} \tag{5-3}$$

The values of the β_1 and β_2 radicals are taken to be either positive real or negative imaginary depending on the magnitude of k in order to ensure outgoing waves from the point source and to satisfy the boundary conditions toward infinity in the z direction. The boundary conditions of continuity of pressure and vertical component of particle velocity along the plane $z = 0$ reduce to

$$\begin{aligned} \rho_1 F_2 &= \rho_2 F_3 \\ \frac{dF_2}{dz} &= \frac{dF_3}{dz} \end{aligned} \qquad \text{at } z = 0 \tag{5-4}$$

In order to satisfy the source condition at $r = 0$, $z = d$ we impose the conditions

$$\left.\begin{aligned} F_1 &= F_2 \\ \frac{dF_2}{dz} - \frac{dF_1}{dz} &= 2k \end{aligned}\right\} \qquad \text{at } z = d \tag{5-5}$$

along the plane $z = d$ and integrate the form (5-1) from 0 to ∞ with respect to k. Substituting Eqs. (5-2) into (5-5) and (5-4) we obtain

$$\begin{aligned} e^{-i\beta_1 d}A - e^{-i\beta_1 d}B - e^{i\beta_1 d}C \qquad\qquad &= 0 \\ e^{-i\beta_1 d}A - e^{-i\beta_1 d}B + e^{i\beta_1 d}C \qquad\qquad &= \frac{2k}{i\beta_1} \\ \rho_1 B + \rho_1 C - \rho_2 D &= 0 \\ -\beta_1 B + \beta_1 C - \beta_2 D &= 0 \end{aligned} \tag{5-6}$$

Solving Eqs. (5-6) for A we obtain

$$A = \frac{\begin{vmatrix} 0 & -e^{-i\beta_1 d} & -e^{i\beta_1 d} & 0 \\ \dfrac{2k}{i\beta_1} & -e^{-i\beta_1 d} & e^{i\beta_1 d} & 0 \\ 0 & \rho_1 & \rho_1 & -\rho_2 \\ 0 & -\beta_1 & \beta_1 & -\beta_2 \end{vmatrix}}{\begin{vmatrix} e^{-i\beta_1 d} & -e^{-i\beta_1 d} & -e^{i\beta_1 d} & 0 \\ e^{-i\beta_1 d} & -e^{-i\beta_1 d} & e^{i\beta_1 d} & 0 \\ 0 & \rho_1 & \rho_1 & -\rho_2 \\ 0 & -\beta_1 & \beta_1 & -\beta_2 \end{vmatrix}}$$

$$= \frac{2k}{i\beta_1}\frac{(\beta_1 - b\beta_2)e^{-i\beta_1 d} + (\beta_1 + b\beta_2)e^{i\beta_1 d}}{2(\beta_1 + b\beta_2)}$$

$$= \frac{k}{i\beta_1}\left(e^{i\beta_1 d} + \frac{\beta_1 - b\beta_2}{\beta_1 + b\beta_2}\,e^{-i\beta_1 d}\right) \tag{5-7}$$

Similarly we obtain for B, C, and D

$$B = \frac{k}{i\beta_1}\left(\frac{\beta_1 - b\beta_2}{\beta_1 + b\beta_2}\right)e^{-i\beta_1 d} \tag{5-8}$$

$$C = \frac{k}{i\beta_1}\,e^{-i\beta_1 d} \tag{5-9}$$

$$D = \frac{k}{i\beta_1}\left(\frac{2b\beta_1}{\beta_1 + b\beta_2}\right)e^{-i\beta_1 d} \tag{5-10}$$

Our final solutions in the three regions are then, substituting (5-7), (5-8), (5-9), (5-10) into (5-2) and (5-1) and integrating from 0 to ∞ with respect to k

$$\Phi_1 = e^{i\omega t}\Psi_1(r, z, \omega) = e^{i\omega t} \int_0^\infty J_0(kr) \frac{k\,dk}{i\beta_1} e^{-i\beta_1(z-d)} + e^{i\omega t} \int_0^\infty J_0(kr) \frac{k\,dk}{i\beta_1} \left(\frac{\beta_1 - b\beta_2}{\beta_1 + b\beta_2}\right) e^{-i\beta_1(z+d)} \tag{5-11}$$

$$\Phi_2 = e^{i\omega t}\Psi_2(r, z, \omega) = e^{i\omega t} \int_0^\infty J_0(kr) \frac{k\,dk}{i\beta_1} e^{-i\beta_1(d-z)} + e^{i\omega t} \int_0^\infty J_0(kr) \frac{k\,dk}{i\beta_1} \left(\frac{\beta_1 - b\beta_2}{\beta_1 + b\beta_2}\right) e^{-i\beta_1(d+z)} \tag{5-12}$$

$$\Phi_3 = e^{i\omega t}\Psi_3(r, z, \omega) = e^{i\omega t} \int_0^\infty J_0(kr) \frac{k\,dk}{i\beta_1} \left(\frac{2b\beta_1}{\beta_1 + b\beta_2}\right) e^{-i(\beta_1 d - \beta_2 z)} \tag{5-13}$$

The relations (5-11) and (5-12) are the same with an interchange of z for d, indicative of whether the depth of the receiver is above or below that of the source. The first terms of (5-11) and (5-12) have been evaluated previously [Eq. (3-74)]. It is simply the direct wave from the source at $(0, d)$ to a receiver at (r, z).

$$\Psi_1' = \int_0^\infty J_0(kr) \frac{k\,dk}{i\beta_1} e^{-i\beta_1(z-d)} = \frac{e^{-ik_1[r^2+(z-d)^2]^{1/2}}}{[r^2 + (z - d)^2]^{1/2}} = \frac{e^{-ik_1R}}{R} \tag{5-14}$$

where from Fig. 5-2, $R = [r^2 + (z - d)^2]^{1/2}$ is the distance between source and receiver. The second terms of (5-11) and (5-12) represent the energy returned to the upper medium from the boundary. From (3-95) and (3-98) we know that for real β_1 and β_2 the quantity in the parentheses of these terms is the Rayleigh reflection coefficient. Relation (5-13) is the expression for the energy transmitted to the lower medium. The quantity in the parentheses of (5-13) is the Rayleigh refraction coefficient [Eq. (2-155)] for real β_1 and β_2.

Evaluation of the resultant integral in the limit of high frequencies

We should like to obtain an evaluation of the integrals of the second terms of (5-11) and (5-12). An approximate evaluation can be given in the limit of high frequencies. Then, since ω is very large, we may consider β_1 and β_2 as real over the limits of integration. The integral may be evaluated by the method of stationary phase. The $J_0(kr)e^{-i\beta_1(z+d)}$ term will be oscillatory and the remaining terms slowly varying with respect to k. We proceed as follows:

$$\Psi_1'' = \int_0^\infty J_0(kr) \frac{k\,dk}{i\beta_1} \left(\frac{\beta_1 - b\beta_2}{\beta_1 + b\beta_2}\right) e^{-i\beta_1(z+d)} \tag{5-15}$$

$$= \frac{1}{2} \int_0^\infty [H_0^{(1)}(kr) + H_0^{(2)}(kr)] \frac{k\,dk}{i\beta_1} \left(\frac{\beta_1 - b\beta_2}{\beta_1 + b\beta_2}\right) e^{-i\beta_1(z+d)} \tag{5-16}$$

$$\Psi_1'' = \frac{1}{2}\int_0^\infty \sqrt{\frac{2}{\pi kr}}\,\frac{k\,dk}{i\beta_1}\left(\frac{\beta_1 - b\beta_2}{\beta_1 + b\beta_2}\right) e^{i(\pi/4)}\{e^{-i[-kr+\beta_1(z+d)]} + e^{-i[kr+\beta_1(z+d)]}\} \tag{5-17}$$

using the relations (3-100) and the asymptotic forms of the Hankel functions. The exponential of the first integral has no stationary value for positive r and z. The exponential of the second integral does have a stationary value, determined as follows:

$$f(k) = kr + \beta_1(z + d) \tag{5-18}$$

$$f'(k) = r - \frac{k(z + d)}{\sqrt{k_1^2 - k^2}} \tag{5-19}$$

$$f''(k) = -\frac{z + d}{\sqrt{k_1^2 - k^2}} - \frac{k^2(z + d)}{(k_1^2 - k^2)^{3/2}} \tag{5-20}$$

From (5-19) the value of k at the stationary value of $f(k)$ is

$$r - \frac{k(z + d)}{\sqrt{k_1^2 - k^2}} = 0$$

$$\alpha \equiv k = k_1 \frac{r}{\sqrt{r^2 + (z + d)^2}} \tag{5-21}$$

The other quantities that are needed for the stationary phase evaluation (2-115) are

$$\beta_1(\alpha) = k_1\sqrt{\frac{(z + d)^2}{r^2 + (z + d)^2}} = k_1\frac{z + d}{R_1} \tag{5-22}$$

$$\sqrt{\frac{2}{\pi\alpha r}}\,\frac{\alpha}{i\beta_1(\alpha)} = \sqrt{\frac{2R_1}{\pi k_1 r^2}}\,\frac{r}{i(z + d)} = \frac{1}{i(z + d)}\sqrt{\frac{2R_1}{\pi k_1}} \tag{5-23}$$

$$\frac{\beta_1(\alpha) - b\beta_2(\alpha)}{\beta_1(\alpha) + b\beta_2(\alpha)} = \frac{\cos\theta - b\delta}{\cos\theta + b\delta} = \kappa \tag{5-24}$$

$$f(\alpha) = k_1R_1 \tag{5-25}$$

$$f''(\alpha) = -\frac{R_1^3}{k_1(z + d)^2} \tag{5-26}$$

where R_1, δ, and θ are defined in Fig. 5-2, the quantity κ being the Rayleigh reflection coefficient (2-154). Substituting the quantities (5-23), (5-24), (5-25), and (5-26) into (5-15) and (2-115) we obtain

$$\Psi_1'' = \kappa\frac{e^{-ik_1R_1}}{R_1} \tag{5-27}$$

Equation (5-27) is simply the expression for the spherical wave which has followed the Snell's law reflection ray path multiplied by the Rayleigh reflection coefficient for the angle θ.

In the stationary phase method of evaluation of an integral the slowly varying term is considered to be constant over the interval for which the stationary value of the oscillating term gives a contribution, and it is factored out of the integration. As a second approximation, we might

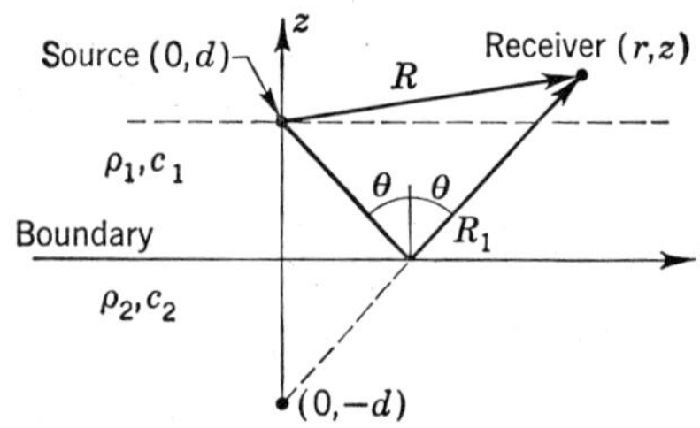

FIG. 5-2. $R = [r^2 + (z - d)^2]^{1/2}$, $R_1 = [r^2 + (z + d)^2]^{1/2}$, $\sin\theta = r/R_1$, $\cos\theta = (z + d)/R_1$, $\delta = [(c_1^2/c_2^2) - \sin^2\theta]^{1/2}$.

wish to consider the variations of this term. Then, in the integral of the form (2-113)

$$u = \int_a^b \varphi(k)e^{if(k)}\,dk \tag{5-28}$$

we presume that we have a stationary value of $f(k)$ between the limits of integration at $k = \alpha$. An expansion of $f(k)$ and $\varphi(k)$ about $k = \alpha$ gives

$$k = \alpha + \xi$$
$$f'(\alpha) = 0$$
$$f(k) = f(\alpha) + \tfrac{1}{2}\xi^2 f''(\alpha) \tag{5-29}$$
$$\varphi(k) = \varphi(\alpha) + \xi\varphi'(\alpha) + \tfrac{1}{2}\xi^2\varphi''(\alpha) + \cdots \tag{5-30}$$

Since the integral has an appreciable value only in the neighborhood of α, we may extend the limits of integration on ξ from $-\infty$ to ∞. We obtain, instead of (2-114),

$$u = e^{if(\alpha)}\left[\varphi(\alpha)\int_{-\infty}^{\infty} e^{i\frac{1}{2}\xi^2 f''(\alpha)}\,d\xi + \varphi'(\alpha)\int_{-\infty}^{\infty} \xi e^{i\frac{1}{2}\xi^2 f''(\alpha)}\,d\xi + \tfrac{1}{2}\varphi''(\alpha)\int_{-\infty}^{\infty} \xi^2 e^{i\frac{1}{2}\xi^2 f''(\alpha)}\,d\xi\right] \tag{5-31}$$

These integrals may all be evaluated,[1] giving

$$u = \frac{\sqrt{2\pi}}{|f''(\alpha)|^{1/2}}\varphi(\alpha)e^{i[f(\alpha)\pm(\pi/4)]} + \frac{\sqrt{\pi/2}}{|f''(\alpha)|^{3/2}}\varphi''(\alpha)e^{i[f(\alpha)\pm(3\pi/4)]} \tag{5-32}$$

The second integral vanishes. The plus or minus sign in the exponentials is taken as $f''(\alpha)$ is negative or positive.

[1] B. O. Pierce, "A Short Table of Integrals," p. 63, Secs. 492, 494, Ginn & Company, Boston, 1929.

In our evaluation we shall consider variations in the reflection-coefficient term of (5-15)

$$\varphi(k) = \frac{\beta_1 - b\beta_2}{\beta_1 + b\beta_2} = \frac{(k_1^2 - k^2)^{1/2} - b\left(\dfrac{c_1^2}{c_2^2}k_1^2 - k^2\right)^{1/2}}{(k_1^2 - k^2)^{1/2} + b\left(\dfrac{c_1^2}{c_2^2}k_1^2 - k^2\right)^{1/2}} \tag{5-33}$$

$$\varphi'(k) = \frac{2bkk_1^2\left(1 - \dfrac{c_1^2}{c_2^2}\right)}{\left[(k_1^2 - k^2)^{1/2} + b\left(\dfrac{c_1^2}{c_2^2}k_1^2 - k^2\right)^{1/2}\right]^2\left(\dfrac{c_1^2}{c_2^2}k_1^2 - k^2\right)^{1/2}(k_1^2 - k^2)^{1/2}} \tag{5-34}$$

$$\begin{aligned}\varphi''(k) = {}& \frac{2bk_1^2\left(1 - \dfrac{c_1^2}{c_2^2}\right)}{\left[(k_1^2 - k^2)^{1/2} + b\left(\dfrac{c_1^2}{c_2^2}k_1^2 - k^2\right)^{1/2}\right]^3\left(\dfrac{c_1^2}{c_2^2}k_1^2 - k^2\right)^{3/2}} \\ & \left[\frac{\left(\dfrac{c_1^2}{c_2^2}k_1^2 - k^2\right)}{(k_1^2 - k^2)^{1/2}}(\cdot\ \cdot\ \cdot\ \cdot) + \frac{k_1^2\left(\dfrac{c_1^2}{c_2^2}k_1^2 - k^2\right)}{(k_1^2 - k^2)^{3/2}}(\cdot\ \cdot\ \cdot\ \cdot)\right. \\ & \quad + \frac{k^2}{(k_1^2 - k^2)^{1/2}}(\cdot\ \cdot\ \cdot\ \cdot) \\ & \left. + \frac{2k^2\left(\dfrac{c_1^2}{c_2^2}k_1^2 - k^2\right)^{1/2}\left[\left(\dfrac{c_1^2}{c_2^2}k_1^2 - k^2\right)^{1/2} + b(k_1^2 - k^2)^{1/2}\right]}{(k_1^2 - k^2)}\right]\end{aligned} \tag{5-35}$$

At $k = \alpha$ the various quantities in (5-35) are given by

$$(k_1^2 - k^2)^{1/2} = k_1 \cos\theta \tag{5-36}$$

$$\left(\frac{c_1^2}{c_2^2}k_1^2 - k^2\right)^{1/2} = k_1\delta \tag{5-37}$$

$$(k_1^2 - k^2)^{1/2} + b\left(\frac{c_1^2}{c_2^2}k_1^2 - k^2\right)^{1/2} = k_1(\cos\theta + b\delta) \tag{5-38}$$

$$k = k_1 \sin\theta \tag{5-39}$$

Substituting (5-36), (5-37), (5-38), and (5-39) into (5-35) we obtain

$$\begin{aligned}\varphi''(\alpha) = \frac{2b[1 - (c_1^2/c_2^2)]}{k_1^2\delta^3(\cos\theta + b\delta)^3\cos^3\theta}\{&\sin^2\theta\cos^3\theta + 3b\delta\sin^2\theta\cos^2\theta \\ &+ \delta^2\cos\theta(1 + 2\sin^2\theta) + b\delta^3\}\end{aligned} \tag{5-40}$$

Substituting (5-35) into (5-32) and (5-15), the final result is

$$\Psi_1'' = \frac{e^{-ik_1R_1}}{R_1}\left[\kappa + i\frac{b[1 - (c_1^2/c_2^2)]}{k_1R_1\delta^3(\cos\theta + b\delta)^3\cos\theta}\{\cdot\ \cdot\ \cdot\ \cdot\}\right] \tag{5-41}$$

where the braces are the same as (5-40).

The first term in the brackets of (5-41) is the plane wave reflection coefficient. The second term gives the conditions under which the plane wave reflection coefficient may be taken as a valid approximation for the reflection of an incident spherical wave. This condition is that

$$\frac{1}{k_1 R_1 \delta^3} = \frac{c_1}{\omega_1 R_1[(c_1^2/c_2^2) - \sin^2\theta]^{3/2}} \ll 1 \tag{5-42}$$

The approximation is valid for high frequencies and for moderate distances away from the source. For a high-speed bottom, $c_2 > c_1$, the plane wave approximation breaks down completely as θ approaches the critical angle, $\theta_c = \sin^{-1}(c_1/c_2)$.

Evaluation of the resultant integral for long ranges and for $c_2 > c_1$

It is apparent from (5-41) and (5-42) that the approximation of a plane wave reflection coefficient is valid for many problems for a low-speed bottom and for a high-speed bottom at angles less than the critical angle. However, for those problems in which angles more grazing than the critical angle are concerned, it will be necessary to make a different evaluation.

Let us make the substitution (3-100) into (5-15) and carry out an integration in the complex k plane similar to that used in Chap. 3. It will be assumed throughout that c_2 is greater than c_1. With reference to Fig. 3-43 and Eqs. (3-101) and (3-102) the $H_0^{(1)}$ integration is taken over the infinite sector of the first quadrant and the $H_0^{(2)}$ integration over the infinite sector of the fourth quadrant. The $H_0^{(1)}$ integral over the infinite quarter circle in the first quadrant and the $H_0^{(2)}$ integral over the infinite quarter circle in the fourth quadrant vanish. The $H_0^{(1)}$ integral along the positive imaginary axis from 0 to $i\infty$ and the $H_0^{(2)}$ integral along the negative imaginary axis from 0 to $-i\infty$ cancel. We have then finally that the sum of the $H_0^{(1)}$ and $H_0^{(2)}$ integrals along the positive real axis from 0 to ∞ is equal to the branch-line integrals along the k_1 and k_2 branch lines. There are no poles. From (3-100) the sum of the $H_0^{(1)}$ and $H_0^{(2)}$ integrals along the positive real axis is also equal to twice the original integral (5-15) that we want to determine.

$$\Psi_1'' = \tfrac{1}{2}\int_{k_1,-i\infty}^{k_1} H_0^{(2)}(kr)k\,dk[F(\beta_1,\beta_2) - F(-\beta_1,\beta_2)] + \tfrac{1}{2}\int_{k_2,-i\infty}^{k_2} H_0^{(2)}(kr)k\,dk[F(\beta_1,\beta_2) - F(\beta_1,-\beta_2)] \tag{5-43}$$

where

$$F(\beta_1,\beta_2) = \frac{1}{i\beta_1}\left(\frac{\beta_1 - b\beta_2}{\beta_1 + b\beta_2}\right)e^{-i\beta_1(z+d)} \tag{5-44}$$

Since $H_0^{(2)}$ vanishes exponentially for large imaginary parts of the argument, the value of the integrands of (5-43) will be appreciable only when the argument of $H_0^{(2)}$ is small. If r is large, we may make an approximate evaluation considering variations of the integrands for small imaginary values of k only. For the first integration let us change the variable of integration to

$$k = k_1\lambda \tag{5-45}$$

where

$$\lambda = 1 - ix \tag{5-46}$$

Then, for small x we shall have

$$k = k_1\lambda \approx k_1 \tag{5-47}$$

$$dk = -ik_1\,dx \tag{5-48}$$

$$\beta_1 = \sqrt{k_1^2 - k^2} = k_1(1 - \lambda^2)^{1/2} \approx k_1(2ix)^{1/2} = k_1(2x)^{1/2}e^{i(\pi/4)} \tag{5-49}$$

$$\begin{aligned}\beta_2 = \sqrt{k_2^2 - k^2} = k_1\left(\frac{c_1^2}{c_2^2} - \lambda^2\right)^{1/2} &\approx k_1\left(\frac{c_1^2}{c_2^2} - 1\right)^{1/2} \\ &= ik_1\left(1 - \frac{c_1^2}{c_2^2}\right)^{1/2} \\ &= ik_1\mu\end{aligned} \tag{5-50}$$

$$H_0^{(2)}(kr) = H_0^{(2)}(k_1r\lambda) \rightarrow \sqrt{\frac{2}{\pi k_1 r}}\, e^{-i[k_1r-(\pi/4)]}e^{-k_1rx} \tag{5-51}$$

where μ is defined as

$$\mu \equiv \left(1 - \frac{c_1^2}{c_2^2}\right)^{1/2} \tag{5-52}$$

$$\begin{aligned}F(\beta_1, \beta_2) - F(-\beta_1, \beta_2) &= \frac{1}{i\beta_1}\left[\frac{\beta_1 - b\beta_2}{\beta_1 + b\beta_2}\, e^{-i\beta_1(z+d)} + \frac{\beta_1 + b\beta_2}{\beta_1 - b\beta_2}\, e^{i\beta_1(z+d)}\right] \\ &= \frac{2}{i\beta_1}\left\{\frac{(\beta_1^2 + b^2\beta_2^2)\cos[\beta_1(z+d)] + i2b\beta_1\beta_2 \sin[\beta_1(z+d)]}{\beta_1^2 - b^2\beta_2^2}\right\} \\ &\approx \frac{-2e^{-i(\pi/4)}}{ik_1(2x)^{1/2}}\left[\cos(mx^{1/2}) + \frac{2(2x)^{1/2}e^{i(\pi/4)}}{b\mu}\sin(mx^{1/2})\right]\end{aligned} \tag{5-53}$$

where m is defined as

$$m \equiv \sqrt{2}\, k_1 e^{i(\pi/4)}(z + d) \tag{5-54}$$

Substituting (5-47), (5-48), (5-49), (5-50), (5-51), and (5-52) into the first integral of (5-43) and making a change in the variable of integration to y where $x = y^2$, we shall have

$$\Psi''_{k_1} = \frac{-2k_1e^{-ik_1r}}{\sqrt{\pi k_1 r}}\int_0^\infty e^{-k_1ry^2}\cos(my)\,dy - \frac{4\sqrt{2}\,k_1e^{-ik_1r}e^{i(\pi/4)}}{b\mu\sqrt{\pi k_1 r}}\int_0^\infty ye^{-k_1ry^2}\sin(my)\,dy \tag{5-55}$$

The first integral is a tabulated form,[1] the second integral may be evaluated through an integration by parts as follows:

$$\int_0^\infty ye^{-k_1ry^2}\sin(my)\,dy = \left[\frac{-1}{2k_1r}e^{-k_1ry^2}\sin(my)\right]_0^\infty + \frac{m}{2k_1r}\int_0^\infty e^{-k_1ry^2}\cos(my)\,dy$$

$$= \frac{m}{2k_1r}\left[\frac{1}{2}\sqrt{\frac{\pi}{k_1r}}\,e^{-(m^2/4k_1r)}\right]$$

The final result for (5-55) then is

$$\Psi''_{k_1} = \frac{-1}{r}\left[1 + i\frac{2(z+d)}{b\mu r}\right]\exp\left\{-ik_1\left[r + \frac{(z+d)^2}{2r}\right]\right\} \tag{5-56}$$

For large r the second term in (5-56) may be neglected, leaving

$$\Psi''_{k_1} = -\frac{1}{r}\exp\left\{-ik_1\left[r + \frac{(z+d)^2}{2r}\right]\right\} \tag{5-57}$$

Following the same procedure for the second integral we shall have

$$k = k_1\eta \tag{5-58}$$

where

$$\eta \equiv \frac{c_1}{c_2} - ix \tag{5-59}$$

Then, for small x we shall have

$$k = k_1\eta \approx k_1\frac{c_1}{c_2} = k_2 \tag{5-60}$$

$$dk = -ik_1\,dx \tag{5-61}$$

$$\beta_1 = \sqrt{k_1^2 - k^2} = k_1(1-\eta^2)^{1/2} \approx k_1\left(1 - \frac{c_1^2}{c_2^2}\right)^{1/2} = k_1\mu \tag{5-62}$$

$$\beta_2 = \sqrt{k_2^2 - k^2} = k_1\left(\frac{c_1^2}{c_2^2} - \eta^2\right)^{1/2} \approx k_1\left(2ix\frac{c_1}{c_2}\right)^{1/2} = k_1\left(2x\frac{c_1}{c_2}\right)^{1/2} e^{i(\pi/4)} \tag{5-63}$$

$$H_0^{(2)}(kr) = H_0^{(2)}(k_1r\eta) \to \sqrt{\frac{2c_2}{\pi k_1rc_1}}\,e^{-i[k_2r-(\pi/4)]}e^{-k_1rx} \tag{5-64}$$

$$F(\beta_1, \beta_2) - F(\beta_1, -\beta_2) = \frac{e^{-i\beta_1(z+d)}}{i\beta_1}\left(\frac{\beta_1 - b\beta_2}{\beta_1 + b\beta_2} - \frac{\beta_1 + b\beta_2}{\beta_1 - b\beta_2}\right) = \frac{4ib\beta_2}{\beta_1^2 - b^2\beta_2^2}e^{-i\beta_1(z+d)} \approx \frac{4ib[2x(c_1/c_2)]^{1/2}e^{i(\pi/4)}}{k_1\mu^2}e^{-ik_1\mu(z+d)} \tag{5-65}$$

[1] B. O. Pierce, "A Short Table of Integrals," p. 64, Sec. 508, Ginn & Company, Boston, 1929.

Substituting (5-60), (5-61), (5-62), (5-63), (5-64), and (5-65) into the second integral of (5-43) we obtain[1]

$$\begin{aligned}\Psi''_{k_2} &= \frac{4ibk_2}{\sqrt{\pi k_1 r}} \frac{e^{-i[k_2 r + k_1 \mu (z+d)]}}{\mu^2} \int_0^\infty x^{\frac{1}{2}} e^{-k_1 r x}\, dx \\ &= \frac{4ibk_2}{\sqrt{\pi k_1 r}} \frac{e^{-i[k_2 r + k_1 \mu (z+d)]}}{\mu^2} \left[\frac{\sqrt{\pi}}{2(k_1 r)^{\frac{3}{2}}}\right] \\ &= \frac{2ibk_2}{\mu^2 k_1{}^2 r^2} e^{-i[k_2 r + k_1 \mu (z+d)]} \end{aligned} \tag{5-66}$$

The generalization of Eqs. (5-57) and (5-66) to an arbitrary pulse by means of a Fourier integral adds nothing significantly new. For Eq. (5-57) there is a frequency dependence only as a multiplying factor. The pulse form of (5-57) is undistorted with respect to the incident pulse. For Eq. (5-66) besides the multiplying factor in the exponential there is a frequency dependence proportional to ω^{-1}. The velocity components of the pulse form of (5-66) will resemble the displacement components of the incident wave. The beginning times for the arrival of the waves represented by (5-57) and (5-66) for a source initiated at $t = 0$ are given, respectively, by the phase factor of the exponentials of each expression.

We see that our analysis has led to expressions for three arrivals [Eqs. (5-14), (5-57), and (5-66)]. Equation (5-14) is the expression for the *direct arrival.* It follows the straight-line path from source to receiver. Its amplitude decreases inversely proportional to the distance between source and receiver R, as expected for spherical wave propagation. Its arrival time at the receiver for a source commenced at $t = 0$ is $t = R/c_1$.

Equation (5-57) is an approximate expression for the *reflection arrival,* the evaluation having been carried out for long ranges. The expression in the brackets of the exponential is

$$R_1 = [r^2 + (z+d)^2]^{\frac{1}{2}} = r\left[1 + \frac{(z+d)^2}{r^2}\right]^{\frac{1}{2}} \approx r\left[1 + \frac{1}{2}\frac{(z+d)^2}{r^2}\right]$$

the first two terms of the series expansion of R_1. The reflection arrival follows the Snell's law reflection ray path. Its amplitude decreases as R^{-1}; in Eq. (5-57) r is an approximation to R_1 for large r. In expression (5-57) there is a phase change of 180 deg, corresponding to the phase change from the Rayleigh reflection coefficient for grazing incidence, large r. Its arrival time for a source commenced at $t = 0$ is $t = R_1/c_1$.

Equation (5-66) is the expression for a new arrival which we have not encountered before. It is referred to as the *refraction arrival.* With regard to terminology the term *the refraction arrival* is not to be confused

[1] B. O. Pierce, "A Short Table of Integrals," p. 63, Sec. 493, Ginn & Company, Boston, 1929.

with the more general term *a refracted arrival* from the ray-theory discussions of the preceding chapters. As we shall see, the refraction arrival is a limiting case of a refracted arrival. Let us examine the expression in the brackets of the exponential of (5-66) to determine what, if any, ray path can be associated with this arrival. Since both the k_1 and k_2 wave numbers enter this expression, we know that part of the path has been traversed with the velocity c_1 of the upper medium and part with the velocity c_2 of the lower medium.

$$\begin{aligned} k_2 r + k_1\mu(z+d) &= \omega\left[\frac{r}{c_2} + \frac{z+d}{c_1}\left(1 - \frac{c_1^2}{c_2^2}\right)^{1/2}\right] \\ &= \omega\left[\frac{r}{c_2} + \frac{z+d}{c_1}\cos\theta_c\right] \\ &= \omega\left[\frac{r-(z+d)\tan\theta_c}{c_2} + \frac{z+d}{c_1}\left(\cos\theta_c + \frac{c_1}{c_2}\tan\theta_c\right)\right] \\ &= \omega\left[\frac{r-(z+d)\tan\theta_c}{c_2} + \frac{z+d}{c_1}\left(\frac{\cos^2\theta_c + \sin^2\theta_c}{\cos\theta_c}\right)\right] \\ &= \omega\left[\frac{r-(z+d)\tan\theta_c}{c_2} + \frac{(z+d)/\cos\theta_c}{c_1}\right] \qquad (5\text{-}67) \end{aligned}$$

The quantity in the brackets of (5-67) is the expression for the arrival time for the ray path of Fig. 5-3. The angle $\theta_c = \sin^{-1}(c_1/c_2)$ is, as

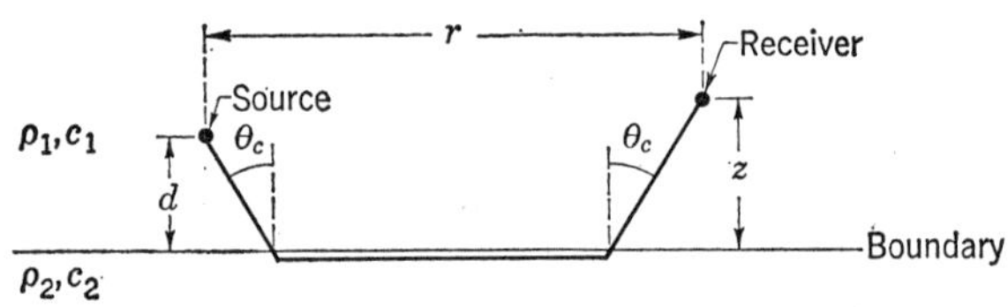

FIG. 5-3.

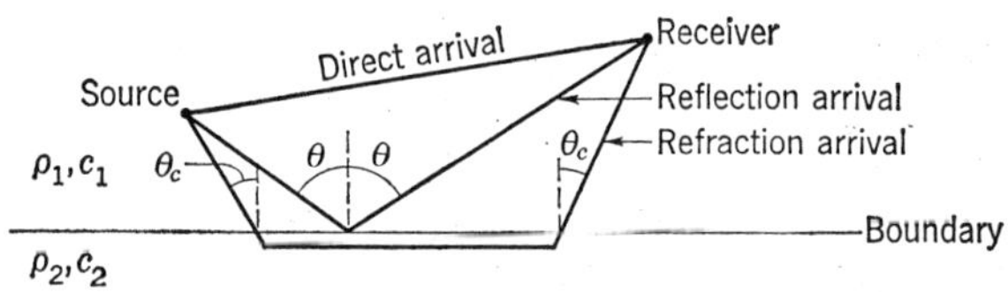

FIG. 5-4.

before, the critical angle. The ray path of Fig. 5-3 is a true Snell's law minimum time path. From ray acoustics alone, however, no energy would be associated with this path; from the more formal solution we see that this is not the case. The amplitude of the refraction arrival decreases as r^{-2}, similar to the Lloyd mirror effect of Chap. 3.

The ray associations for these three arrivals [(5-14), (5-57), and (5-66)] are summarized in Fig. 5-4.

It is sometimes helpful to a physical understanding of these three arrivals to consider the ray and wave surface constructions of Fig. 5-5. The only energy that can be returned to the upper medium from propagation in the second medium is that which propagates adjacent to the boundary in the lower medium and reradiates back into the upper medium. From Snell's law this reradiation will be at the critical angle. This is in addition to the direct transmission and the reflected transmission from the boundary between the two media.

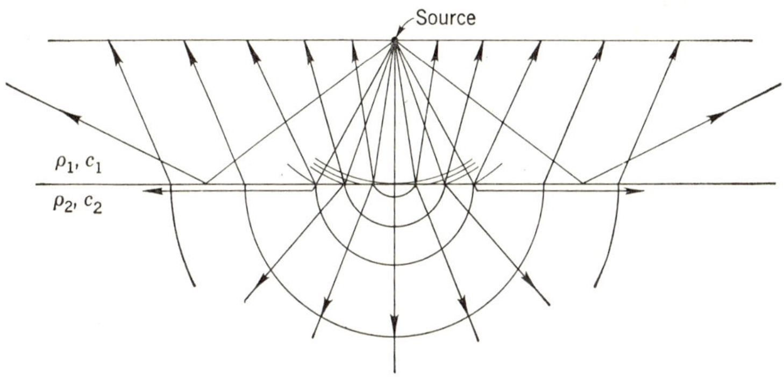

Fig. 5-5.

Another concept which is helpful to a physical understanding of the refraction arrival is a Huygens construction for the effects in the upper medium from a disturbance propagating parallel to the boundary in the lower medium with the velocity c_2 of the lower medium. If the velocity c_2 is greater than c_1, the wavelets will overlap and form an envelope as shown in Fig. 5-6. The envelope forms a plane wave front, and the ray

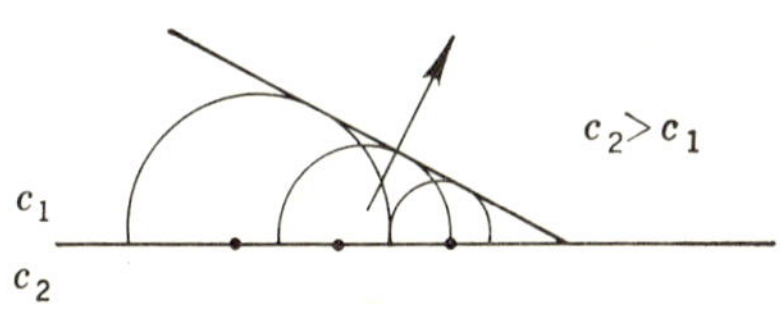

Fig. 5-6.

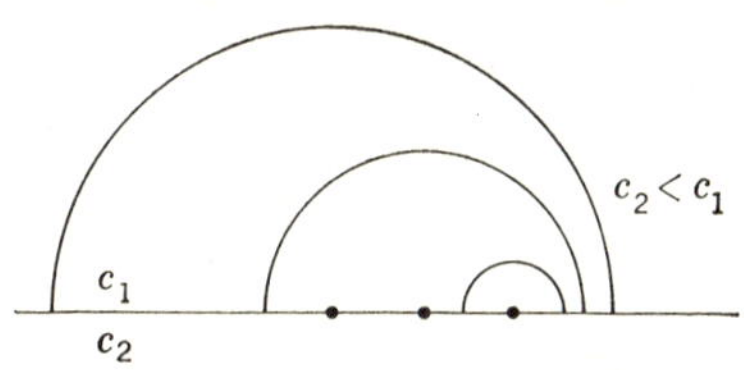

Fig. 5-7.

direction is defined by the angle whose sine is the ratio of the two velocities, c_1/c_2. If the velocity c_2 is less than c_1, the wavelets will not overlap and will not form an envelope as shown in Fig. 5-7. There is no propagation into the upper medium but only an exponential decrease away from the boundary into the upper medium. In general if the Huygens wavelets do not form an envelope, there will be no propagation. If the wavelets do form an envelope, there will be propagation. The envelope is the wave surface, and the normals to the wave surfaces define the ray directions. The propagation will be associated with a finite amplitude. If the rays themselves form an envelope, the propagation will be associated with an infinite amplitude along this caustic.

Evaluation of the branch-line integral from the normal-mode solution of Chap. 3

Returning to the branch-line integral from the formal solution for two superposed liquids of Chap. 3, its evaluation has a special meaning for long-range propagation. It is the refraction arrival corresponding to the refraction arrival obtained above for propagation in the vicinity of a single boundary. From Eqs. (3-104) and (3-103) the branch-line integral is

$$\begin{aligned}
\Psi_2'' &= \int_{k_2,-i\infty}^{k_2} H_0^{(2)}(kr)[F(\beta_1, \beta_2) - F(\beta_1, -\beta_2)]k\,dk \\
&= \int_{k,-i\infty}^{k_2} H_0^{(2)}(kr)\frac{\sin(\beta_1 d)}{\beta_1}\left\{\frac{\beta_1\cos[\beta_1(H-z)] + ib\beta_2\sin[\beta_1(H-z)]}{\beta_1\cos(\beta_1 H) + ib\beta_2\sin(\beta_1 H)}\right. \\
&\qquad\left. - \frac{\beta_1\cos[\beta_1(H-z)] - ib\beta_2\sin[\beta_1(H-z)]}{\beta_1\cos(\beta_1 H) - ib\beta_2\sin(\beta_1 H)}\right\}k\,dk \\
&= \int_{k_2,-i\infty}^{k_2} H_0^{(2)}(kr)\frac{\sin(\beta_1 d)}{\beta_1}\left(\frac{\begin{array}{c}2ib\beta_1\beta_2\{\sin[\beta_1(H-z)]\cos(\beta_1 H) \\ -\cos[\beta_1(H-z)]\sin(\beta_1 H)\}\end{array}}{\beta_1^2\cos^2(\beta_1 H) + b^2\beta_2^2\sin^2(\beta_1 H)}\right)k\,dk \\
&= -2ib\int_{k_2,-i\infty}^{k_2} H_0^{(2)}(kr)\frac{\beta_2\sin(\beta_1 d)\sin(\beta_1 z)}{\beta_1^2\cos^2(\beta_1 H) + b^2\beta_2^2\sin^2(\beta_1 H)}k\,dk
\end{aligned} \tag{5-68}$$

Since $H_0^{(2)}$ vanishes exponentially for large imaginary parts of the argument, the value of the integrand will be appreciable only when the argument of $H_0^{(2)}$ is small. If r is large, we may make an approximate evaluation considering variations of the integrand for small imaginary values of k only. Proceeding as before we change the variable of integration to

$$k = k_1\eta \tag{5-69}$$

where

$$\eta \equiv \frac{c_1}{c_2} - ix \tag{5-70}$$

Then, for small x, we shall have expressions (5-60), (5-61), (5-62), (5-63), and (5-64). For the denominator of (5-66) we shall have

$$\begin{aligned}
\beta_1^2\cos^2(\beta_1 H) + b^2\beta_2^2\sin^2(\beta_1 H) &\approx \beta_1^2\cos^2(\beta_1 H) \\
&= k_1^2\mu^2\cos^2(k_1\mu H)
\end{aligned} \tag{5-71}$$

which will be valid generally except in the neighborhood of zeros of

$\cos(\beta_1 H)$. Substituting into (5-68) gives[1]

$$\Psi_2'' = 2ib \int_0^\infty \sqrt{\frac{2c_2}{\pi k_1 r c_1}}\, e^{i[(\pi/4)-k_2 r]} e^{-k_1 r x} \frac{k_1[2x(c_1/c_2)]^{1/2} e^{i(\pi/4)} \sin(k_1\mu d)\sin(k_1\mu z)}{k_1{}^2\mu^2\cos^2(k_1\mu H)} (-ik_1k_2)\,dx$$

$$= \frac{4ibk_2}{\sqrt{\pi k_1 r}} \frac{\sin(k_1\mu d)\sin(k_1\mu z)}{\mu^2\cos^2(k_1\mu H)} e^{-ik_2 r}\int_0^\infty x^{1/2} e^{-k_1 r x}\,dx$$

$$= \frac{4ibk_2}{\sqrt{\pi k_1 r}} \frac{\sin(k_1\mu d)\sin(k_1\mu z)}{\mu^2\cos^2(k_1\mu H)} e^{-ik_2 r}\left[\frac{\sqrt{\pi}}{2(k_1 r)^{3/2}}\right]$$

$$= \frac{2ibk_2}{(k_1 r)^2} \frac{\sin(k_1\mu d)\sin(k_1\mu z)}{\mu^2\cos^2(k_1\mu H)} e^{-ik_2 r} \tag{5-72}$$

The case of particular interest is the generalization of the above to a point-source pulse. As representative we shall take an exponential pulse (3-112) whose Fourier transform is (3-113). Then, the resultant disturbance at long ranges will be, following the forms of (3-116) and (3-120),

$$R(r, z, t) = \frac{1}{\pi}\int_0^\infty \frac{2ibk_2}{(k_1 r)^2} \frac{\sin(k_1\mu d)\sin(k_1\mu z)}{\mu^2\cos^2(k_1\mu H)} \frac{\cos[\omega t - k_2 r - \tan^{-1}(\omega/\lambda)]}{(\lambda^2+\omega^2)^{1/2}}\,d\omega \tag{5-73}$$

This integral will have its major contributions at values of ω satisfying the equation

$$\cos(k_1\mu H) = 0 \tag{5-74}$$

However, at these points the approximation (5-71) will no longer be valid, and that expression will be given by

$$\beta_1{}^2\cos^2(\beta_1 H) + b^2\beta_2{}^2\sin^2(\beta_1 H) \approx b^2\beta_2{}^2\sin^2(\beta_1 H) = b^2 k_1{}^2\left(2x\frac{c_1}{c_2}\right) e^{i(\pi/2)}\sin^2(k_1\mu H) \tag{5-75}$$

Substituting (5-75) for (5-71) in (5-68) gives[1]

$$\Psi_2'' = \frac{2k_2c_2}{bc_1\sqrt{\pi k_1 r}} \frac{\sin(k_1\mu d)\sin(k_1\mu z)}{\sin^2(k_1\mu H)} e^{-ik_2 r}\int_0^\infty x^{-1/2} e^{-k_1 r x}\,dx$$

$$= \frac{2k_2c_2}{bc_1\sqrt{\pi k_1 r}} \frac{\sin(k_1\mu d)\sin(k_1\mu z)}{\sin^2(k_1\mu H)} e^{-ik_2 r}\left[\frac{\sqrt{\pi}}{(k_1 r)^{1/2}}\right]$$

$$= \frac{2}{br}\frac{\sin(k_1\mu d)\sin(k_1\mu z)}{\sin^2(k_1\mu H)} e^{-ik_2 r} \tag{5-76}$$

[1] B. O. Pierce, "A Short Table of Integrals," p. 63, Sec. 493, Ginn & Company, Boston, 1929.

We may then say that the integration (5-73) over all frequencies has contributions only for those values of $\omega = \omega_n$ satisfying Eq. (5-74). The propagation is nondispersive of constant frequency ω_n. The velocity of propagation in the r direction is

$$V = \frac{\omega}{k_2} = c_2 \tag{5-77}$$

the velocity of the lower medium. The range dependence for the velocity potential is r^{-1}, the same as that for the dispersive normal-mode train.

The quantity $R_n(r, z, t)$ is the expression for the refraction arrival; it travels with the velocity of the lower medium. In contrast to the single boundary solution it is frequency-dependent with frequencies, $\omega = \omega_n$; and its intensity decreases as spherical spreading r^{-2}, rather than Lloyd mirror spreading r^{-4}. Its incidence is at the same time as the beginning of the dispersive normal-mode train; but it is a train of constant frequency.

From (5-74) the discrete frequencies for the refraction arrival are

$$\begin{aligned} k_1\mu H &= (2n - 1)\frac{\pi}{2} \\ \frac{\omega}{c_1}\sqrt{1 - \frac{c_1^2}{c_2^2}}\, H &= (2n - 1)\frac{\pi}{2} \end{aligned} \tag{5-78}$$

or simply

$$f_n = \frac{(2n - 1)c_1c_2}{4H\sqrt{c_2^2 - c_1^2}}, \quad n = 1, 2, 3, \ldots \tag{5-79}$$

This set of frequencies is the same as the cutoff frequencies for the normal-mode train.

The physical significance of these results can be seen from the following. For the refraction arrival we are concerned only with those rays which are incident at the critical angle θ_c where

$$\sin\theta_c = \frac{c_1}{c_2} \tag{5-80}$$

From this particular set of multiply reflected rays there will be a contribution to the refraction path upon each multiple reflection within the overlying channel. The resultant refraction arrival will have a significant contribution only for those frequencies for which the multiply reflected contributions are always added in phase. From Fig. 5-8 this condition, similar to that for the normal modes described previously, will be

$$\frac{\omega}{c_1}(\overline{AC} + \overline{CB}) - \pi = 2(n - 1)\pi \tag{5-81}$$

The phase change upon reflection from the bottom at the critical angle is

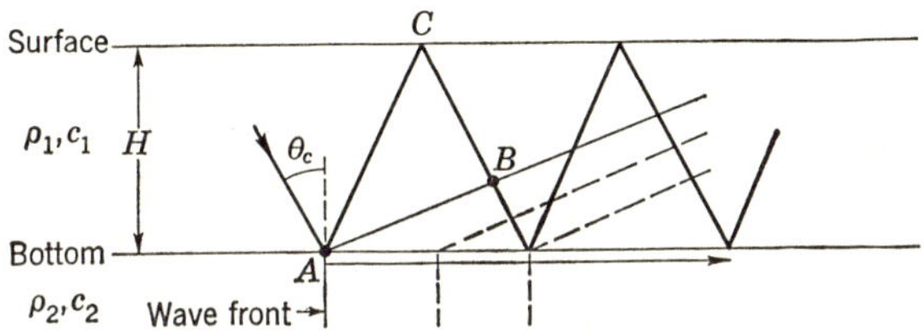

FIG. 5-8.

zero, and the phase change upon reflection from the surface is π. Rewriting Eq. (5-81) in terms of the angle θ_c and the velocities c_1 and c_2 we obtain

$$\frac{\omega}{c_1}(2H \cos \theta_c) = (2n - 1)\pi$$

$$f_n = \frac{(2n - 1)c_1}{4H \cos \theta_c}$$

$$= \frac{(2n - 1)c_1c_2}{4H \sqrt{c_2^2 - c_1^2}} \tag{5-82}$$

the same as Eq. (5-79).

Normal incidence to a rapid change in velocity

Up to now we have considered the boundary between two media to be a finite discontinuity in velocity and density. It is of interest to consider what the effects will be for incidence to a zone of rapid change in velocity. For simplicity we shall consider only normal incidence of a simple harmonic plane wave. Further we shall consider a medium of constant density whose velocity changes linearly with depth in the transition region II of Fig. 5-9. For the regions I and III we shall have the usual one-dimensional wave equations

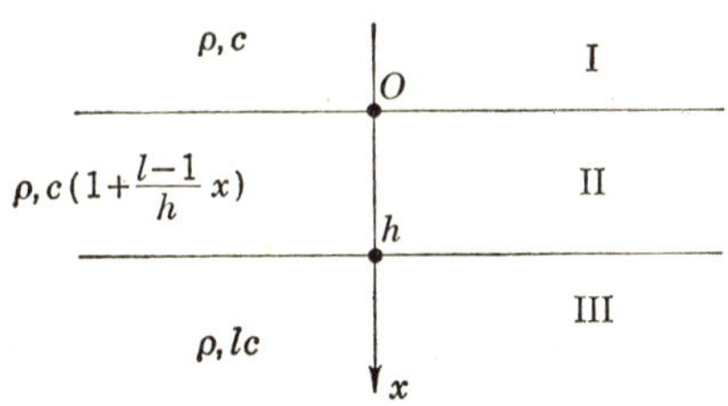

FIG. 5-9.

$$\frac{\partial^2 \varphi_1}{\partial t^2} = c^2 \frac{\partial^2 \varphi_1}{\partial x^2}$$

and

$$\frac{\partial^2 \varphi_3}{\partial t^2} = (lc)^2 \frac{\partial^2 \varphi_3}{\partial t^2} \tag{5-83}$$

where φ is the velocity potential. Their solutions will be

$$\varphi_1 = e^{i\omega t}\bar{\varphi}_1 = Ae^{i\omega[t-(x/c)]} + Be^{i\omega[t+(x/c)]} \tag{5-84}$$

and

$$\varphi_3 = e^{i\omega t}\bar{\varphi}_3 = Ee^{i\omega[t-(x/lc)]} \tag{5-85}$$

where A is the amplitude of the incident wave, B the amplitude of the wave returned to the upper region, and E the amplitude of the wave transmitted to the lower region. For the equation defining propagation in the transition region II we must return to Eqs. (1-28). The only nonvanishing components of stress and displacement are σ_{xx} and u. Equations (1-28) reduce to

$$\begin{aligned}
\rho \frac{\partial^2 u_2}{\partial t^2} &= \frac{\partial \sigma_{xx}}{\partial x} \\
\rho \frac{\partial^2 u_2}{\partial t^2} &= \frac{\partial}{\partial x}\left[\lambda(x)\frac{\partial u_2}{\partial x}\right] \\
\frac{\partial^2 u_2}{\partial t^2} &= \frac{\partial}{\partial x}\left[V^2(x)\frac{\partial u_2}{\partial x}\right] \\
\frac{\partial^2 \varphi_2}{\partial t^2} &= \frac{\partial}{\partial x}\left[V^2(x)\frac{\partial \varphi_2}{\partial x}\right]
\end{aligned} \tag{5-86}$$

where from Fig. 5-9 V is given by

$$V = c\left(1 + \frac{l-1}{h}x\right) \tag{5-87}$$

For a simple harmonic plane wave the solution to (5-86) will be of the form

$$\varphi_2 = e^{i\omega t}\bar{\varphi}_2(x) \tag{5-88}$$

Substituting into (5-86) we obtain the differential equation

$$c^2\left(1 + \frac{l-1}{h}x\right)^2 \frac{d^2\varphi_2}{dx^2} + 2c^2\left(\frac{l-1}{h}\right)\left(1 + \frac{l-1}{h}x\right)\frac{d\varphi_2}{dx} + \omega^2\varphi_2 = 0 \tag{5-89}$$

Let us make a change of variables to z where z is defined by

$$z \equiv \left(1 + \frac{l-1}{h}x\right) \tag{5-90}$$

Equation (5-89) becomes

$$z^2 \frac{d^2\varphi_2}{dz^2} + 2z \frac{d\varphi_2}{dz} + \frac{\omega^2 h^2}{c^2(l-1)^2} \varphi_2 = 0 \tag{5-91}$$

Equation (5-91) is a recognizable form. Its solution is

$$\varphi_2 = z^m \tag{5-92}$$

where m is given by

$$m(m+1) + \frac{\omega^2 h^2}{c^2(l-1)^2} = 0 \tag{5-93}$$

upon substituting (5-92) back into (5-91). The solutions for (5-93) are

$$\begin{aligned} m_1 &= -\frac{1}{2} + \sqrt{\frac{1}{4} - \frac{\omega^2 h^2}{c^2(l-1)^2}} \\ m_2 &= -\frac{1}{2} - \sqrt{\frac{1}{4} - \frac{\omega^2 h^2}{c^2(l-1)^2}} \end{aligned} \tag{5-94}$$

The velocity potential in the transition region is then

$$\varphi_2 = e^{i\omega t}\bar{\varphi}_2 = C\left(1 + \frac{l-1}{h} x\right)^{m_1} e^{i\omega t} + D\left(1 + \frac{l-1}{h} x\right)^{m_2} e^{i\omega t} \tag{5-95}$$

At the boundaries between the regions I and II and the regions II and III we must satisfy the usual conditions of continuity of pressure and vertical component of particle velocity,

$$\begin{aligned} \varphi_1 &= \varphi_2 \\ \frac{\partial \varphi_1}{\partial x} &= \frac{\partial \varphi_2}{\partial x} \qquad \text{at } x = 0 \end{aligned}$$

and

$$\begin{aligned} \varphi_2 &= \varphi_3 \\ \frac{\partial \varphi_2}{\partial x} &= \frac{\partial \varphi_3}{\partial x} \qquad \text{at } x = 0 \end{aligned} \tag{5-96}$$

Substituting Eqs. (5-84), (5-95), and (5-85) into (5-96) we obtain

$$\begin{aligned} -B + C + D &= A \\ B - \frac{m_1 c(l-1)}{i\omega h} C - \frac{m_2 c(l-1)}{i\omega h} D &= A \\ l^{m_1} C + l^{m_2} D - E e^{-i(\omega h/lc)} &= 0 \\ \frac{m_1 c(l-1)}{i\omega h} l^{m_1} C + \frac{m_2 c(l-1)}{i\omega h} l^{m_2} D + E e^{-i(\omega h/lc)} &= 0 \end{aligned} \tag{5-97}$$

Solving Eqs. (5-97) for the ratio of the amplitudes of the wave returned to the upper medium B to the incident wave A gives

$$R \equiv \frac{B}{A} = \frac{\begin{vmatrix} 1 & 1 & 1 & 0 \\ 1 & -\frac{m_1}{\sigma} & -\frac{m_2}{\sigma} & 0 \\ 0 & l^{m_1} & l^{m_2} & -1 \\ 0 & \frac{m_1}{\sigma} l^{m_1} & \frac{m_2}{\sigma} l^{m_2} & 1 \end{vmatrix}}{\begin{vmatrix} -1 & 1 & 1 & 0 \\ 1 & -\frac{m_1}{\sigma} & -\frac{m_2}{\sigma} & 0 \\ 0 & l^{m_1} & l^{m_2} & -1 \\ 0 & \frac{m_1}{\sigma} l^{m_1} & \frac{m_2}{\sigma} l^{m_2} & 1 \end{vmatrix}}$$

$$= \frac{-\left(\frac{m_1}{\sigma}+1\right)\left(\frac{m_2}{\sigma}+1\right) l^{m_2} + \left(\frac{m_1}{\sigma}+1\right)\left(\frac{m_2}{\sigma}+1\right) l^{m_1}}{\left(\frac{m_1}{\sigma}-1\right)\left(\frac{m_2}{\sigma}+1\right) l^{m_2} - \left(\frac{m_1}{\sigma}+1\right)\left(\frac{m_2}{\sigma}-1\right) l^{m_1}}$$

$$= \frac{[m_1 m_2 + \sigma(m_2 + m_1) + \sigma^2](l^{m_1} - l^{m_2})}{[m_1 m_2 - \sigma(m_2 - m_1) - \sigma^2] l^{m_2} - [m_1 m_2 + \sigma(m_2 - m_1) - \sigma^2] l^{m_1}}$$

$$= \frac{-(l^\gamma - l^{-\gamma})}{2\sigma(l^\gamma - l^{-\gamma}) + 2\gamma(l^\gamma + l^{-\gamma})}$$

$$= \frac{-1}{2\sigma + 2\gamma \dfrac{l^\gamma + l^{-\gamma}}{l^\gamma - l^{-\gamma}}}$$

$$= \frac{-1}{2\sigma + 2\gamma \coth(\gamma \log l)} \tag{5-98}$$

where σ and γ are defined by

$$\sigma \equiv \frac{i\omega h}{c(l-1)} \tag{5-99}$$

$$\gamma \equiv \sqrt{\frac{1}{4} - \frac{\omega^2 h^2}{c^2(l-1)^2}} = \sqrt{\frac{1}{4} + \sigma^2} \tag{5-100}$$

In general the reflection coefficient R is complex. Both its amplitude and phase are dependent on frequency. In the limit as $\omega \to 0$ or $h \to 0$, R becomes

$$R \to \frac{-(l^{1/2} - l^{-1/2})}{(l^{1/2} + l^{-1/2})} = \frac{1-l}{1+l} \tag{5-101}$$

Either limit corresponds to making the transition region II appear small to the incident wave. From Eq. (2-156) the reflection coefficient for no transition layer is

$$R = \frac{\rho_2 c_2 - \rho_1 c_1}{\rho_2 c_2 + \rho_1 c_1} = \frac{l - 1}{l + 1} \tag{5-102}$$

The two are not equal; the difference in sign between them is discussed for the similar problem of reflection from a sequence of three layers later in

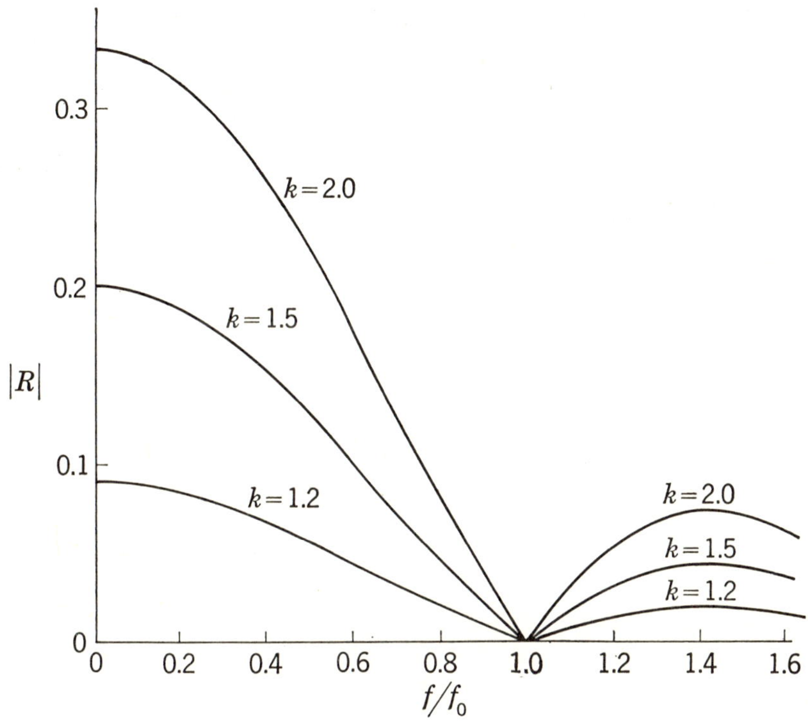

FIG. 5-10. (*From A. Wolf, Geophysics, vol. 2, 1937.*)

this chapter. In Eq. (5-98) as ω increases from zero, the amplitude of the reflection coefficient decreases. It becomes zero at

$$\gamma \coth (\gamma \log l) = -\bar{\gamma} \cot (\bar{\gamma} \log l) = \infty$$

$$\sqrt{\frac{\omega_0{}^2 h^2}{c^2(l - 1)^2} - \frac{1}{4}} \log l = \pi$$

$$f_0 = \frac{c(l - 1)}{2h \log l} \left[1 + \left(\frac{\log l}{2\pi} \right)^2 \right]^{1/2} \tag{5-103}$$

It then goes through a series of zeros for integral values of $n\pi$ with decreasing maximum amplitude between the zeros. Figure 5-10 is a graph of the amplitude of the reflection coefficient as a function of f/f_0 for various values of l.

In many problems of moderate complexity where the velocity of propagation is a function of one or more space coordinates it is often convenient to use the approximate wave equation

$$V^2 \frac{\partial^2 \varphi}{\partial x^2} = \frac{\partial^2 \varphi}{\partial t^2} \tag{5-104}$$

Equation (5-104) is an approximation to Eq. (5-86) under the conditions that the velocity is a slowly varying function of x. Let us see what the reflection coefficient would be if we had used this approximate equation. As before, we want a simple harmonic solution of the form

$$\varphi_2 = e^{i\omega t}\bar{\varphi}_2(x) \tag{5-105}$$

Substituting into Eq. (5-104) we obtain

$$c^2 \left(1 + \frac{l-1}{x}\right)^2 \frac{d^2\varphi_2}{dx^2} + \omega^2\varphi_2 = 0 \tag{5-106}$$

Making the change of variable (5-90), we obtain

$$z^2 \frac{d^2\varphi_2}{dz^2} + \frac{\omega^2 l^2}{c^2(l-1)^2}\varphi_2 = 0 \tag{5-107}$$

whose solution is also of the form (5-92) where here m is given by the solution to the equation

$$m(m-1) + \frac{\omega^2 h^2}{c^2(l-1)^2} = 0 \tag{5-108}$$

or

$$\begin{aligned} m_1 &= \frac{1}{2} + \sqrt{\frac{1}{4} - \frac{\omega^2 h^2}{c^2(l-1)^2}} \\ m_2 &= \frac{1}{2} + \sqrt{\frac{1}{4} - \frac{\omega^2 h^2}{c^2(l-1)^2}} \end{aligned} \tag{5-109}$$

The velocity potential in the transition region is then

$$\varphi_2 = e^{i\omega t}\bar{\varphi}_2 = C\left(1 + \frac{l-1}{h}x\right)^{m_1} e^{i\omega t} + D\left(1 + \frac{l-1}{h}x\right)^{m_2} e^{i\omega t} \tag{5-110}$$

Following through the derivation of Eq. (5-99) with these values for m_1 and m_2 we obtain

$$R = \frac{1}{2\sigma + 2\gamma \coth(\gamma \log k)} \tag{5-111}$$

the negative of Eq. (5-98). The amplitude of the reflection coefficient is the same for the approximate solution, but the phase is incorrect by 180 deg.

It is of further interest to see that as the velocity gradient $(l - 1)/h$ approaches zero, the solutions (5-95) and (5-110) reduce to $e^{\pm i(\omega/c)x}$, the correct solution for a constant velocity. As $(l - 1)/h = a$ approaches zero,

$$
\begin{aligned}
\left(1 + \frac{l-1}{h} x\right)^m &\to (1 + ax)^{\pm \frac{iw}{ac}} \\
&= 1 + \frac{\pm i\omega}{ac}(ax) + \frac{\pm(i\omega/ac)[\pm(i\omega/ac) - 1]}{2!}(ax)^2 \\
&\qquad + \cdots \\
&\to 1 + \left(\pm \frac{i\omega}{c} x\right) + \frac{[\pm(i\omega/c)x]^2}{2!} + \frac{[\pm(i\omega/c)x]^3}{3!} \\
&\qquad + \cdots \\
&= e^{\pm i(\omega/c)x}
\end{aligned}
$$

Reflection from an elastic medium

In Chap. 1 the reflection coefficient for plane wave incidence to a plane boundary between two fluid media was derived. Here we shall derive the reflection coefficient for a fluid medium underlain by an elastic medium. The incident wave is taken to be a simple harmonic wave propagating in the fluid medium in the xz plane (Fig. 5-11). For plane waves the displacements will be functions of x and z only. The problem will be solved in terms of the displacement potentials of (1-86) and (1-107).

$$
\begin{aligned}
u_1 &= \frac{\partial \varphi_1}{\partial x} & w_1 &= \frac{\partial \varphi_1}{\partial z} \\
u_2 &= \frac{\partial \varphi_2}{\partial x} + \frac{\partial \psi_2}{\partial z} & w_2 &= \frac{\partial \varphi_2}{\partial z} - \frac{\partial \psi_2}{\partial x}
\end{aligned}
\tag{5-112}
$$

The displacement v_2 in the lower medium will be zero; for in order to satisfy the boundary condition that $(\sigma_{xy})_2 = 0$, v_2 must be zero. There will be no transverse vibration in the lower medium parallel to the y axis.

For plane wave propagation in an elastic medium the component of motion in the direction of propagation is called the P component. It is the portion of the displacements u_2 and w_2 determined by the φ_2 solution and propagates with the velocity α_2. The component of motion transverse to the direction of propagation and parallel to the reflecting boundary is called the SH component. In our case this is v_2. The third component of motion also transverse to the direction of propagation and normal to the SH component is called the SV component. It is the portion of the displacements u_2 and w_2 determined by the ψ_2 solution. Both the SH and SV components propagate with the velocity β_2.

The displacement potentials φ and ψ satisfy the wave equations

$$\begin{aligned} c_1{}^2\nabla^2\varphi_1 &= \frac{\partial^2\varphi_1}{\partial t^2} \\ \alpha_2{}^2\nabla^2\varphi_2 &= \frac{\partial^2\varphi_2}{\partial t^2} \\ \beta_2{}^2\nabla^2\psi_2 &= \frac{\partial^2\psi_2}{\partial t^2} \end{aligned} \tag{5-113}$$

The simple harmonic plane wave solutions to these equations will be

$$\begin{aligned} \varphi_1 &= \varphi_{\text{in}} + \varphi_{\text{re}} \\ \varphi_{\text{in}} &= e^{i(\omega t + a_1 x + bz)} \\ \varphi_{\text{re}} &= Ae^{i(\omega t - a_1 x + bz)} \\ \varphi_2 &= Be^{i(\omega t + a_2 x + bz)} \\ \psi_2 &= Ce^{i(\omega t + d_2 x + bz)} \end{aligned} \tag{5-114}$$

for an incident wave, φ_{in}, of unit amplitude. In order that the boundary conditions may be met for all t and all z we obtain, as before, the relations

$$\begin{aligned} a_1 &= \frac{\omega}{c_1}\cos\theta_1 = k_1\cos\theta_1 = k_1\sqrt{1-\sin^2\theta_1} \\ a_2 &= \frac{\omega}{\alpha_2}\cos\theta_2 = k_1\sqrt{\frac{c_1{}^2}{\alpha_2{}^2} - \sin^2\theta_1} \\ d_2 &= \frac{\omega}{\beta_2}\cos\eta_2 = k_1\sqrt{\frac{c_1{}^2}{\beta_2{}^2} - \sin^2\theta_1} \\ b &= \frac{\omega}{c_1}\sin\theta_1 = \frac{\omega}{\alpha_2}\sin\theta_2 = \frac{\omega}{\beta_2}\sin\eta_2 \end{aligned} \tag{5-115}$$

where the angles θ_1, θ_2, and η_2 are, respectively, the angles of the incident and reflected P waves, the refracted P wave, and the refracted SV wave.

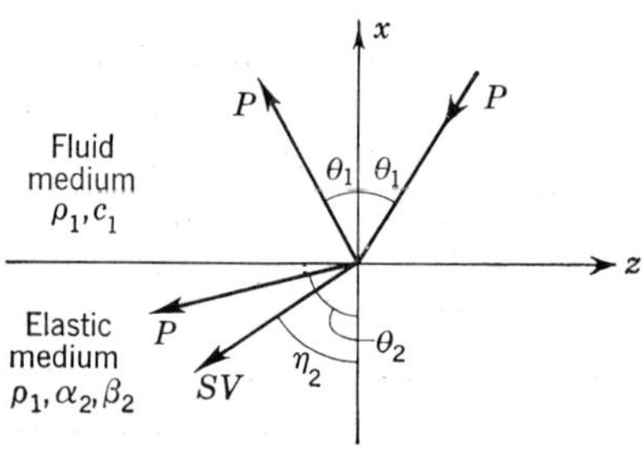

FIG. 5-11.

There are three conditions which must be met at the interface between the two media. These three conditions allow us to determine the coefficients A, B, and C in terms of the incident amplitude. Two of the boundary conditions are the usual ones, that of equality of normal component of displacement and normal component of stress. The third boundary condition is that the tangential component of stress in the lower medium must vanish at the interface. Since the upper medium cannot support a shear stress, it is a free boundary for shear stresses. From Eqs. (1-12) and (5-114) the stresses are given by

$$\begin{aligned}
(\sigma_{xx})_1 &= -\rho_1\omega^2\varphi_1 \\
(\sigma_{xx})_2 &= \lambda_2\Delta_2 + 2\mu_2 \frac{\partial u_2}{\partial x} = \lambda_2\nabla^2\varphi_2 + 2\mu_2\left(\frac{\partial^2\varphi_2}{\partial x^2} + \frac{\partial^2\psi_2}{\partial x\,\partial z}\right) \\
&= (\lambda_2 + 2\mu_2)\left(-\frac{\omega^2}{\alpha_2{}^2}\varphi_2\right) + 2\mu_2\left(\frac{\partial^2\varphi_2}{\partial x^2} + \frac{\omega^2}{\alpha_2{}^2}\varphi_2 + \frac{\partial^2\psi_2}{\partial x\,\partial z}\right) \\
&= \rho_2\left[-\omega^2\varphi_2 + 2\beta_2{}^2\left(-\frac{\partial^2\varphi_2}{\partial z^2} + \frac{\partial^2\psi_2}{\partial x\,\partial z}\right)\right] \qquad (5\text{-}116) \\
(\sigma_{xz})_2 &= \mu_2\left(\frac{\partial u_2}{\partial z} + \frac{\partial w_2}{\partial x}\right) = \mu_2\left(2\frac{\partial^2\varphi_2}{\partial x\,\partial z} - \frac{\partial^2\psi_2}{\partial x^2} + \frac{\partial^2\psi_2}{\partial z^2}\right)
\end{aligned}$$

Substituting Eqs. (5-116) and (5-114) into the boundary conditions

$$u_1 = u_2,\ (\sigma_{xx})_1 = (\sigma_{xx})_2,\ (\sigma_{xz})_2 = 0 \qquad (5\text{-}117)$$

we obtain the following

$$\begin{aligned}
\bar{a}_1 - \bar{a}_1 A &= \bar{a}_2 B + \bar{b} C \\
-\rho_1(1 + A) &= \rho_2[-B + 2\beta_2{}^2(\bar{b}^2 B - \bar{d}_2\bar{b}C)] \qquad (5\text{-}118) \\
\mu_2[-2\bar{a}_2\bar{b}B + \bar{d}_2{}^2C - \bar{b}^2C] &= 0
\end{aligned}$$

where $\bar{a}_1$, $\bar{a}_2$, $\bar{b}$, and $\bar{d}_2$ are defined by

$$a_1 \equiv \omega\bar{a}_1,\ a_2 \equiv \omega\bar{a}_2,\ b \equiv \omega\bar{b},\ d_2 \equiv \omega\bar{d}_2 \qquad (5\text{-}119)$$

Equations (5-118) reduce to

$$\begin{aligned}
\bar{a}_1 A + \bar{a}_2 B + \bar{b} C &= \bar{a}_1 \\
-q^2 A - \delta e^2 B + 2\bar{d}_2\bar{b}\delta C &= q^2 \qquad (5\text{-}120) \\
2\bar{a}_2\bar{b}B + e^2 C &= 0
\end{aligned}$$

where δ, e^2, and q^2 are defined by

$$\delta \equiv \frac{\rho_2}{\rho_1},\ e^2 \equiv 2\bar{b}^2 - \frac{1}{\beta_2{}^2},\ q^2 \equiv \frac{1}{\beta_2{}^2} \qquad (5\text{-}121)$$

Solving Eq. (5-120) for A, the reflection coefficient, we obtain

$$\begin{aligned}
R \equiv A &= \frac{\bar{a}_1(-\delta e^4 - 4\bar{a}_2\bar{d}_2\bar{b}^2\delta) - q^2(\bar{a}_2 e^2 - 2\bar{a}_2\bar{b}^2)}{\bar{a}_1(-\delta e^4 - 4\bar{a}_2\bar{d}_2\bar{b}^2\delta) + q^2(\bar{a}_2 e^2 - 2\bar{a}_2\bar{b}^2)} \\
&= \frac{\bar{a}_1\delta(e^4 + 4\bar{a}_2\bar{d}_2\bar{b}^2) - \bar{a}_2 q^4}{\bar{a}_1\delta(e^4 + 4\bar{a}_2\bar{d}_2\bar{b}^2) + \bar{a}_2 q^4} \qquad (5\text{-}122)
\end{aligned}$$

The quantities $\bar{a}_1$, $\bar{b}$, δ, e^2, and q^2 will always be real. The quantities $\bar{a}_2$ and $\bar{d}_2$ may be real or imaginary. Following the same argument as given in Chap. 1, we want to take the positive square root for $\bar{a}_1$, $\bar{b}$, $\bar{a}_2$, and $\bar{d}_2$ real. For $\bar{a}_2$ or $\bar{d}_2$ imaginary we want to take the negative root to agree with the sense of the x axis as chosen in Fig. 5-11 and produce a decreasing exponential away from the boundary into the lower medium. From (5-115) and (5-119) $\bar{a}_2$ and $\bar{d}_2$ are given by

$$\bar{a}_2 = \frac{1}{c_1}\sqrt{\frac{c_1^2}{\alpha_2^2} - \sin^2\theta_1}$$

or

$$\bar{a}_2 = \frac{-i}{c}\sqrt{\sin^2\theta_1 - \frac{c_1^2}{\alpha_2^2}} = -i\bar{a}_2'$$

and (5-123)

$$\bar{d}_2 = \frac{1}{c_1}\sqrt{\frac{c_1^2}{\beta_2^2} - \sin^2\theta_1}$$

or

$$\bar{d}_2 = \frac{-i}{c_1}\sqrt{\sin^2\theta_1 - \frac{c_1^2}{\beta_2^2}} = -i\bar{d}_2'$$

For the normal conditions encountered in nature α_2 will be greater than c_1. The quantity $\bar{a}_2$ will then be imaginary over some values of the incident angle θ_1. The angle at which $\bar{a}_2$ becomes imaginary,

$$\theta_\alpha = \sin^{-1}\frac{c_1}{\alpha_2} \tag{5-124}$$

is called the critical angle for P waves. The velocity β_2 may be greater or less than c_1. If β_2 is less than c_1, the quantity $\bar{d}_2$ will be real for all values of θ_1. If β_2 is greater than c_1, the quantity $\bar{d}_2$ will then be imaginary over some values of θ_1. The angle at which $\bar{d}_2$ becomes imaginary,

$$\theta_\beta = \sin^{-1}\frac{c_1}{\beta_2} \tag{5-125}$$

is called the critical angle for SV waves. Since α_2 is always greater than β_2, the angle θ_β will be greater than θ_α. The critical angle for P will always be reached before the critical angle for SV. Beyond the critical angle θ_α there will be no refracted P wave into the lower medium. Beyond the critical angle θ_β there will be no refracted SV wave. Beyond the critical angle θ_β there will be total reflection to the upper medium.

Let us express the reflection coefficient R in the form me^{in} where m is the amplitude of the reflected wave with respect to the incident wave and n the phase change of the reflected wave with respect to the incident wave. For the case of β_2 less than c_1 we shall obtain

$$R = \frac{\bar{a}_1\delta(e^4 + 4\bar{a}_2\bar{d}_2\bar{b}) - \bar{a}_2q^4}{\bar{a}_1\delta(e^4 + 4\bar{a}_2\bar{d}_2\bar{b}) + \bar{a}_2q^4}\,e^{i0} \qquad 0 < \theta \leqslant \theta_\alpha \tag{5-126}$$

and

$$R = \frac{\bar{a}_1\delta e^4 + i\bar{a}_2'(q^4 - 4\bar{d}_2\bar{b}\delta)}{\bar{a}_1\delta e^4 - i\bar{a}_2'(q^4 + 4\bar{d}_2\bar{b}\delta)} = \frac{P + i(Q - R)}{P - i(Q + R)}$$
$$= \left(\frac{P^2 + Q^2 + R^2 - 2QR}{P^2 + Q^2 + R^2 + 2QR}\right)^{1/2} e^{i[(\lambda/2)+(m/2)]} \qquad \theta_\alpha < \theta \leqslant \frac{\pi}{2} \tag{5-127}$$

where

$$\cot\frac{\lambda}{2} = \frac{P}{Q - R}, \cot\frac{m}{2} = \frac{P}{Q + R}$$

The amplitude of the reflected wave will always be less than unity. There will be a phase change for angles of incidence greater than θ_α. For the case of β_2 greater than c_1, Eq. (5-126) will remain the same for the angular range defined. Equation (5-127) will be valid for an angular range defined by $\theta_\alpha < \theta \leqslant \theta_\beta$. For angles greater than θ_β, we shall have

$$R = \frac{\bar{a}_1\delta(e^4 - 4\bar{a}_2'\bar{d}_2'\bar{b}^2) + i\bar{a}_2'q^4}{\bar{a}_1\delta(e^4 - 4\bar{a}_2'\bar{d}_2'\bar{b}^2) - i\bar{a}_2'q^4} = \frac{M + iN}{M - iN}$$
$$= e^{i\epsilon} \qquad \theta_\beta < \theta \leqslant \frac{\pi}{2} \tag{5-128}$$

where

$$\cot\frac{\epsilon}{2} = \frac{M}{N} = \frac{\bar{a}_1\delta(e^4 - 4\bar{a}_2'\bar{d}_2'\bar{b}^2)}{\bar{a}_2'q^4}$$

The amplitude of the reflected wave will be unity for angles of incidence greater than θ_β. There will be a phase change for angles greater than θ_α.

Normal incidence to a corrugated boundary

In all the problems up to now we have considered the discontinuity between two media to be plane. It is of interest to examine what effects will be produced by a roughness in the discontinuity. In particular we shall consider normal incidence to a corrugated reflector. The corrugations are taken to be periodic and are represented by a simple cosine term,

$$z = \zeta = a \cos (px) \tag{5-129}$$

From Fig. 5-12 the velocity potential of a simple harmonic plane wave of unit amplitude incident normal to the boundary will be

$$\varphi_{\text{in}} = e^{i\omega t}e^{ik_1 z} \tag{5-130}$$

where

$$k_1 = \frac{\omega}{c_1} \tag{5-131}$$

The energy returned to the upper medium will not be entirely along the z axis. There will be energy reflected back at various angles to the upper medium. Since the corrugations of the boundary are periodic in x and the incident plane wave is normal to the boundary, the x dependence of the energy returned to the upper medium must also be periodic in x. The form of the x solution may then be represented by a Fourier series. The z

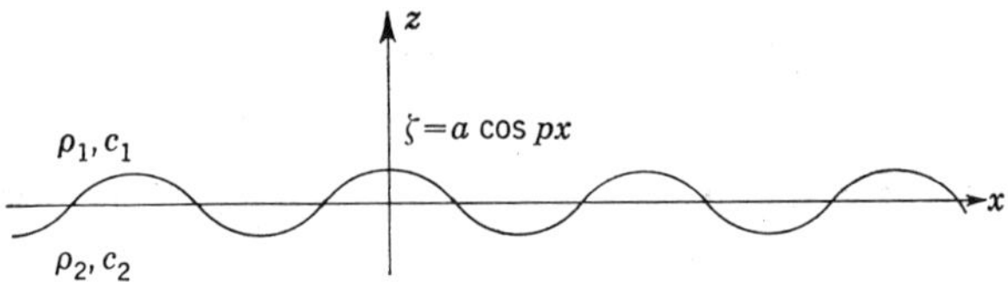

FIG. 5-12.

dependence will be taken as the usual exponential for propagation back in the direction of positive z, giving for the form of φ_{re},

$$\varphi_{re} = e^{i\omega t} \sum_n A_n \cos(npx) e^{-i\mu_n z} \tag{5-132}$$

Substituting into the appropriate wave equation

$$c_1^2 \nabla^2 \varphi_1 = \frac{\partial^2 \varphi_1}{\partial t^2} \tag{5-133}$$

we obtain the conditional equation determining μ_n,

$$\mu_n^2 = k_1^2 - n^2p^2, \; n = 0, 1, 2, \ldots . \tag{5-134}$$

We then have for φ_1

$$\varphi_1 = e^{i\omega t}\bar{\varphi}_1 = e^{i\omega t}[e^{ik_1 z} + A_0 e^{-ik_1 z} + A_1 e^{-i\mu_1 z} \cos(px) + A_2 e^{-i\mu_2 z} \cos(2px) + \cdots] \tag{5-135}$$

From a similar derivation for φ_2, we obtain

$$\varphi_2 = e^{i\omega t}\bar{\varphi}_2 = e^{i\omega t}[B_0 e^{ik_2 z} + B_1 e^{i\nu_1 z} \cos(px) + B_2 e^{i\nu_2 z} \cos(2px) + \cdots] \tag{5-136}$$

where

$$k_2 = \frac{\omega}{c_2} \tag{5-137}$$

$$\nu_n^2 = k_2^2 - n^2p^2, \; n = 0, 1, 2, \ldots . \tag{5-138}$$

From Eq. (5-134) we see that μ_n can be a real or imaginary quantity. As we have done previously, we take the positive square root for μ_n real to give waves propagating back into the upper medium and the negative square root for μ_n imaginary to give a decreasing exponential toward infinity in the z direction. For a real value of μ_n, say μ_m, the mth term will be

$$A_m e^{-i\mu_m z} \cos(mpx) = \tfrac{1}{2} A_m [e^{-i(\mu_m z - mpx)} + e^{-i(\mu_m z + mpx)}] \tag{5-139}$$

The mth term represents two plane waves each inclined at an angle θ_m on either side of the z axis. From Eq. (5-139) the angle θ_m is given by

$$\cos \theta_m = \frac{\mu_m}{k_1} = \frac{\mu_m}{(\mu_m{}^2 + m^2p^2)^{1/2}} \tag{5-140}$$

In the expression (5-135) the zeroth term, corresponding to the Snell's law path for a plane boundary, is called the *specular reflection*. The other terms with μ_n real are called the *scattered reflections* or scattered waves. For an imaginary value of μ_n, say μ_r, the rth term will be

$$A_r e^{-\mu_r' z} \cos (rpx) \tag{5-141}$$

where

$$\mu_r = -i\mu_r' \tag{5-142}$$

The rth term represents a vibration confined to the neighborhood of the boundary. Its amplitude decreases exponentially away from the boundary into the upper medium.

If the wavelength of the incident plane wave is greater than the wavelength of the corrugations, $p > k$, all the μ_n except μ_0 are imaginary. At a sufficient distance away from the boundary the only sensible term in the reflected wave will be the zeroth,

$$A_0 e^{-ik_1 z} \tag{5-143}$$

all others decreasing exponentially. Only the specular reflection is important. The reflection coefficient A_0 is less than the Rayleigh reflection coefficient by the amount that has gone into the vibrations confined to the neighborhood of the boundary, represented by the higher terms.

If the wavelength of the incident plane wave is much smaller than the wavelength of the corrugations, $p \ll k$, the μ_n of Eq. (5-134) may be approximated by k_1. The scattered reflections are nearly along the direction of the specular reflection. The boundary conditions to be satisfied are as before. The pressure and normal components of particle velocity must be equal along the boundary. The normal components of velocity should be taken normal to the corrugations; but under the approximation that $p \ll k$, it may be approximated by the vertical component. The boundary conditions are then

$$\begin{aligned} \rho_1\varphi_1 &= \rho_2\varphi_2 \\ \frac{\partial \varphi_1}{\partial z} &= \frac{\partial \varphi_2}{\partial z} \qquad \text{at } z = \zeta \end{aligned} \tag{5-144}$$

From Eqs. (5-144) and (5-136) we obtain

$$\frac{\partial \varphi_1}{\partial z} = \frac{\partial \varphi_2}{\partial z} = ik_2\varphi_2 = ik_2 \frac{\rho_1}{\rho_2} \varphi_1 \tag{5-145}$$

under the approximation that $\nu_n \approx k_2$. Substituting (5-135) into (5-145) gives

$$\begin{aligned} ik_2\rho_1[e^{ik_1\zeta} + A_0 e^{-ik_1\zeta} + A_1 e^{-ik_1\zeta}\cos(px) + A_2 e^{-ik_1\zeta}\cos(2px) + \cdots] \\ = ik_1\rho_2[e^{ik_1\zeta} - A_0 e^{-ik_1\zeta} - A_1 e^{-ik_1\zeta}\cos(px) \\ - A_2 e^{-ik_1\zeta}\cos(2px) + \cdots] \end{aligned} \quad (5\text{-}146)$$

Collecting terms we obtain

$$\frac{k_2\rho_1 - k_1\rho_2}{k_2\rho_1 + k_1\rho_2} e^{i2k_1\zeta} + A_0 + A_1\cos(px) + A_2\cos(2px) + \cdots = 0 \quad (5\text{-}147)$$

The coefficients $A_0, A_1, A_2, \ldots$ are determined from Eq. (5-147) with the known form of the corrugations, ζ. If $\zeta = 0$, a plane boundary, Eq. (5-147) reduces to

$$A_0 = \frac{k_1\rho_2 - k_2\rho_1}{k_1\rho_2 + k_2\rho_1} = \frac{\rho_2 c_2 - \rho_1 c_1}{\rho_2 c_2 + \rho_1 c_1} \quad (5\text{-}148)$$

the Rayleigh reflection coefficient. The coefficients $A_1, A_2, \ldots$ vanish. In order to determine the coefficients for $\zeta \neq 0$, we need to express the exponential of (5-147) in terms of a Fourier series in px. For the form (5-129) this is a known series and is given by

$$\begin{aligned} e^{i2k_1\zeta} = e^{i2k_1 a\cos(px)} = J_0(2k_1a) - 2J_2(2k_1a)\cos(2px) \\ + 2J_4(2k_1a)\cos(4px) + \cdots + i[2J_1(2k_1a)\cos(px) \\ - 2J_3(2k_1a)\cos(3px) + 2J_5(2k_1a)\cos(5px) + \cdots] \end{aligned} \quad (5\text{-}149)$$

where the $J_0, J_1, \cdots$ are the Bessel functions of the various orders. Equating (5-149) to (5-147) gives

$$\begin{array}{ll} A_0 = RJ_0(2k_1a) & A_1 = i2RJ_1(2k_1a) \\ A_2 = -2RJ_2(2k_1a) & A_3 = -i2RJ_3(2k_1a) \\ A_4 = 2RJ_4(2k_1a) & A_5 = i2RJ_5(2k_1a) \\ \cdots\cdots\cdots & \cdots\cdots\cdots \end{array} \quad (5\text{-}150)$$

Reflection from a porous medium

In Chap. 6 the propagation of sound in a porous medium is discussed. We shall borrow from those results to discuss in general terms the reflection of sound from a boundary between a fluid medium and a highly porous medium (Fig. 5-13). For a highly porous medium the velocity of propagation may be less than that of the contained fluid at low frequencies. The velocity will increase with frequency and eventually

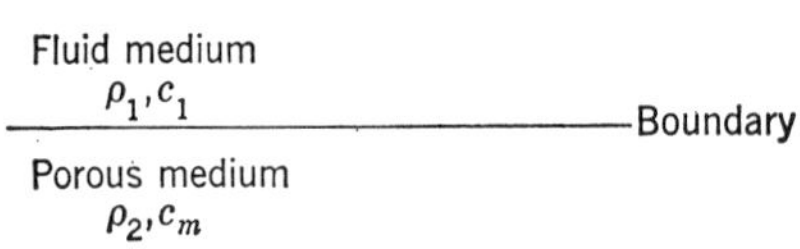

FIG. 5-13.

become greater than that of the contained fluid. Such a relation is shown in Fig. 5-14 where c_m is the velocity of the porous medium and c_0 the velocity of the fluid. If the overlying fluid medium is the same as the fluid contained in the porous medium, $c_1 = c_0$, the velocity of the fluid medium will be less than that of the underlying porous medium at low frequencies and greater than that of the porous medium at high frequencies. The density ρ_m of the porous medium will be greater than that of the overlying fluid medium ρ_1; and in general $\rho_m c_m$ will be greater than $\rho_1 c_1$ for all frequencies. From Eq. (2-154) and Figs. 2-29*d* and 2-29*e* the Rayleigh

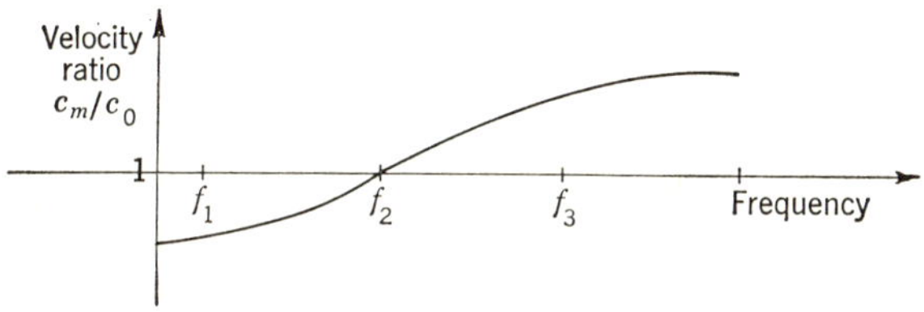

FIG. 5-14.

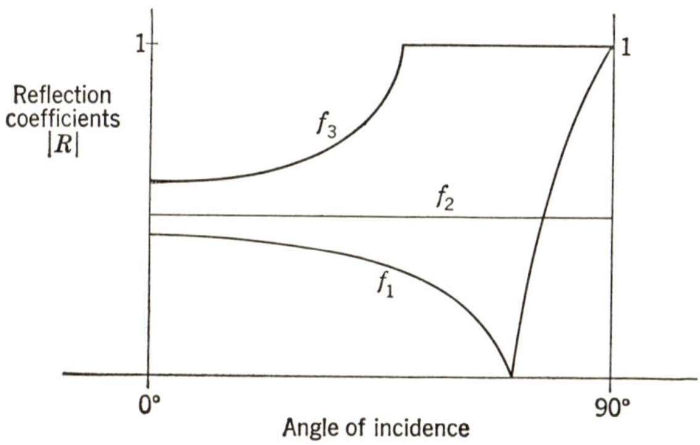

FIG. 5-15.

reflection coefficient will be as shown in Fig. 5-15 as a function of angle of incidence and frequency. The boundary between the fluid and highly porous medium will be in general a good reflector of sound at high frequencies and a good transmitter of sound at low frequencies.

Reflection from a boundary for which the boundary condition is continuity of acoustic impedance

For completeness we should consider an approximate boundary condition that is applied to a number of architectural acoustic problems. It is often used to describe the reflection of sound in a fluid medium, air or water, from an absorbing wall. Its usefulness in sound-transmission problems is limited.

It states that the acoustic effects at a boundary are independent of the angle of incidence and that the acoustic properties of a boundary are given entirely by its *specific acoustic impedance.* The specific acoustic impedance is defined as the ratio of the acoustic impedance of the wall to that of the fluid. This condition is valid in the limit that there is complete local interaction at the surface. The motion in the lower medium at any point along the boundary is assumed to be independent of the motion at adjacent points along the boundary. The particle motion and direction of propagation in the lower medium must be then essentially normal to the boundary. These conditions are approximated by a highly absorbent wall material.

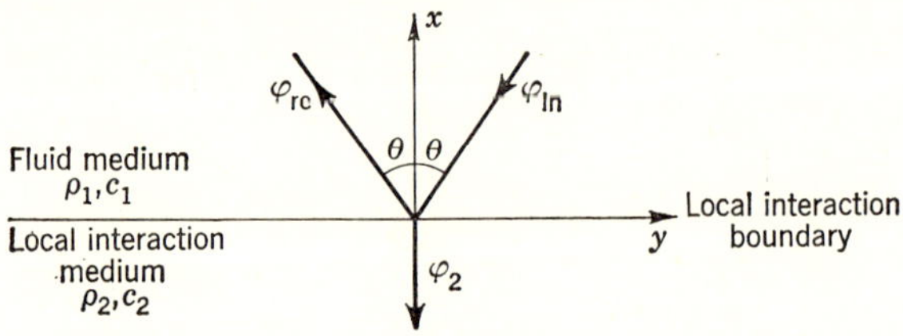

FIG. 5-16.

Under these conditions the reflection coefficient for simple harmonic plane waves may be derived as follows. With reference to Fig. 5-16, the velocity potential φ for the fluid medium is as before

$$\begin{aligned} \varphi_1 &= \varphi_{in} + \varphi_{re} \\ &= e^{i(\omega t + k_1 x \cos\theta + k_1 y \sin\theta)} + Ae^{i(\omega t - k_1 x \cos\theta + k_1 y \sin\theta)} \end{aligned} \tag{5-151}$$

where the amplitude of the incident wave φ_{in} is taken as unity. The velocity potential for the lower medium will be

$$\varphi_2 = Be^{i\omega t}e^{i(k_2 - il_2)x} \tag{5-152}$$

where l_2 is the absorption coefficient of the lower medium. The boundary conditions are as before

$$\begin{aligned} \rho_1\varphi_1 &= \rho_2\varphi_2 \\ \frac{\partial\varphi_1}{\partial x} &= \frac{\partial\varphi_2}{\partial x} \end{aligned} \qquad \text{at } x = 0 \tag{5-153}$$

It is immediately apparent from the forms of (5-151) and (5-152) that the boundary conditions cannot be met uniquely everywhere along the boundary. This is a consequence of the violation of Snell's law in Eq. (5-152). The coefficient B will be dependent on position along the boundary. This result is within our condition of local interaction at the surface.

From (5-153) and (5-152) we obtain

$$\frac{\partial \varphi_1}{\partial x} = \frac{\partial \varphi_2}{\partial x} = i(k_2 - il_2)\varphi_2 = i(k_2 - il_2)\frac{\rho_1}{\rho_2}\varphi_1 \tag{5-154}$$

Substituting (5-151) into (5-154) we obtain for the reflection coefficient A

$$A = R = \frac{\zeta \cos \theta - 1}{\zeta \cos \theta + 1} \tag{5-155}$$

where ζ is defined as

$$\zeta \equiv \frac{k_1\rho_2}{(k_2 - il_2)\rho_1} = \frac{\rho_2\omega/(k_2 - il_2)}{\rho_1 c_1} \equiv \frac{Z_2}{\rho_1 c_1} \tag{5-156}$$

If l_2 were zero, Eqs. (5-156) and (5-155) would reduce to

$$\zeta = \frac{\rho_2 c_2}{\rho_1 c_1} \tag{5-157}$$

and

$$R = \frac{\rho_2 c_2 \cos \theta - \rho_1 c_1}{\rho_2 c_2 \cos \theta + \rho_1 c_1} = \frac{(\rho_2/\rho_1) - [(c_1/c_2)/\cos \theta]}{(\rho_2/\rho_1) + [(c_1/c_2)/\cos \theta]} \tag{5-158}$$

Equation (5-158) is different from the Rayleigh reflection coefficient (2-154) by the substitution of the term (c_1/c_2) for $[(c_1^2/c_2^2) - \sin^2 \theta]^{1/2}$.

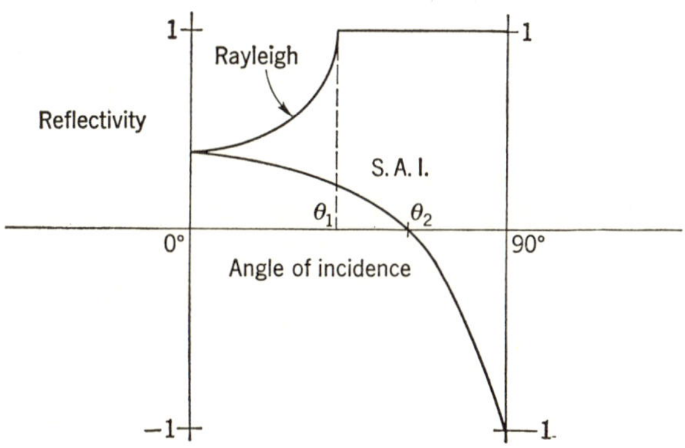

FIG. 5-17.

Figure 5-17 is a plot of (5-158) and the Rayleigh reflection coefficient for the same values of density and velocity.

From Eq. (1-77) we have that

$$\frac{p_{\text{in}}}{\dot{u}'_{\text{in}}} = \rho_1 c_1, \quad \frac{p_{\text{re}}}{\dot{u}''_{\text{re}}} = \rho_1 c_1 \tag{5-159}$$

where $\dot{u}'_{\text{in}}$ is the particle velocity component along the direction of the incident ray and $\dot{u}''_{\text{re}}$ is the particle velocity component along the direction of the reflected ray. The ratio of the pressure to particle velocity along the direction of propagation has been defined as the acoustic impedance,

$$Z_1 \equiv \rho_1 c_1 = \frac{\rho_1 \omega}{k_1} \tag{5-160}$$

This definition is analogous to that used in mechanical and electrical problems. For the lower medium we shall have a similar relation,

$$\frac{p_2}{\dot{u}_2'''} = \frac{\rho_2 \omega}{k_2 - il_2} = Z_2 \tag{5-161}$$

Rewriting the boundary condition (5-154) in this notation we obtain

$$\frac{i\rho_1 \omega \varphi_1}{\partial \varphi_1/\partial x} = \frac{\rho_2 \omega}{k_2 - il_2}$$

$$\frac{p_1}{\dot{u}_{1x}} = Z_2 \qquad \text{at } x = 0 \tag{5-162}$$

Equation (5-162) states that the ratio of pressure to normal component of particle velocity of the fluid must be equal to the wall impedance at the boundary. This is the boundary condition to be satisfied for this type of problem.

Reflection from three layers

This section is simply an extension to three layers of the solution for the Rayleigh reflection coefficient of Chap. 2. It introduces a number of new effects, most notable of which is the frequency dependence of the energy returned to the upper medium. We shall carry through the derivation first for normal incidence and then extend it to oblique incidence.

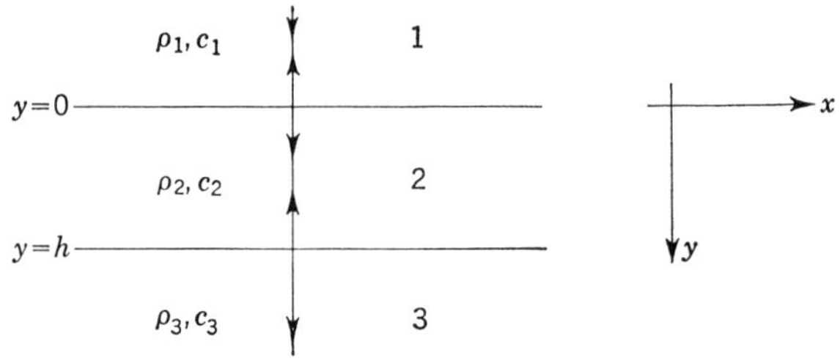

FIG. 5-18.

With reference to Fig. 5-18 consider a simple harmonic plane wave incident normal to the sequence of three layers from the upper medium. There will be energy returned upward to the first medium along the y direction from reflection at the first boundary and from internal reflections within the second medium refracted back into the first medium. The sum total of all the contributions returned to the upper medium will form, in the steady-state solution, a wave propagating in the negative y direction whose amplitude and phase will be the convergent sum of all the individual contributions. Similarly, there will be both downgoing and upgoing waves in the second medium; and there will be downgoing energy refracted into the lower medium along the y direction. The resultant steady-state solution may then be written in general form for the three media in terms of the velocity potential as, respectively,

$$\begin{aligned} \varphi_1 &= A_1 e^{i(\omega t - k_1 y)} + B_1 e^{i(\omega t + k_1 y)} \\ \varphi_2 &= A_2 e^{i(\omega t - k_2 y)} + B_2 e^{i(\omega t + k_2 y)} \\ \varphi_3 &= A_3 e^{i(\omega t - k_3 y)} \end{aligned} \tag{5-163}$$

where $A_1 e^{i(\omega t - k_1 y)}$ is the incident wave.

The boundary conditions that must be satisfied are as before that there be continuity of pressure and vertical component of particle velocity at each boundary.

$$\left.\begin{aligned} \rho_1 \frac{\partial \varphi_1}{\partial t} &= \rho_2 \frac{\partial \varphi_2}{\partial t} \\ \frac{\partial \varphi_1}{\partial y} &= \frac{\partial \varphi_2}{\partial y} \end{aligned}\right\} \quad \text{at } y = 0$$

and

$$\left.\begin{aligned} \rho_2 \frac{\partial \varphi_2}{\partial t} &= \rho_3 \frac{\partial \varphi_3}{\partial t} \\ \frac{\partial \varphi_2}{\partial y} &= \frac{\partial \varphi_3}{\partial y} \end{aligned}\right\} \quad \text{at } y = h \tag{5-164}$$

Substituting (5-163) into (5-164) gives

$$\begin{aligned} \rho_1 A_1 + \rho_1 B_1 &= \rho_2 A_2 + \rho_2 B_2 \\ k_1 A_1 - k_1 B_1 &= k_2 A_2 - k_2 B_2 \\ \rho_2 A_2 e^{-ik_2 h} + \rho_2 B_2 e^{ik_2 h} &= \rho_3 A_3 e^{-ik_3 h} \\ k_2 A_2 e^{-ik_2 h} - k_2 B_2 e^{ik_2 h} &= k_3 A_3 e^{-ik_3 h} \end{aligned} \tag{5-165}$$

Rewriting Eqs. (5-165) in terms of the amplitude of the incident wave A_1 and substituting for the k's their values

$$k_1 = \frac{\omega}{c_1}, \; k_2 = \frac{\omega}{c_2}, \; k_3 = \frac{\omega}{c_3} \tag{5-166}$$

we obtain

$$\begin{aligned} -\rho_1 B_1 + \rho_2 A_2 \quad + \rho_2 B_2 \qquad\qquad\qquad &= \rho_1 A_1 \\ c_2 B_1 + c_1 A_2 \quad - c_1 B_2 \qquad\qquad\qquad &= c_2 A_1 \\ \rho_2 e^{-ik_2 h} A_2 + \rho_2 e^{ik_2 h} B_2 - \rho_3 e^{-ik_3 h} A_3 &= 0 \\ c_3 e^{-ik_2 h} A_2 - c_3 e^{ik_2 h} B_2 - c_2 e^{-ik_3 h} A_3 &= 0 \end{aligned} \tag{5-167}$$

Equations (5-167) are four simultaneous linear equations in terms of the complex amplitudes B_1, A_2, B_2, and A_3 and may be solved for the amplitude of any one of these in terms of the incident amplitude A_1. We are particularly interested in the amplitude and phase of the downgoing wave transmitted into the lower medium A_3 and the upgoing wave returned to the upper medium B_1. Solving Eqs. (5-167) for the reflection coefficient B_1/A_1 we obtain

$$\frac{B_1}{A_1} = \frac{\begin{vmatrix} \rho_1 & \rho_2 & \rho_2 & 0 \\ c_2 & c_1 & -c_1 & 0 \\ 0 & \rho_2 e^{-ik_2h} & \rho_2 e^{ik_2h} & -\rho_3 \\ 0 & c_3 e^{-ik_2h} & -c_3 e^{ik_2h} & -c_2 \end{vmatrix}}{\begin{vmatrix} -\rho_1 & \rho_2 & \rho_2 & 0 \\ c_2 & c_1 & -c_1 & 0 \\ 0 & \rho_2 e^{-ik_2h} & \rho_2 e^{ik_2h} & -\rho_3 \\ 0 & c_3 e^{-ik_2h} & -c_3 e^{ik_2h} & -c_2 \end{vmatrix}}$$

$$= \frac{(\rho_1c_1 - \rho_2c_2)(-\rho_2c_2 - \rho_3c_3)e^{ik_2h} + (\rho_1c_1 + \rho_2c_2)(-\rho_2c_2 + \rho_3c_3)e^{-ik_2h}}{(-\rho_1c_1 - \rho_2c_2)(-\rho_2c_2 - \rho_3c_3)e^{ik_2h} + (-\rho_1c_1 + \rho_2c_2)(-\rho_2c_2 + \rho_3c_3)e^{-ik_2h}}$$

$$= \frac{(-1 + m_1m_2)\cos(k_2h) + i(m_1 - m_2)\sin(k_2h)}{(1 + m_1m_2)\cos(k_2h) + i(m_1 + m_2)\sin(k_2h)}$$

$$= \left[\frac{(-1 + m_1m_2)^2\cos^2(k_2h) + (m_1 - m_2)^2\sin^2(k_2h)}{(1 + m_1m_2)^2\cos^2(k_2h) + (m_1 + m_2)^2\sin^2(k_2h)}\right]^{1/2} e^{i(\delta-\epsilon)} \tag{5-168}$$

where m_1, m_2, δ, and ϵ are defined as

$$m_1 \equiv \frac{\rho_2c_2}{\rho_1c_1} \tag{5-169}$$

$$m_2 \equiv \frac{\rho_3c_3}{\rho_2c_2} \tag{5-170}$$

$$\delta \equiv \tan^{-1}\frac{(m_1 - m_2)\sin(k_2h)}{(-1 + m_1m_2)\cos(k_2h)} \tag{5-171}$$

$$\epsilon \equiv \tan^{-1}\frac{(m_1 + m_2)\sin(k_2h)}{(1 + m_1m_2)\cos(k_2h)} \tag{5-172}$$

Similarly, we obtain for the transmission coefficient A_3/A_1

$$\frac{A_3}{A_1} = \frac{2c_3e^{ik_3h}}{c_1[(1 + m_1m_2)\cos(k_2h) + i(m_1 + m_2)\sin(k_2h)]}$$

$$= \frac{2c_3e^{i(k_3h-\epsilon)}}{c_1[(1 + m_1m_2)^2\cos^2(k_2h) + (m_1 + m_2)^2\sin^2(k_2h)]^{1/2}} \tag{5-173}$$

There are a number of things to be gathered from an examination of Eqs. (5-168) and (5-173). Both amplitudes depend not only on the acoustic resistances of the three media, ρ_1c_1, ρ_2c_2, ρ_3c_3, but on the thickness of the intermediate medium and the frequency of the incident wave. The frequency dependence is repetitive in the cos (k_2h) and sin (k_2h) terms. The amplitudes go through a series of alternate maxima and minima as a function of frequency.

If the condition existed that

$$m_1m_2 = 1$$

so that

$$\rho_1c_1 = \rho_3c_3 \tag{5-174}$$

there would be no energy returned to the upper medium for frequencies such that

$$\sin^2(k_2h) = 0$$

$$f = 0, \frac{c_2}{2h}, \frac{2c_2}{2h}, \frac{3c_2}{2h}, \ldots \tag{5-175}$$

For these frequencies all the energy would be transmitted to the lower medium. The condition (5-174) is that for a plate or a layer of undefined acoustic resistance ρ_2c_2 immersed in a liquid. At the frequencies (5-175) the intensity of the transmitted wave is equal to that of the incident wave.

If the condition existed that

$$m_1 = m_2$$

so that

$$(\rho_1c_1)(\rho_3c_3) = (\rho_2c_2)^2 \tag{5-176}$$

there would be no energy returned to the upper medium for frequencies such that

$$\cos^2(k_2h) = 0$$

$$f = \frac{c_2}{4l}, \frac{3c_2}{4l}, \frac{5c_2}{4l}, \ldots \tag{5-177}$$

At these frequencies the intermediate layer gives perfect matching for transmission of energy from the first medium to the third medium.

It is instructive to derive the relation (5-168) directly from a summation of all the individual contributions returned back to the upper medium. This is a somewhat more physical approach to the problem than the general formulations of (5-163) and (5-164). From Fig. 5-19 we see that the total effect returned back to the upper medium consists of the summation of an infinite number of terms of decreasing amplitude, the first being the wave reflected from the 1-2 interface, the second being the wave refracted from 1 to 2 reflected from the 2-3 interface and refracted back into 1, the third being the wave refracted from 1 to 2 reflected twice from the 2-3 interface, once from the 2-1 interface, and refracted back into 1, and so on for the other terms. If A_1 is the ampli-

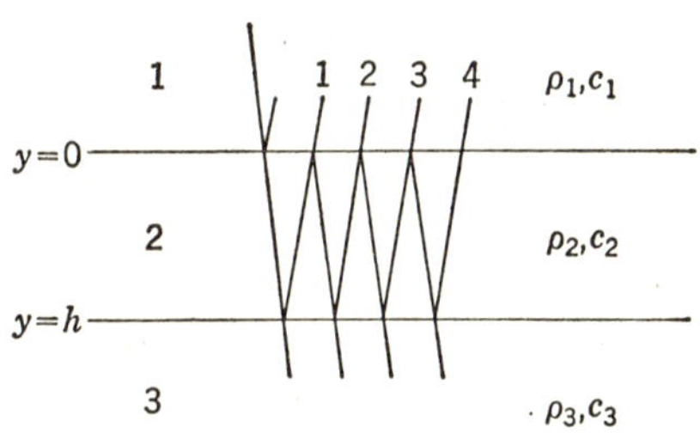

FIG. 5-19.

tude of the incident wave, the total return, $B_1e^{i(\omega t+k_1y)}$, to the upper medium will be

$$B_1e^{i(\omega t+k_1y)} = A_1(R_{12} + T_{12}R_{23}T_{21}e^{-i2k_2h} + T_{12}R_{21}R_{23}{}^2T_{21}e^{-i4k_2h} + T_{12}R_{21}{}^2R_{23}{}^3T_{21}e^{-i6k_2h} + \cdots)e^{i(\omega t+k_1y)}$$

$$\frac{B_1}{A_1} = R_{12} + T_{12}T_{21}(R_{23}e^{-i2k_2h} + R_{21}R_{23}{}^2e^{-i4k_2h} + R_{21}{}^2R_{23}{}^3e^{-i6k_2h} + \cdots) \qquad (5\text{-}178)$$

where R and T are the respective reflection and refraction coefficients. From (2-156) and (2-157) they may be written as

$$\begin{aligned} R_{12} &= \frac{\rho_2c_2 - \rho_1c_1}{\rho_2c_2 + \rho_1c_1} = \frac{m_1 - 1}{m_1 + 1} \\ R_{21} &= \frac{\rho_1c_1 - \rho_2c_2}{\rho_1c_1 + \rho_2c_2} = \frac{1 - m_1}{1 + m_1} \\ R_{23} &= \frac{\rho_3c_3 - \rho_2c_2}{\rho_3c_3 + \rho_2c_2} = \frac{m_2 - 1}{m_2 + 1} \\ T_{12}T_{21} &= \frac{2\rho_1c_2}{\rho_2c_2 + \rho_1c_1}\,\frac{2\rho_2c_1}{\rho_2c_2 + \rho_1c_1} = \frac{4m_1}{(1 + m_1)^2} \end{aligned} \qquad (5\text{-}179)$$

Substituting (5-179) into (5-178) and expressing the quantity in the parentheses of (5-178) by its sum as an infinite geometric series gives

$$\begin{aligned} \frac{B_1}{A_1} &= \frac{m_1 - 1}{m_1 + 1} + \frac{4m_1}{(m_1 + 1)^2}\left\{\frac{[(m_2 - 1)/(m_2 + 1)]e^{-i2k_2h}}{1 - [(1 - m_1)/(1 + m_1)][(m_2 - 1)/(m_2 + 1)]e^{-i2k_2h}}\right\} \\ &= \frac{(m_1 - 1)(m_1 + 1)(m_2 + 1)e^{ik_2h} + [(m_1 - 1)^2(m_2 - 1) + 4m_1(m_2 - 1)]e^{-ik_2h}}{(m_1 + 1)^2(m_2 + 1)e^{ik_2h} + (m_1 + 1)(m_1 - 1)(m_2 - 1)e^{-ik_2h}} \\ &= \frac{(-1 + m_1m_2)\cos(k_2h) + i(m_1 - m_2)\sin(k_2h)}{(1 + m_1m_2)\cos(k_2h) + i(m_1 + m_2)\sin(k_2h)} \end{aligned} \qquad (5\text{-}180)$$

the same as Eq. (5-168) from the formal solution. A similar derivation can be given for the refraction coefficient. From Eq. (5-178) we may see how the condition of no energy return can be obtained; for although there is a definite contribution from the reflection at the 1-2 interface, at those frequencies for which there is no energy return the total effect of all the upward refractions caused by the internal reflections within the second medium is such as to nullify that of the 1-2 reflection.

In the limit of no second layer, $h \to 0$, the reflection coefficient becomes

$$\frac{B_1}{A_1} \to \frac{-\rho_1c_1 + \rho_3c_3}{\rho_1c_1 + \rho_3c_3} \qquad (5\text{-}181)$$

the negative of the Rayleigh reflection (2-156) for reflection from a 1-3 interface. From the derivation of (5-178) we can see why this reduction does not agree with Eq. (2-156). The internal phase change from the 2-1 reflections is not removed in the limit of $h \to 0$. For the derivation of Eq. (5-178) we cannot describe the transmission effects within the second medium by rays for h small. For the derivation of Eq. (5-168) we cannot treat the problem as two separate, unconnected boundary conditions for h small.

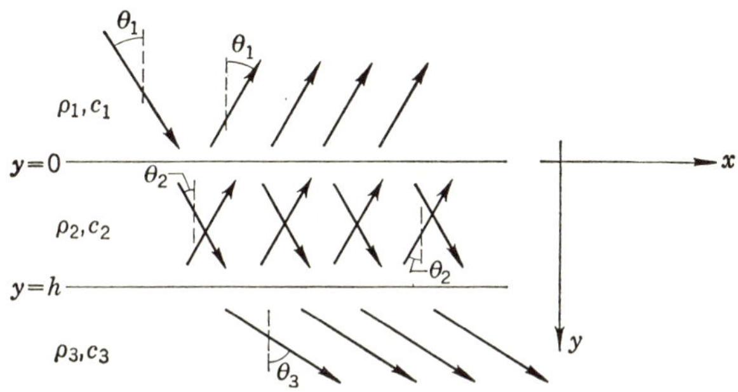

FIG. 5-20.

Let us extend our solution for normal incidence to the general case of oblique incidence. From Snell's law we are assured that all the contributions returned to the first medium will be propagating in parallel directions and similarly for the second and third media. Thus, as before, we may write our formal solutions in the form

$$\begin{aligned} \varphi_1 &= A_1 e^{i(\omega t - ax - b_1 y)} + B_1 e^{i(\omega t - ax + b_1 y)} \\ \varphi_2 &= A_2 e^{i(\omega t - ax - b_2 y)} + B_2 e^{i(\omega t - ax + b_2 y)} \\ \varphi_3 &= A_3 e^{i(\omega t - ax - b_3 y)} \end{aligned} \tag{5-182}$$

where from Fig. 5-20 the quantities b_1, b_2, b_3, and a are defined by

$$\begin{aligned} b_1 &= k_1 \cos\theta_1 = \frac{2\pi f}{c_1}\cos\theta_1 = \frac{2\pi f}{s_1} \qquad & s_1 &\equiv \frac{c_1}{\cos\theta_1} \\ b_2 &= k_2 \cos\theta_2 = \frac{2\pi f}{s_2} & s_2 &\equiv \frac{c_2}{\cos\theta_2} \\ b_3 &= k_3 \cos\theta_3 = \frac{2\pi f}{s_3} & s_3 &\equiv \frac{c_3}{\cos\theta_3} \\ a &= k_1 \sin\theta_1 = \frac{2\pi f}{c_1}\sin\theta_1 = \frac{2\pi f}{c_2}\sin\theta_2 = \frac{2\pi f}{c_3}\sin\theta_3 \end{aligned} \tag{5-183}$$

Substituting (5-182) and (5-183) into (5-164) we obtain

$$\begin{aligned} -\rho_1 B_1 + \rho_2 A_2 + \rho_2 B_2 &= \rho_1 A_1 \\ s_2 B_1 + s_1 A_2 - s_1 B_2 &= s_2 A_1 \\ \rho_2 e^{-ib_2 h} A_2 + \rho_2 e^{ib_2 h} B_2 - \rho_3 e^{-ib_3 h} A_3 &= 0 \\ s_3 e^{-ib_2 h} A_2 - s_3 e^{ib_2 h} B_2 - s_2 e^{-ib_3 h} A_3 &= 0 \end{aligned} \tag{5-184}$$

Equations (5-184) are of the same form as (5-167) where c has been replaced by s and k by b. We may then write down our solutions immediately from (5-168) and (5-173).

$$\frac{B_1}{A_1} = \left[\frac{(-1 + m_1m_2)^2 \cos^2 (b_2h) + (m_1 - m_2)^2 \sin^2 (b_2h)}{(1 + m_1m_2)^2 \cos^2 (b_2h) + (m_1 + m_2)^2 \sin^2 (b_2h)}\right]^{1/2} e^{i(\delta-\epsilon)} \tag{5-185}$$

and

$$\frac{A_3}{A_1} = \frac{2s_3e^{i(b_3h-\epsilon)}}{s_1[(1 + m_1m_2)^2 \cos^2 (b_2h) + (m_1 + m_2)^2 \sin^2 (b_2h)]^{1/2}} \tag{5-186}$$

where m_1 and m_2 are now defined as

$$\begin{aligned} m_1 &\equiv \frac{\rho_2s_2}{\rho_1s_1} = \frac{\rho_2c_2}{\rho_1c_1}\frac{\cos\theta_1}{\cos\theta_2} \\ m_2 &\equiv \frac{\rho_3s_3}{\rho_2s_2} = \frac{\rho_3c_3}{\rho_2c_2}\frac{\cos\theta_2}{\cos\theta_3} \end{aligned} \tag{5-187}$$

The expressions (5-184) and (5-185) are slightly more difficult to interpret than the corresponding expressions for normal incidence. The argument of the sine and cosine terms is a function of the frequency and the incident angle. The amplitudes of these terms are not simply constants with respect to the three media but functions of the incident angle as well.

If the third medium is the same as the first, i.e., the second medium a plate within the first, m_1m_2 will equal unity. The reflected wave will be zero for values of $\sin^2 (b_2h)$ given by

$$\begin{aligned} \sin^2 (b_2h) &= 0 \\ \frac{2\pi fh}{c_2} \cos\theta_2 &= n\pi \\ f\cos\theta_2 &= \frac{c_2}{2h}\, n \\ f\left[1 - \left(\frac{c_2}{c_1}\right)^2 \sin^2\theta_1\right]^{1/2} &= 0, \frac{c_2}{2h}, \frac{2c_2}{2h}, \frac{3c_2}{2h}, \ \ldots \end{aligned} \tag{5-188}$$

For any given angle of incidence up to the critical angle, there will be a set of discrete frequencies given by (5-188) for which the amplitude of the reflected wave is zero. For increasing angles of incidence from normal incidence, the set of frequencies of zero reflectivity will migrate toward higher frequencies. Similarly, the reflected wave will be a maximum for values of $\sin^2 (b_2h)$ equal to unity. Another condition for no reflection in this case is met if

$$\begin{aligned} m_1 = m_2 &= 1 \\ \frac{\rho_2c_2\cos\theta_1}{\rho_1c_1\cos\theta_2} &= 1 \end{aligned} \tag{5-189}$$

For those angles satisfying (5-189) there will be no reflection at all frequencies. Normally if we have the conditions

$$\rho_2 c_2 > \rho_1 c_1$$

we shall also have

$$c_2 > c_1$$

so that

$$\frac{1}{\cos \theta_2} > \frac{1}{\cos \theta_1}$$

The condition (5-189) then cannot be met for any angle of incidence. The only way in which a solution to (5-189) will exist is for the acoustic resistance of one medium to be greater than that for the other and the sound velocity of the second to be greater than that for the first.

In the general case, Eqs. (5-185) and (5-186) will show a series of maxima and minima as a function of the frequency at any particular angle of incidence. The frequencies and amplitudes of the maxima and minima will change with the angle of incidence. For an expression of the form (5-185) the maxima and minima as a function of frequency at any particular angle of incidence will be given by

$$\frac{d}{dx}\left[\frac{\alpha \cos^2 x + \beta \sin^2 x}{\gamma \cos^2 x + \delta \sin^2 x}\right]$$

$$= \frac{(\gamma \cos^2 x + \delta \sin^2 x)(-2\alpha \sin x \cos x + 2\beta \sin x \cos x) - (\alpha \cos^2 x + \delta \sin^2 x)(-2\gamma \sin x \cos x + 2\delta \sin x \cos x)}{(\gamma \cos^2 x + \delta \sin^2 x)^2}$$

$$= \frac{\sin x \cos x[(-2\alpha + 2\beta)(\gamma \cos^2 x + \delta \sin^2 x) - (2\gamma + 2\delta)(\alpha \cos^2 x + \beta \sin^2 x)]}{(\gamma \cos^2 x + \delta \sin^2 x)^2} = 0$$

The above expression will be zero if

$$\sin x = 0 \qquad \text{or} \qquad \cos x = 0$$

Depending on the absolute value of the relative magnitudes of α/γ and β/δ one set will give the maxima and the other the minima. Let us examine these ratios. From (5-185) and (5-187)

$$\left|\frac{\beta}{\delta}\right| = \left|\frac{m_1 - m_2}{m_1 + m_2}\right| = \left|\frac{\dfrac{r_2}{r_1} - \dfrac{r_3}{r_2}}{\dfrac{r_2}{r_1} + \dfrac{r_3}{r_2}}\right| = \left|\frac{r_2{}^2 - r_1 r_3}{r_2{}^2 + r_1 r_3}\right| \tag{5-190}$$

$$\left|\frac{\alpha}{\gamma}\right| = \left|\frac{-1 + m_1 m_2}{1 + m_1 m_2}\right| = \left|\frac{-1 + \dfrac{r_2}{r_1}\cdot\dfrac{r_3}{r_2}}{1 + \dfrac{r_2}{r_1}\cdot\dfrac{r_3}{r_2}}\right| = \left|\frac{r_3 - r_1}{r_3 + r_1}\right| = \left|\frac{r_3{}^2 - r_1 r_3}{r_3{}^2 + r_1 r_3}\right|$$

$$= \left|\frac{r_1 - r_3}{r_1 + r_3}\right| = \left|\frac{r_1{}^2 - r_1 r_3}{r_1{}^2 + r_1 r_3}\right| \tag{5-191}$$

where the substitutions

$$r_1 = \frac{\rho_1 c_1}{\cos \theta_1}, \; r_2 = \frac{\rho_2 c_2}{\cos \theta_2}, \; r_3 = \frac{\rho_3 c_3}{\cos \theta_3} \tag{5-192}$$

have been made. For (5-191) to be greater than (5-190) ($\cos x = 0$, minima) we must have either

$$r_3 > r_2 > r_1 \tag{5-193}$$

or

$$r_1 > r_2 > r_3 \tag{5-194}$$

For all other combinations (5-190) will be greater than (5-191) ($\sin x = 0$, minima). In general if we have

$$\rho_3 c_3 > \rho_2 c_2 > \rho_1 c_1$$

we also have

$$c_3 > c_2 > c_1$$

so that from Snell's law

$$\frac{1}{\cos \theta_3} > \frac{1}{\cos \theta_2} > \frac{1}{\cos \theta_1}$$

For this case we may be assured that

$$r_3 > r_2 > r_1$$

for all angles of incidence. If the velocity inequalities are not in the same direction as the impedance inequalities, it is possible for the conditions of

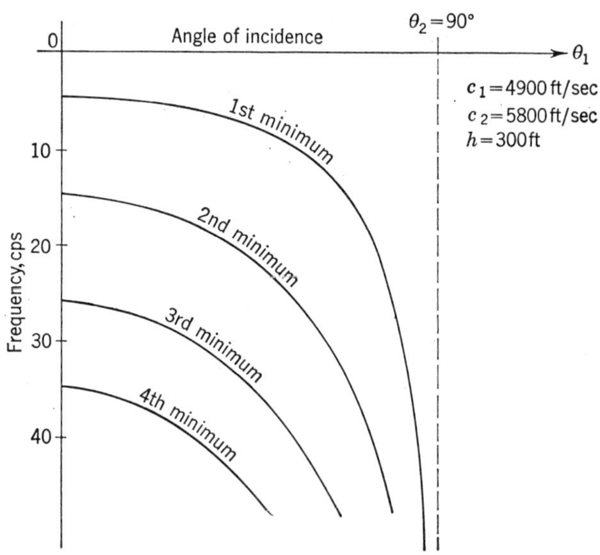

FIG. 5-21.

maxima to become conditions of minima as a function of the angle of incidence. For the conditions (5-193) or (5-194), which are the ones usually encountered in nature, the reflection minima will occur at

$$\frac{2\pi fh}{c_2}\cos\theta_2 = (2n+1)\frac{\pi}{2}$$

$$f\cos\theta_2 = \frac{c_2}{4h}(2n+1)$$

$$f\left[1-\left(\frac{c_2}{c_1}\right)^2\sin^2\theta_1\right]^{1/2} = \frac{c_2}{4h}, \frac{3c_2}{4h}, \frac{5c_2}{4h}, \ldots \qquad (5\text{-}195)$$

Figure 5-21 is a graph of Eq. (5-195) for particular values of c_1, c_2, and h.

Horizontal coupling to a wave guide

Let us examine briefly the problem shown in Fig. 5-22. This resembles the normal-mode problem of Chap. 2. The free surface has been replaced by an upper medium of density ρ_0 and sound velocity c_0, and the source is located in the upper medium. We are particularly interested in transmission under the conditions that $c_2 > c_0 > c_1$.

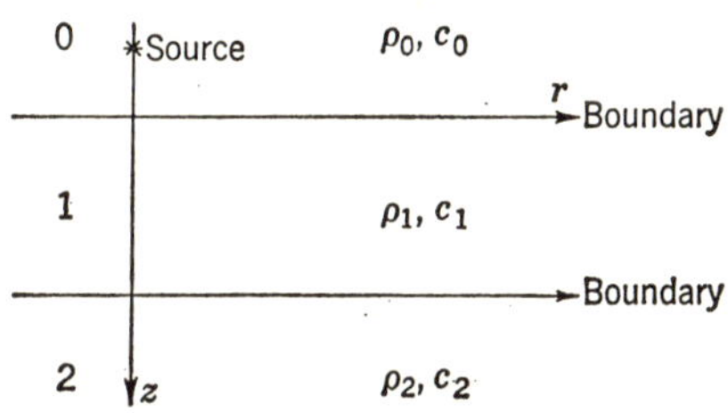

FIG. 5-22.

The formal solution is similar to that given in Chap. 2. There will be a normal-mode solution with phase velocities as measured in the r direction extending from $V = \infty$ to $V = c_1$. In terms of ray directions the phase velocity is given by $V = \sin\theta/c_1$ where θ is the angle the ray makes with the z direction. For $V = \infty$, the propagation is vertical; for $V = c_1$, the propagation is horizontal and confined to the wave guide, layer 1. For $V > c_2$, the wave-guide propagation is attenuated with range; energy is lost from the wave guide by partial refraction to both the upper and lower media. The velocity $V = c_2$ corresponds to critical angle incidence with the lower medium. For $c_2 > V > c_0$, the wave-guide propagation is attenuated with range because of partial refraction of energy into the upper medium only. The velocity $V = c_0$ corresponds to critical angle incidence with the upper medium. For $c_0 > V > c_1$, the wave-guide propagation is unattenuated. The excitation function for each mode and for the various parts of each mode will be different for the source in the upper medium than for the source in the wave guide. In particular the excitation of the portion of modes corresponding to $c_0 > V > c_1$ will be considerably less and will decrease exponentially with distance of the source away from the upper boundary. As before there will be a branch-line integral solution for $V = c_2$, corresponding at long ranges to the refraction arrival for the lower medium.

In addition there will be a third solution for wave-guide transmission. At long ranges and in the vicinity of the upper boundary the disturbance in the upper medium may be considered to be propagating parallel to the boundary with the velocity c_0. The third solution is that for the coupling of the horizontal transmission in the upper medium to the wave guide. It corresponds in a reciprocal manner to the refraction arrival.

Consider a disturbance moving along the upper boundary with the velocity c_0. It will set up successive wave trains in the underlying wave guide. Those portions of the interfering wave trains will predominate which are in phase with the moving disturbance. The dominant contributions will be those for which there is always constructive interference with the moving disturbance and reinforcement to the resultant wave-guide transmission. This condition defines a set of discrete frequencies whose phase velocity is equal to the velocity of the upper medium, $V = c_0$. The wave-guide transmission will be a set of constant frequency waves. If the velocity c_0 of the upper medium is less than the velocity c_1 of the layer, the condition $V = c_0$ cannot be met; and there would be no coupling.

Interpretation of echo sounding records

Two important and in principle straightforward uses of transmitted sound in the earth sciences are echo sounding measurements and geophysical reflection and refraction profiles. Echo sounding measurements are used to determine the configuration of the bottom in oceanic areas. Geophysical reflection and refraction profiles are used to determine subsurface geologic structure under continental or oceanic areas. The principles and methods of interpretation involved in the first will be discussed in this section and those of the latter in the next section.

The standard method for the determination of marine topography is echo sounding. A piezoelectric or magnetostriction transducer is located in the hull of the echo sounding ship or towed over the side on a strut, wire, or chain at a modest depth. The transducer converts electrical energy to mechanical (sound) energy and vice versa. A high-frequency ping of short duration is emitted by the transducer. The pulse travels outward from the source. A portion is reflected from the bottom and received back at the transducer, as illustrated in Fig. 5-22′. The received signal is suitably amplified and recorded. The record is made by a stylus or other marking device which moves across the recording paper at a constant rate. It makes a mark on the paper when a signal is received. The recording stylus and outgoing ping are controlled together in such a manner that the outgoing ping always occurs at the same place on the record, usually near the top. The echo time can then be read directly and conveniently from the record. The recording paper is generally graduated in horizontal time lines to aid in reading. The outgoing pings are repeated at a high repetition rate, and the recording paper progresses out horizontally so that a nearly continuous record of the bottom echoes is obtained for the bottom over which the echo sounding ship has moved. An echo sounding record is shown schematically in Fig. 5-23.

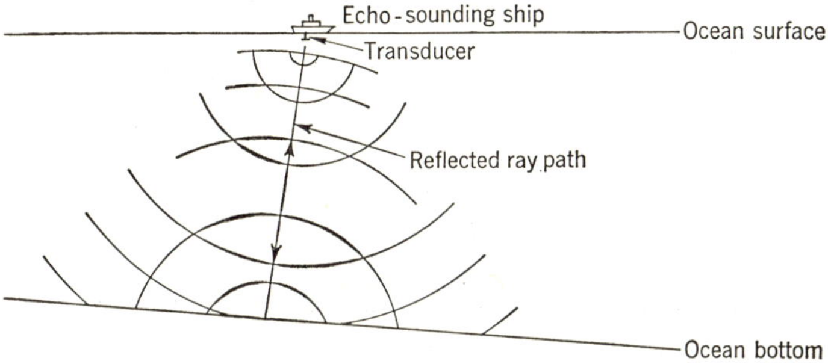

FIG. 5-22′.

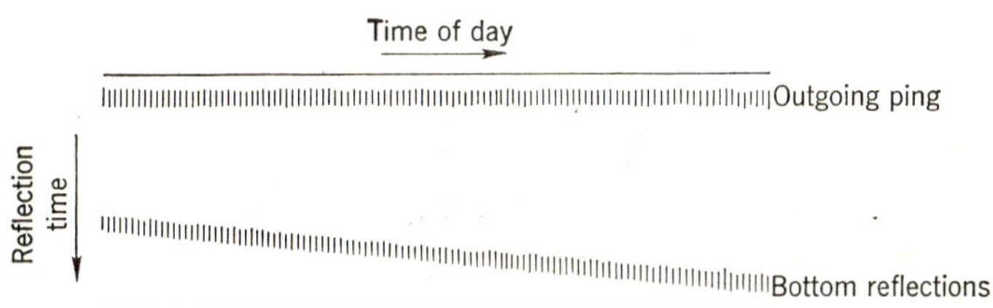

FIG. 5-23.

If the average sound velocity $\bar{c}$ from the ocean surface to the reflector is known, the distance to the reflector H can be determined by the simple formula

$$H = \bar{c}\,\frac{\Delta t}{2} \tag{5-196}$$

where Δt is the time interval between the outgoing ping and the reception of the bottom reflection. The velocity $\bar{c}$ is not the velocity given by

$$\bar{c} = \frac{1}{H}\int_0^H c(h)\,dh$$

It is the velocity which when multiplied by the time $\Delta t/2$ will give the distance to the reflector. It is a time average, not a straight velocity average, and is given by the relation

$$\frac{\Delta t}{2} = \frac{H}{\bar{c}} = \int_0^H \frac{dh}{c(h)}$$
$$\bar{c} = \frac{H}{\int_0^H [dh/c(h)]} \qquad (5\text{-}197)$$

Fortunately the sound velocity in the ocean is a sufficiently slowly varying function that an average velocity may be defined over various depth ranges and areas of the ocean. Tables of such values are available for computations.

It is important to emphasize that H is the distance to the reflector and not necessarily the depth under the echo sounding ship. Only on a flat

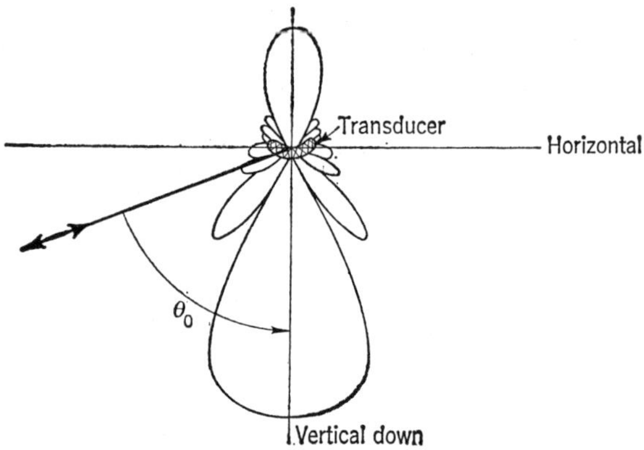

FIG. 5-24.

bottom will H also be the depth. As shown in Fig. 5-22′, the reflection occurs at an angle to the vertical; the ray path distance is shorter than the vertical depth under the ship.

In most cases the echo path will be the specular reflection from the bottom, i.e., a ray path. In some cases, especially in the vicinity of steep slopes, this may not be the case. The transducer itself has a directivity pattern for transmitting and receiving. It transmits and receives best in a near vertical direction, and its characteristics will decrease with angles away from the vertical. These characteristics may resemble that shown in the polar plot of Fig. 5-24. For a specular reflection from a steep slope at an angle θ_0 as shown in Fig. 5-24 the transmitted and received signal may be too weak to give a detectable signal on the record. Then only the scattered reflections from steeper angles will be observed. It is often possible to discern this situation when it occurs: a specular reflection will give a fairly good reproduction of the outgoing ping. The scattered reflections will be more diffuse and more prolonged. An illustration of such a case is given in Fig. 5-25. It is possible to reach a limit in which no detectable signals are received from a sufficiently precipitous slope. The scattered reflections are of limited use. Their incidence and duration are dependent on a number of things such as the transducer characteristics, gain of the receiving system, inclination and irregularity of the reflecting slope, reflection and scattering properties of the bottom and subbottom materials, and background noise, all of which are difficult to include in a quantitative form.

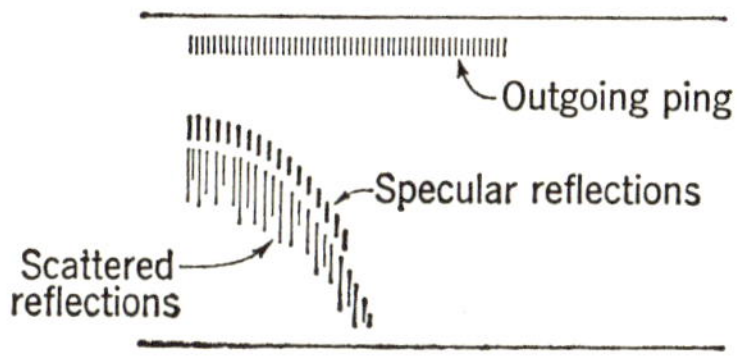

FIG. 5-25.

It should be obvious that in an area of varying topography it may be possible to obtain more than one specular reflection for a given position of the echo sounding ship. An example of an echo sounding traverse across a submarine canyon with a flat bottom is shown in Figs. 5-26 and 5-27. Problems of focusing and defocusing due to bottom curvature are also important. A knob or highlight on the ocean bottom will be received at diminished intensity and over an extended range. A depression in the ocean bottom will be detected over a limited range and at increased intensity.

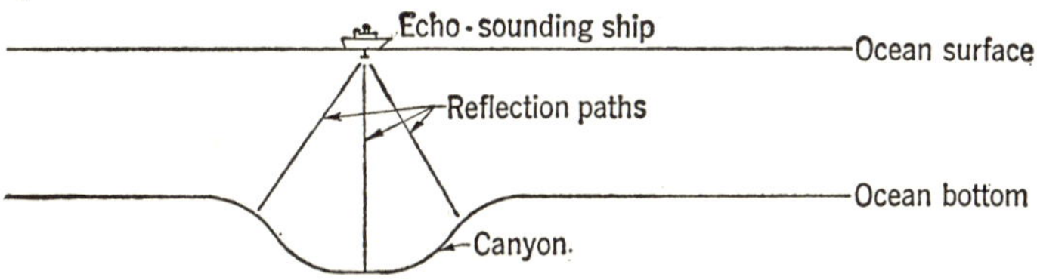

FIG. 5-26.

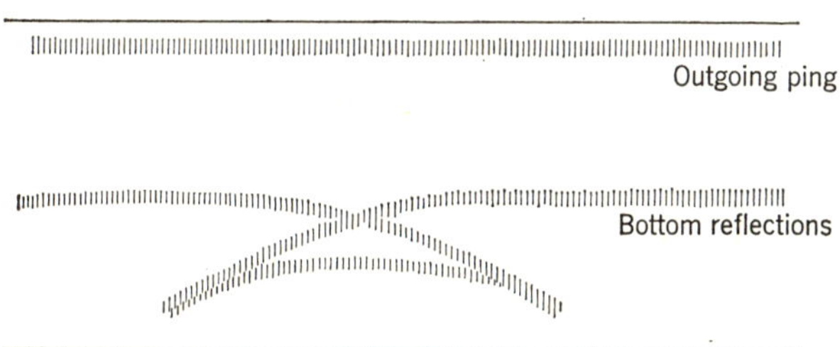

FIG. 5-27.

The final result that is desired from a series of echo sounding traverses is generally a map of the ocean bottom. The distances H determined from Eq. (5-194) are the depths of water under the echo sounding ship only in the limit of a flat bottom. Over much of the ocean bottom the inclinations are quite modest so that a map of H will be a reasonable approximation to a map of the bottom. For those areas where the bottom inclinations are steep or in general where a more accurate map is desired a convenient method to use is the following. Take an echo sounding traverse or a line from a map of H which is approximately normal to the bottom contours. The bottom reflections from this traverse or line will then be in vertical plane including the traverse line. For selected points along this line draw circular arcs in the vertical plane with radii H and centers at the ship's location for that recording of H. The envelope of the circular arcs will then outline the bottom topography. This is true only for specular reflection paths. From a number of such lines a corrected map can be drawn. For example, if in Fig. 5-26 we presume that the ship's track is normal to the contours of the canyon slopes and that the canyon floor and ocean bottom adjacent to the canyon are nearly flat, we may make such a construction for the record of Fig. 5-27. Figure 5-28 is illustrative of the result that would be obtained.

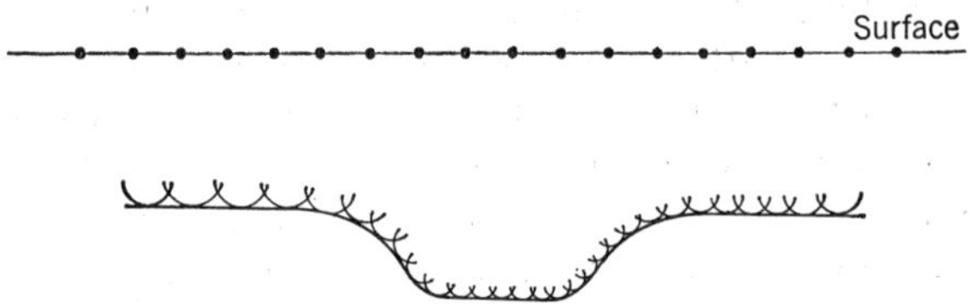

Fig. 5-28.

It should be mentioned that almost without exception all the bathymetric charts that have been published are not true maps of the submarine topography. First, most maps are based on an average sound velocity $\bar{c}$ of 4,800 ft/sec. Most echo sounder records have depth lines for convenience in reading the record which are also based on $\bar{c}$ equal to 4,800 ft/sec. Second, most maps do not include angularity corrections such as that shown in Fig. 5-28 for the reduction of the record of Fig. 5-27. One of the primary reasons why the charts are made in this way is that the sparsity of good echo sounding records and often the lack of good navigational control of the echo sounding ship do not warrant even such gross refinements as the above. To this should be added the statement that, if the primary purpose of the chart is an aid to navigation, it would not be desirable to make these corrections. For such a purpose the chart should duplicate what a mariner would read from his own echo sounding record as he crossed the area.

Interpretation of geophysical reflection and refraction profiles

Two of the standard methods of outlining subsurface geologic structure are the geophysical reflection and refraction profiles. They are based on the two arrivals, the reflection arrival and the refraction arrival, that can be returned from a subsurface change in acoustic properties (Fig. 5-4).

Both methods have received considerable attention in the exploration for subsurface structures related to oil and gas accumulations. There is considerable variation in the actual procedures, methods, and equipment that have been used for various problems and in various areas. In principle they all reduce to the following. First, a controlled source of acoustic energy is needed. The frequencies that give usable information extend from about 5 up to 500 cps. Generally a chemical explosive, such as dynamite, is used as the source. Sometimes various mechanical or other types of sources are used, and sometimes multiple explosive sources are used. The explosive charges are generally detonated at a modest depth in the ground in a hole drilled for that purpose. Second, a detector for the detection of the subsurface reflection or refraction arrivals is needed. The detector may be sensitive to displacement u, velocity $\dot{u}$, or pressure p. In general a number of detectors are used for the reception of the signals from a given source initiation. The detectors may be arranged in various patterns on the ground or in a deep hole. Sometimes the received signals from a number of closely spaced detectors are combined to give one piece of information. This tends to accentuate arrivals incident at near vertical angles against those incident at more grazing angles. The sum from the set of closely spaced detectors will always be additive, constructive interference, for near vertical incidence. The arrivals from a given source to a set of detectors at various ranges or depths constitute a reflection or refraction profile. In some applications the procedure is reversed; a single detector with a series of separate sources initiated individually is used to make a profile. Third, a recording system for the signals received by the detectors is needed. The system generally includes a set of electronic amplifiers, one or more for each detector or set of closely spaced detectors. The amplifiers often contain a controlled means for frequency filtering and automatic gain control. The amplified signals are recorded either directly on a record which can be examined visually or on magnetic tape or other storage media for later reduction to a visual record or computer. Sometimes corrections such as that for the depth of the source, variations in the elevations of the detectors, or variations in the near-surface velocities or layer thicknesses are applied to the received information by various electronic or mechanical means before the visual record is made. Sometimes other functional operations are also applied to the received information before the final visual record is made. One is to apply to the reflection times the velocity as a function of depth so that the final record will be a function of depth rather than time for near vertical reflections. Another is to apply separately to each detector on a reflection profile a function which will make a reflection from a horizontal surface at a particular depth appear with a horizontal time correlation from record to record at succeeding ranges rather than with a curved time correlation (hyperbolic for a constant velocity) on the normal record. All these corrections or operations are applied as an aid to the interpretation of the final visual record. A variety of final visual record presentations are also used. All have time, corrected time, or depth along one axis and distance from source to detector on the other axis. Usually the information from a single profile is presented. In some applications the information from a single detector or from a series of profiles is presented. The records may show the changes in the amplitudes of the received signals to each detector by variations in a wiggly line about a zero position or by differences in darkness of a rectified signal from zero. One of the more common systems is to record the movements of the light spots reflected from the mirrors of a series of galvanometers on a

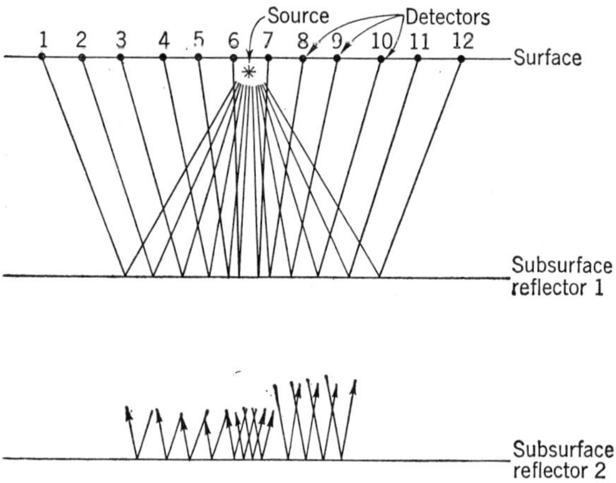

FIG. 5-29.

moving photosensitive paper. This gives a wiggly trace record. In normal operations each trace will correspond to the signals received by detectors at successive distances away from the source. Figure 5-29 is a diagram of a common arrangement of source and detectors for a reflection profile. Figure 5-30 is illustrative of the galvanometer camera record that might be obtained for such an arrangement.

The geometry of source and detectors is somewhat different for reflection and refraction profiles. Reflections from a subsurface horizon can be received from normal incidence out to the critical angle for a previous overlying high velocity contrast. Refractions from a subsurface horizon will not be received until the range at which critical angle incidence for that horizon has been reached. In general then the detectors will be closer to the source and the total distance of the detector spread will be less for a reflection profile than that for a refraction profile of the same problem. In some cases where detail on a particular subsurface horizon is desired a short detector spread at distance from the source may be used in a refraction profile. These geometrical relations are illustrated in Fig. 5-31 for the simple case of two subsurface horizons.

The problem in interpreting geophysical reflection or refraction profile records amounts essentially to transforming the observed arrival times and shot to detector distances to the desired velocity of the transmitting

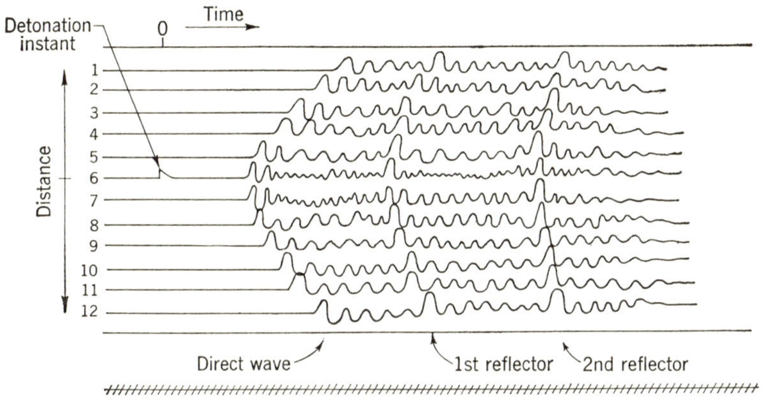

FIG. 5-30.

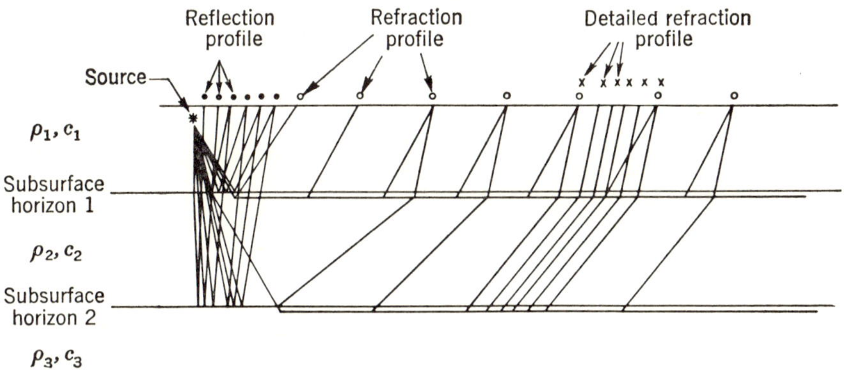

FIG. 5-31.

medium and the depths of the subsurface horizons. We might represent this by the operation

$$f(t, x) \rightarrow g(c, h) \tag{5-198}$$

It is not always possible from a known $f(t, x)$ to obtain a unique $g(c, h)$. Generally, however, the most difficult part of the interpretation is the recognition and correlation of distinct events on reflection profiles or in the case of refraction profiles the determination of the beginnings of the refraction events. We shall proceed to derive in the following paragraphs the relations (5-198) for a few simple cases.

Consider a single horizontal reflector at a depth h with a constant velocity medium above it (Fig. 5-32). The equation for the reflection times as a function of range will be simply

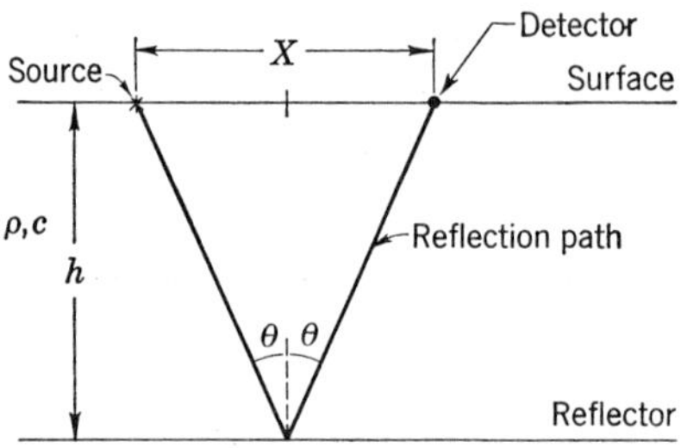

Fig. 5-32.

$$T = \frac{2[(X/2)^2 + h^2]^{1/2}}{c}$$
$$T^2 = \frac{X^2}{c^2} + \frac{4h^2}{c^2} \tag{5-199}$$

The equation for the reflection time versus range is a hyperbola. If, as is the usual case, the velocity c is a function of depth $c(z)$, the ray paths will be curved and the simple relation (5-199) for the reflection times will not hold. However, if the detectors are near the source, the reflected ray incidences will be nearly vertical; and the ray paths may be approximated by straight lines. Equation (5-199) may be used where $\bar{c}$, defined by the relation (5-197), is substituted for c. From Eq. (5-199) we see that for a given increment of distance away from the source the reflection time will increase by a corresponding increment in time. For reflections from successively deeper horizons the increment in reflection time will be less for

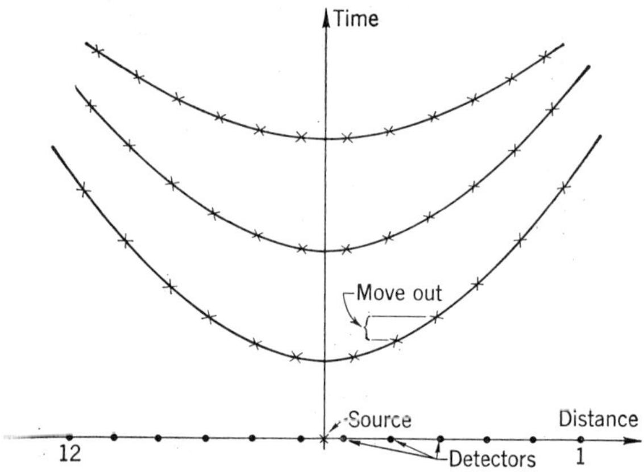

Fig. 5-33.

the same increment in distance. The progressive increase of reflection time with distance for a reflection from a given horizon is referred to as *move-out*. If the horizon is horizontal, the increase is referred to as normal move-out. These relations are illustrated by the hyperbolas of Fig. 5-33.

Consider a single plane reflector inclined at an angle ω to the horizontal surface (Fig. 5-34). The equation for the reflection time as a function of range can be derived as follows.

$$T = \frac{(a^2 + b^2)^{1/2}}{c}$$
$$= \frac{1}{c}[(X - 2h \sin \omega)^2 + (2h \cos \omega)^2]^{1/2}$$
$$= \frac{1}{c}(X^2 + 4h^2 - 4hX \sin \omega)^{1/2} \qquad (5\text{-}200)$$

This formula may be used in a variable velocity medium for near vertical incidence where $\bar{c}$ is substituted for c. There are three quantities to be determined in this problem, $\bar{c}$, h, and ω. A value of $\bar{c}$ can be estimated

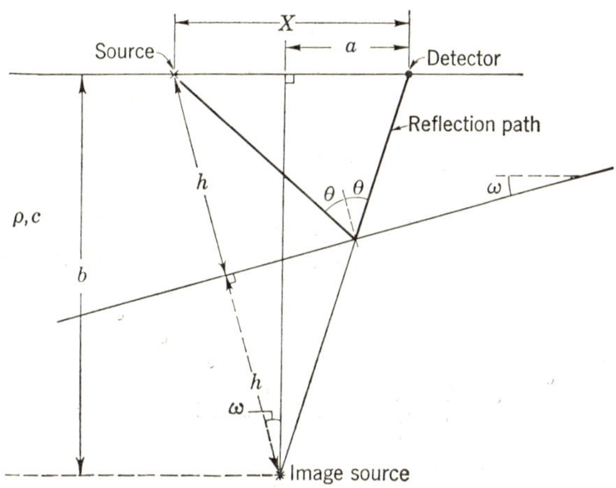

FIG. 5-34.

from the observed reflection times in a manner such as that shown in Fig. 3-9. However, usually the reflection profiles are not carried out to a sufficiently long range to allow a good value of $\bar{c}$ to be determined. A general method that is used is to determine $c(z)$ incrementally from measurements in a neighboring deep hole or by an interpolation from measurements in a series of neighboring deep holes to the profile location. Even when this method is available, inaccuracies can ensue from variations of $c(z)$ within the interpolation grid. With a value of $\bar{c}$ the quantities h and ω can be determined. For a detector spread as shown in Fig. 5-33 let T_0 be the extrapolated time for a particular reflection at zero distance. Then from Eq. (5-200)

$$h = \frac{cT_0}{2} \qquad (5\text{-}201)$$

Let T_1 and T_{12} be the reflection times to the most distant detectors, presumed to be equidistant on each side of the source. Then from Eq. (5-200)

$$T_{12} = \frac{1}{c}(X_1^2 + 4h^2 + 4hX_1 \sin \omega)^{1/2}$$

$$T_1 = \frac{1}{c}(X_1^2 + 4h^2 - 4hX_1 \sin \omega)^{1/2}$$

Squaring and subtracting we obtain

$$c^2(T_{12}^2 - T_1^2) = 8hX_1 \sin \omega$$

$$\sin \omega = \frac{c^2(T_{12} + T_1)(T_{12} - T_1)}{8hX_1}$$

$$\approx \frac{c(T_{12} - T_1)}{2X_1} \tag{5-202}$$

since

$$\frac{T_{12} + T_1}{2} \approx T_0$$

The angle of inclination ω of Fig. 5-34 is measured in the plane of the reflections, i.e., the plane containing the line of the source and receivers

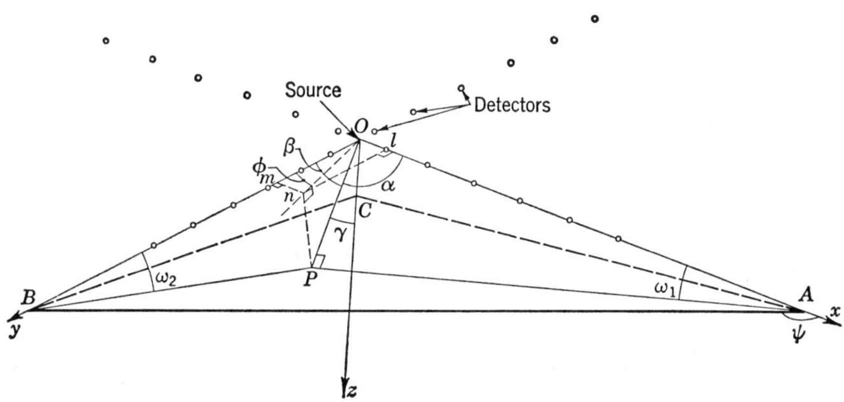

FIG. 5-35.

normal to the reflector. In order to determine the spatial relations of the reflecting surface it is necessary to make two profiles at an angle to each other. A convenient arrangement is two profiles at right angles to each other as shown in Fig. 5-35. The xy plane is the horizontal surface, and z axis is vertical down. The plane ABC is the plane of the reflector. The line OP is normal to the plane ABC. The plane OPA is the plane of the reflections for the profile along the x axis, and the angle ω_1 is the inclination angle measured in this plane. The plane OPB is the plane of the reflections for the profile along the y axis, and the angle ω_2 is the inclination angle measured in this plane. The geometry of a subsurface geologic horizon is usually designated by its dip and strike. The strike is the line of intersection of the inclined plane with the horizontal. It is designated by its bearing with a prescribed direction, generally north. The dip is the angle between the inclined plane and the horizontal measured in a vertical plane normal to the strike. In Fig. 5-35 the angle γ is the dip; and if we presume Ox to be a north direction, the angle ψ is the strike bearing. From triangles OPA and OPB we have

$$\begin{aligned} \sin \omega_1 &= \cos \alpha \\ \sin \omega_2 &= \cos \beta \end{aligned}$$

Since α, β, γ are the directions angles of the line OP, we have

$$\begin{aligned} \cos^2 \alpha + \cos^2 \beta + \cos^2 \gamma &= 1 \\ \sin^2 \gamma = 1 - \cos^2 \gamma &= \cos^2 \alpha + \cos^2 \beta \\ \sin \gamma &= (\sin^2 \omega_1 + \sin^2 \omega_2)^{1/2} \end{aligned} \tag{5-203}$$

for γ in terms of ω_1 and ω_2. From triangles OlP and OmP we have

$$\begin{aligned} Ol &= OP \cos \alpha \\ Om &= OP \cos \beta \end{aligned}$$

From triangle Omn we have

$$\tan \varphi = \frac{mn}{Om} = \frac{Ol}{Om} = \frac{\cos \alpha}{\cos \beta}$$

Since the line On is normal to the line AB, the angle φ will be the complement of the angle ψ so that

$$\tan \psi = \cot \varphi = \frac{\cos \beta}{\cos \alpha}$$

$$\tan \psi = \frac{\sin \omega_2}{\sin \omega_1} \tag{5-204}$$

for ψ in terms of ω_1 and ω_2. The line OP is the distance h of Eq. (5-201) to the reflector. The projection n of the point P on the horizontal surface is offset from the source by the distance

$$On = h \sin \gamma \tag{5-205}$$

in the direction On.

For the interpretation of a refraction profile we shall begin with the simplest case of a single velocity discontinuity at a depth h (Fig. 5-36).

The equation for the refraction time as a function of range will be

$$
\begin{aligned}
T &= 2\frac{a}{c_1} + \frac{b}{c_2} \\
&= \frac{2h}{c_1 \cos \theta_c} + \frac{X - 2h \tan \theta_c}{c_2} \\
&= \frac{X}{c_2} + \frac{2h}{c_1}\left(\frac{1}{\cos \theta_c} - \frac{\sin^2 \theta_c}{\cos \theta_c}\right) \\
&= \frac{X}{c_2} + \frac{2h}{c_1} \cos \theta_c \\
&= \frac{X}{c_2} + \frac{2h\sqrt{c_2^2 - c_1^2}}{c_1 c_2}
\end{aligned}
\tag{5-206}
$$

since

$$\sin \theta_c = \frac{c_1}{c_2}$$

The equation for the refraction time versus range is a straight line whose inverse slope is the velocity of the lower medium c_2 and whose time

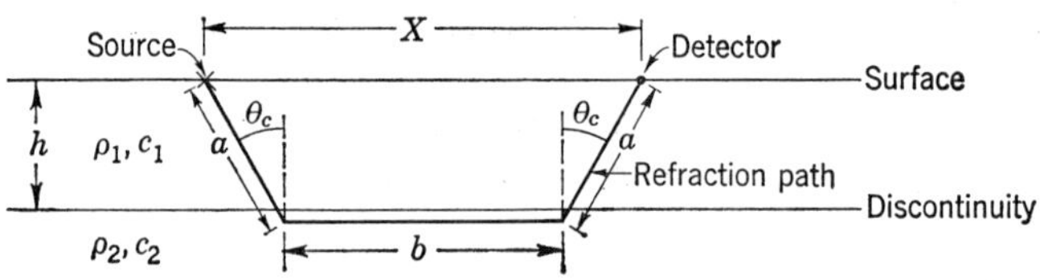

FIG. 5-36.

intercept at $x = 0$ is the second term on the right-hand side of Eq. (5-206). On a refraction profile the velocities are determined directly from the slope of the refraction line. The depth h can be determined conveniently from the extrapolated time intercept T_0 at $x = 0$ by

$$h = \frac{c_1 c_2 T_0}{2\sqrt{c_2^2 - c_1^2}} \tag{5-207}$$

The velocity c_1 needed in the calculation (5-207) can be determined from the direct arrival whose equation is

$$T = \frac{X}{c_1} \tag{5-208}$$

The equation for the direct arrival is also a straight line. Its inverse slope is the velocity c_1 and its time intercept is zero. The refraction arrival will not exist until a range X_c given by

$$X_c = 2h \tan \theta_c \tag{5-209}$$

has been reached. If the velocity c_1 is greater than c_2, there will be no critical angle θ_c to the lower medium; and therefore there will be no refraction arrival.

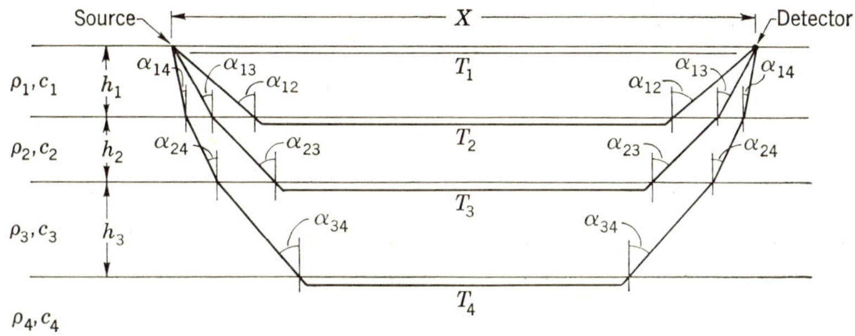

FIG. 5-37.

Next consider a sequence of horizontal layers for which the velocity c_n is greater than the velocity c_{n-1} of the next overlying layer. With reference to Fig. 5-37 the equations will be

$$
\begin{aligned}
T_1 &= \frac{X}{c_1} \\
T_2 &= \frac{X}{c_2} + \frac{2h_1}{c_1}\cos\alpha_{12} \\
T_3 &= \frac{X}{c_3} + \frac{2h_1}{c_1}\cos\alpha_{13} + \frac{2h_2}{c_2}\cos\alpha_{23} \\
T_4 &= \frac{X}{c_4} + \frac{2h_1}{c_1}\cos\alpha_{14} + \frac{2h_2}{c_2}\cos\alpha_{24} + \frac{2h_3}{c_3}\cos\alpha_{34} \\
&\cdots\cdots\cdots\cdots\cdots\cdots\cdots \\
&\cdots\cdots\cdots\cdots\cdots\cdots\cdots \\
T_m &= \frac{X}{c_m} + 2\sum_{n=1}^{m-1}\frac{h_n}{c_n}\cos\alpha_{nm}
\end{aligned}
\tag{5-210}
$$

From Snell's law we have the relations between the angles α,

$$
\begin{aligned}
&\frac{\sin\alpha_{12}}{c_1} = \frac{1}{c_2} \\
&\frac{\sin\alpha_{13}}{c_1} = \frac{\sin\alpha_{23}}{c_2} = \frac{1}{c_3} \\
&\frac{\sin\alpha_{14}}{c_1} = \frac{\sin\alpha_{24}}{c_2} = \frac{\sin\alpha_{34}}{c_3} = \frac{1}{c_4} \\
&\cdots\cdots\cdots\cdots\cdots\cdots\cdots \\
&\cdots\cdots\cdots\cdots\cdots\cdots\cdots \\
&\frac{\sin\alpha_{nm}}{c_n} = \frac{1}{c_m}, \quad n = 1, 2, \ldots, m-1
\end{aligned}
\tag{5-211}
$$

Substituting (5-211) into (5-210) we obtain

$$
\begin{aligned}
T_1 &= \frac{X}{c_1} \\
T_2 &= \frac{X}{c_2} + \frac{2h_1\sqrt{c_2^2 - c_1^2}}{c_1c_2} \\
T_3 &= \frac{X}{c_3} + \frac{2h_1\sqrt{c_3^2 - c_1^2}}{c_1c_3} + \frac{2h_2\sqrt{c_3^2 - c_2^2}}{c_2c_3} \\
T_4 &= \frac{X}{c_4} + \frac{2h_1\sqrt{c_4^2 - c_1^2}}{c_1c_4} + \frac{2h_2\sqrt{c_4^2 - c_2^2}}{c_2c_4} + \frac{2h_3\sqrt{c_4^2 - c_3^2}}{c_3c_4} \\
&\cdots\cdots\cdots\cdots\cdots\cdots\cdots\cdots\cdots\cdots \\
&\cdots\cdots\cdots\cdots\cdots\cdots\cdots\cdots\cdots\cdots \\
T_m &= \frac{X}{c_m} + 2\sum_{n=1}^{m-1} \frac{h_n\sqrt{c_m^2 - c_n^2}}{c_nc_m}
\end{aligned}
\tag{5-212}
$$

The calculations of layer velocities and thicknesses can then be made stepwise. From the T_1 equation we obtain c_1. From the T_2 equation, knowing c_1, we obtain c_2 and h_1. From the T_3 equation, knowing c_1, c_2, and h, we obtain c_3 and h_2, and so on. Figure 5-38 is illustrative of the

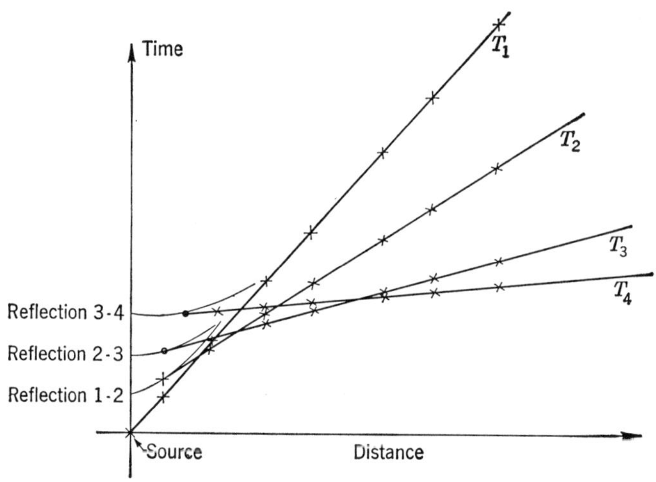

FIG. 5-38.

time-distance plot that would be obtained for the refraction profile of Fig. 5-37. The reflection hyperbolas have been included for reference. For the velocity structure as shown the reflection hyperbolas will cross each other at range. The refraction lines start out tangent to their respective hyperbolas at distances X_c. In Fig. 5-38 T_1 is the first arrival from zero out to a certain range. Then T_2 is the first arrival over the next range interval followed at succeeding range intervals by T_3 and T_4 as first arrivals. This successive progression is not always the case. If a given layer is sufficiently thin, its refraction arrival will never appear as a first arrival. For example, if we diminish h_2, the T_2 line will disappear as a first arrival; T_1 will be a first arrival followed by T_3. When such a condition exists, the arrival from that layer is referred to as a *masked arrival.* In general it is fairly difficult to pick a second refraction arrival behind the first when the time interval is short. At the longer ranges where there is a greater time separation between arrivals second arrivals can be followed more easily. If a masked layer is missed in the interpretation, the depth calculations will be in error by neglect of this velocity change. The error will be dependent on the quantity $h_2(c_2 - c_1)$.

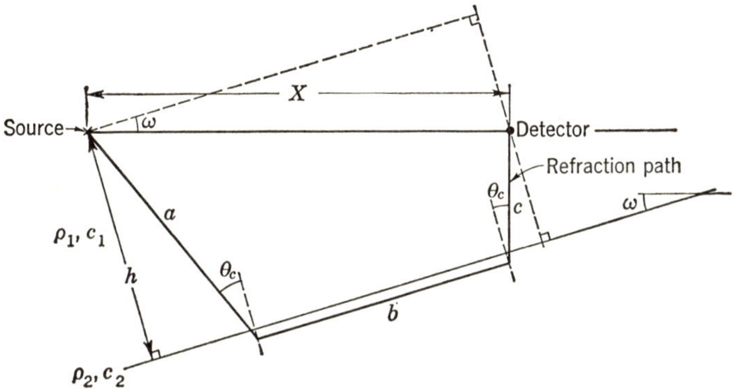

FIG. 5-39.

For a single inclined layer (Fig. 5-39) we shall have the following for the equation of the refraction arrival.

$$\begin{aligned} T &= \frac{a + c}{c_1} + \frac{b}{c_2} \\ &= \frac{2h - X \sin \omega}{c_1 \cos \theta_c} + \frac{X \cos \omega - (2h - X \sin \omega) \tan \theta_c}{c_2} \\ &= \frac{X \cos \omega}{c_2} + \frac{2h - X \sin \omega}{c_1} \cos \theta_c \\ &= \frac{X \sin (\theta_c - \omega)}{c_1} + \frac{2h}{c_1} \cos \theta_c \\ &= \frac{X \sin (\theta_c - \omega)}{c_2 \sin \theta_c} + \frac{2h}{c_1} \cos \theta_c \end{aligned} \tag{5-213}$$

where as before,

$$\sin \theta_c = \frac{c_1}{c_2}$$

The equation for the refraction time versus range for an inclined layer is also a straight line. The time intercept of Eq. (5-213) is the same as that for Eq. (5-206). However, the inverse slope of the refraction line is not the velocity c_2 but c_2 times the ratio of the sine of the angle θ_c to $\theta_c - \omega$. There are three unknowns in Eq. (5-213), c_2, h, and ω. The three unknowns cannot be determined from the two parameters defining a straight line.

This difficulty may be resolved by making a second profile in the direction opposite to that of Fig. 5-39. There are two conventional methods. In one a second set of detectors is used to the left of the source with the same source location for both sets of detectors. This is referred to as an

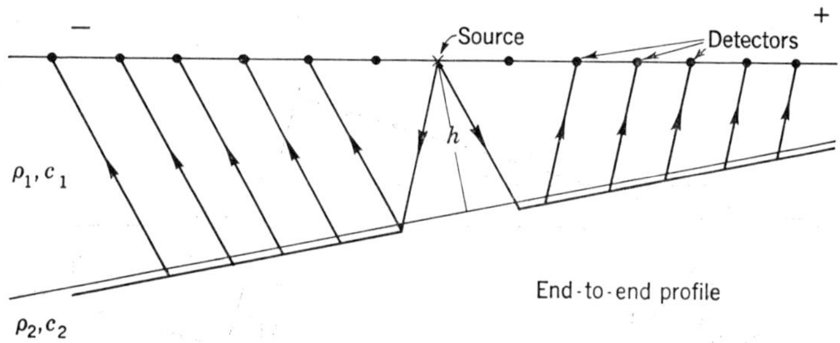

FIG. 5-40. End-to-end profile.

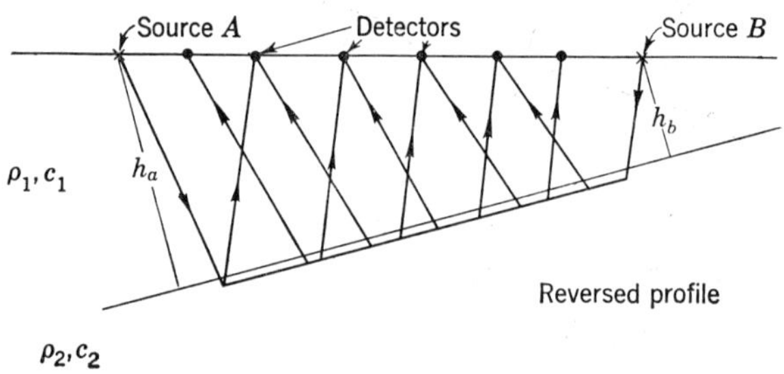

FIG. 5-41. Reversed profile.

end-to-end profile (Fig. 5-40). In the other a second source location at the far end of the original detector spread is used with the same set of detectors. This is referred to as a reversed profile (Fig. 5-41). An end-to-end profile has the disadvantage that each half of the profile measures a different portion of the subsurface layer. The inclination ω may change from one half to the other. Calculations on the assumption of the same ω for both halves would give an error. From (5-213) the equations for an end-to-end profile will be

$$
\begin{aligned}
T^{+} &= \frac{X \sin(\theta_c - \omega)}{c_1} + \frac{2h}{c_1}\cos\theta_c \\
T^{-} &= \frac{X \sin(\theta_c + \omega)}{c_1} + \frac{2h}{c_1}\cos\theta_c
\end{aligned}
\tag{5-214}
$$

The equations for a reversed profile will be

$$T_a = \frac{X \sin(\theta_c - \omega)}{c_1} + \frac{2h_a}{c_1}\cos\theta_c$$
$$T_b = \frac{X \sin(\theta_c + \omega)}{c_1} + \frac{2h_b}{c_1}\cos\theta_c \tag{5-215}$$

where T_a is the refraction time to the detectors from source A and T_b from source B. From Eqs. (5-215), or (5-214), the velocity of the lower medium c_2 and the angle of inclination ω can be determined from the two apparent velocities c_{2a} and c_{2b}.

$$c_{2b} = \frac{c_1}{\sin(\theta_c + \omega)}$$
$$c_{2a} = \frac{c_1}{\sin(\theta_c - \omega)}$$

so that

$$\theta_c = \frac{\sin^{-1}(c_1/c_{2b}) + \sin^{-1}(c_1/c_{2a})}{2}$$
$$\omega = \frac{\sin^{-1}(c_1/c_{2b}) - \sin^{-1}(c_1/c_{2a})}{2} \tag{5-216}$$

and

$$c_2 = \frac{c_1}{\sin\theta_c}$$

If ω is small, the velocity c_2 may be approximated by the following.

$$\sin(\theta_c + \omega) = \sin\theta_c \cos\omega + \cos\theta_c \sin\omega = \frac{c_1}{c_{2b}}$$
$$\sin(\theta_c - \omega) = \sin\theta_c \cos\omega - \cos\theta_c \sin\omega = \frac{c_1}{c_{2a}}$$
$$2\sin\theta_c \cos\omega = c_1 \frac{c_{2b} + c_{2a}}{c_{2b}c_{2a}}$$
$$c_2 = \frac{1}{2\cos\omega}\left(\frac{c_{2a} + c_{2b}}{c_{2a}c_{2b}}\right) \approx \frac{1}{2}\left(\frac{c_{2a} + c_{2b}}{c_{2a}c_{2b}}\right) \tag{5-217}$$

For a detector spread updip from the source the apparent velocity will be greater than the actual velocity of the lower medium. For a detector spread downdip from the source the apparent velocity will be less than the actual velocity. From the value of θ_c calculated from Eqs. (5-216) the depths h_{2a} and h_{2b} can be determined from the time intercepts of Eqs. (5-215).

The last case that we shall consider will be a sequence of inclined layers. An abbreviated derivation of the time-distance relations is given using some of the results of the previous sections on ray theory. The measured apparent velocity for any refraction line is related to the incidence angle of the refracted ray and the velocity of the medium underlying the detectors by the relation

$$c_n = \frac{c_1}{\sin \theta} \tag{5-218}$$

as can be seen from the geometry of Fig. 5-42. From (5-218) and Snell's

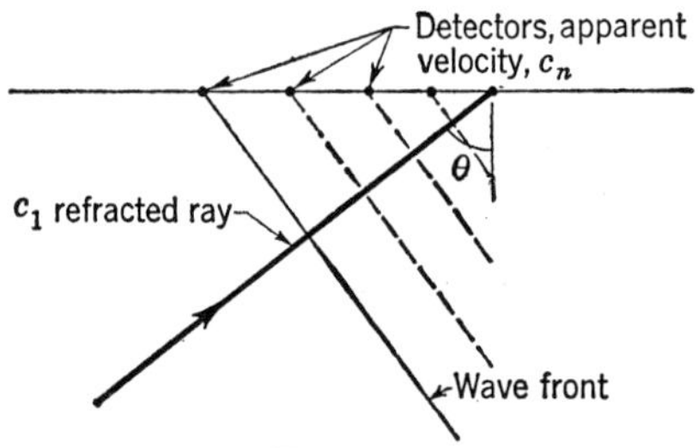

FIG. 5-42.

law we shall then have for the reversed refraction profile paths of Fig. 5-43

$$
\begin{aligned}
&\frac{c_1}{c_{2a}} = \sin (i_{12} - \omega_{12}) \qquad \frac{c_1}{c_{2b}} = \sin (i_{12} + \omega_{12}) \\
&\frac{c_1}{c_2} = \sin i_{12} \\
&\frac{c_1}{c_{3a}} = \sin (\alpha_{13} - \omega_{12}) \qquad \frac{c_1}{c_{3b}} = \sin (\beta_{13} + \omega_{12}) \\
&\frac{c_1}{c_2} = \frac{\sin \alpha_{13}}{\sin (i_{23} - \omega_{23})} \qquad \frac{c_1}{c_2} = \frac{\sin \beta_{13}}{\sin (i_{23} + \omega_{23})} \\
&\frac{c_2}{c_3} = \sin i_{23} \\
&\frac{c_1}{c_{4a}} = \sin (\alpha_{14} - \omega_{12}) \qquad \frac{c_1}{c_{4b}} = \sin (\beta_{14} + \omega_{12}) \\
&\frac{c_1}{c_2} = \frac{\sin \alpha_{14}}{\sin (\alpha_{24} - \omega_{23})} \qquad \frac{c_1}{c_2} = \frac{\sin \beta_{14}}{\sin (\beta_{24} + \omega_{23})} \\
&\frac{c_2}{c_3} = \frac{\sin \alpha_{24}}{\sin (i_{34} - \omega_{34})} \qquad \frac{c_2}{c_3} = \frac{\sin \beta_{24}}{\sin (i_{34} + \omega_{34})} \\
&\frac{c_3}{c_4} = \sin i_{34}
\end{aligned} \tag{5-219}
$$

The velocities c_{na} and c_{nb} are the measured apparent velocities from the refraction lines for the sources A and B, respectively. The inclination angles $\omega_{n-1,n}$ are the inclinations of a given horizon with respect to the next overlying horizon. The actual velocities and the inclinations of the respective media can be calculated stepwise from Eqs. (5-219). From the first set the angle ω_{12} and the velocity c_2 can be determined. From the second set, knowing ω_{12} and c_2, the angle ω_{23} and velocity c_3 can be determined, and so on for succeeding layers. The equations for the intercept times can be obtained from the following consideration. The

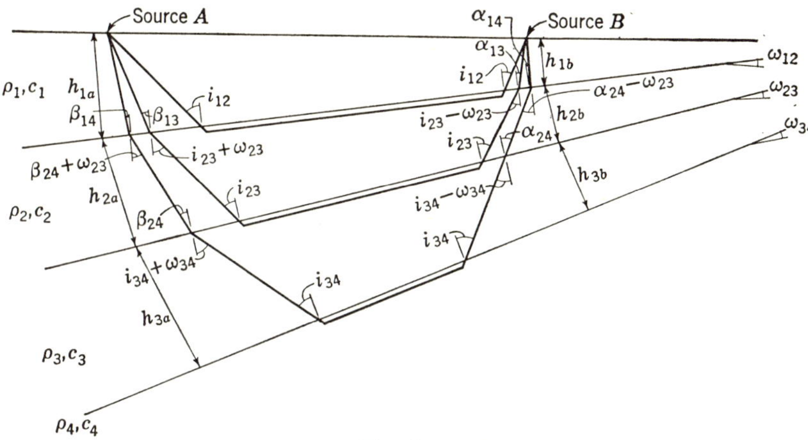

FIG. 5-43.

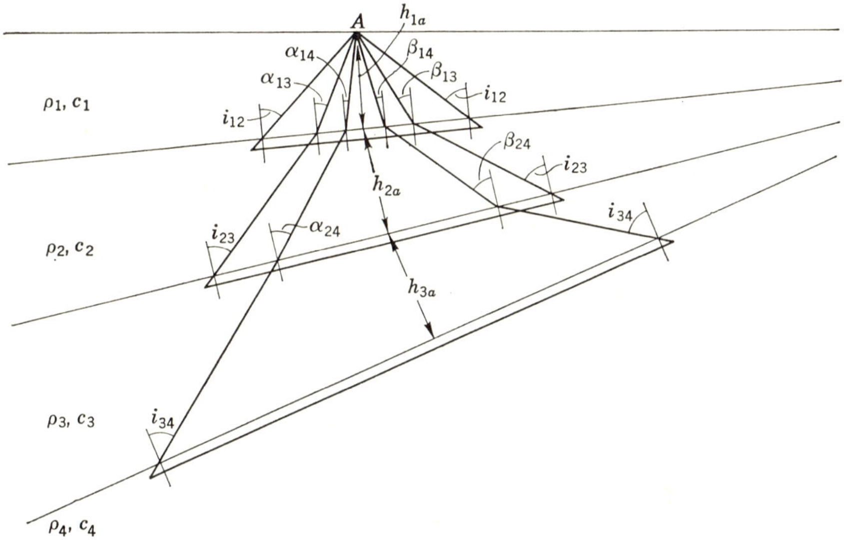

FIG. 5-44.

intercept times are the times extrapolated back from the measured refraction line to $X = 0$. The extrapolated ray paths would be those shown in Fig. 5-44. The times along these paths would be the sum of the times along the inclined paths out from A minus the time along the path parallel to the refraction horizon. For each refraction path these will be

$$\begin{aligned} t_{1a} &= \frac{2h_{1a}}{c_1}\cos i_{12} \\ t_{2a} &= \frac{2h_{2a}}{c_2}\cos i_{23} + \frac{h_{1a}}{c_1}(\cos\alpha_{13} + \cos\beta_{13}) \\ t_{3a} &= \frac{2h_{3a}}{c_3}\cos i_{34} + \frac{h_{2a}}{c_2}(\cos\alpha_{24} + \cos\beta_{24}) + \frac{h_{1a}}{c_1}(\cos\alpha_{14} + \cos\beta_{14}) \end{aligned} \tag{5-220}$$

The times t_{na} are the measured intercept times at A. From the calculations (5-220) the thicknesses h_{na} may be calculated stepwise. From the first equation h_{1a} can be determined. From the second equation, knowing h_{1a}, h_{2a} can be determined, and so on. A similar set of equations exists for point B for the calculations of h_{nb}.

BIBLIOGRAPHY

Bullen, K. E.: "Introduction to the Theory of Seismology," Cambridge University Press, New York, 1947.

Ewing, M., G. P. Woolard, and A. C. Vine: Geophysical Investigations in the Emerged and Submerged Atlantic Coastal Plain, *Bull. Geol. Soc. Amer.*, vol. 50, pp. 257–296, 1939.

Jardetsky, W. S., and F. Press: Rayleigh Wave Coupling to Atmospheric Compression Waves, *Bull. Seismological Soc. Amer.*, vol. 42, pp. 135–144, 1952.

Jeffreys, H.: On Compressional Waves in Two Superposed Layers, *Proc. Cambridge Phil. Soc.*, vol. 23, pp. 472–481, 1926.

Lamb, H.: "Hydrodynamics," Cambridge University Press, New York, 1932, and Dover Publications, New York, 1945.

Morse, P. M.: "Vibration and Sound," 2d ed., McGraw-Hill Book Company, Inc., New York, 1948.

Morse, P. M., and R. H. Bolt: Sound Waves in Rooms, *Revs. Mod. Phys.*, vol. 16, pp. 69–150, 1944.

Muskat, M.: The Theory of Refraction Shooting, *Physics*, vol. 4, pp. 14–28, 1933.

Nettleton, L. L.: "Geophysical Prospecting for Oil," McGraw-Hill Book Company, Inc., New York, 1940.

Officer, C. B.: The Refraction Arrival in Water Covered Areas, *Geophysics*, vol. 18, pp. 805–819, 1953.

Officer, C. B.: A Deep Sea Seismic Reflection Profile, *Geophysics*, vol. 20, pp. 270–282, 1955.

Pekeris, C. L.: Theory of Propagation of Explosive Sound in Shallow Water, *in* Propagation of Sound in the Ocean, *Geol. Soc. Amer., Mem.* 27, 1948.

Stewart, G. W., and R. B. Lindsay: "Acoustics," D. Van Nostrand Company, Inc., Princeton, N.J., 1930.

Strutt, J. W. (Lord Rayleigh): "The Theory of Sound," The Macmillan Company, London, 1894, and Dover Publications, New York, 1945.

Wolf, A.: The Reflection of Elastic Waves from Transition Layers of Variable Velocity, *Geophysics*, vol. 2, pp. 357–363, 1937.

6. *Attenuation*

In the previous chapters we have assumed that there have been no inhomogeneities and no attenuation processes associated with the transmitting medium itself. The primary purpose there has been the description of the transmission phenomena; attenuation effects have been neglected. Here we shall consider the influence of some of these attenuation processes on sound propagation.

The subjects which are discussed in this chapter divide into two groups. The first group is concerned with the *absorption losses* and associated *material dispersion* dependent on the physical properties of a homogeneous fluid medium. The first section deals with propagation in a viscous fluid. The equations of motion and wave equations are derived, and the plane wave viscous absorption coefficient is obtained. In the second section losses due to thermal conduction are considered. The equations of motion are derived, and the plane wave solution is obtained including the thermal conduction coefficient. In the third section propagation in a porous medium of two components is discussed. The velocity of propagation under static conditions is derived, and the losses as a function of frequency are discussed.

The second group is concerned with *scattering* and *diffraction*. The general solution for scattering and diffraction problems is derived in the first section. In the second section the expression for the secondary waves scattered from a small region of anomalous physical properties in a normal medium is derived. The incident primary wave is assumed to be plane. In the third section an approximate solution for diffraction from a plane obstacle is given. In the last section the solution for the reception to a linear array of detectors is obtained. This problem is simply the inverse of that for a simple diffraction grating.

Absorption, viscous fluid

In Chap. 1 we derived the stress-strain relation for a perfect fluid (1-26). We found that for a perfect fluid only hydrostatic stresses could be maintained. A distortional stress component would produce an instantaneous infinite distortion. Equation (1-26) was used to derive the wave equation (1-34) which has been the basis of all the subsequent discussions.

Any real fluid will exhibit some resistance to distortion. This is known as *viscosity* or *internal friction*. It is expressed by a relation, similar to Hooke's law, between stress and rate of change of strain. For infinite time the distortional strain is still infinite, but the rate of change of strain is finite. The hydrostatic strain components remain finite as before. This empirical law gives a good description of the relations for a viscous fluid for small rates of change of strain. The derivation of the stress–rate-of-change-of-strain equations follows the same arguments as those given in Chap. 1 for the stress-strain relations. The resulting equations will be the same as (1-12) or (1-18) where rate of change of strain is substituted for strain. Generally the hydrostatic viscous coefficient, corresponding to k of (1-18), can be neglected; the finite volume changes are assumed to occur instantaneously. The final relations are then, from (1-23),

$$\begin{aligned}
\sigma_{xx} &= -p - \frac{2}{3}\nu\frac{\partial\Delta}{\partial t} + 2\nu\frac{\partial^2 u}{\partial x\,\partial t}\\
\sigma_{yy} &= -p - \frac{2}{3}\nu\frac{\partial\Delta}{\partial t} + 2\nu\frac{\partial^2 v}{\partial y\,\partial t}\\
\sigma_{zz} &= -p - \frac{2}{3}\nu\frac{\partial\Delta}{\partial t} + 2\nu\frac{\partial^2 w}{\partial z\,\partial t}\\
\sigma_{yz} &= \nu\left(\frac{\partial^2 w}{\partial y\,\partial t} + \frac{\partial^2 v}{\partial z\,\partial t}\right)\\
\sigma_{zx} &= \nu\left(\frac{\partial^2 u}{\partial z\,\partial t} + \frac{\partial^2 w}{\partial x\,\partial t}\right)\\
\sigma_{xy} &= \nu\left(\frac{\partial^2 v}{\partial x\,\partial t} + \frac{\partial^2 u}{\partial y\,\partial t}\right)
\end{aligned} \tag{6-1}$$

where

$$-p \equiv \sigma_H = \frac{\sigma_{xx} + \sigma_{yy} + \sigma_{zz}}{3} \tag{6-2}$$

and ν is the distortional viscosity coefficient. As before, we have the same hydrostatic stress-strain relation (1-26)

$$-p = k\Delta \tag{6-3}$$

Substituting the above equations into the equations of motion (1-28), we obtain

$$\rho\frac{\partial^2 u}{\partial t^2} = \frac{\partial}{\partial x}\left(-p - \frac{2}{3}\nu\frac{\partial\Delta}{\partial t} + 2\nu\frac{\partial^2 u}{\partial x\,\partial t}\right) + \frac{\partial}{\partial y}\left[\nu\left(\frac{\partial^2 v}{\partial x\,\partial t} + \frac{\partial^2 u}{\partial y\,\partial t}\right)\right] + \frac{\partial}{\partial z}\left[\nu\left(\frac{\partial^2 w}{\partial x\,\partial t} + \frac{\partial^2 u}{\partial z\,\partial t}\right)\right]$$

which reduces to

$$\rho \frac{\partial^2 u}{\partial t^2} = -\frac{\partial p}{\partial x} + \frac{1}{3}\nu \frac{\partial^2 \Delta}{\partial x\, \partial t} + \nu \nabla^2 \frac{\partial u}{\partial t}$$

$$\rho \frac{\partial^2 u}{\partial t^2} = \left(k + \frac{1}{3}\nu \frac{\partial}{\partial t}\right) \frac{\partial \Delta}{\partial x} + \nu \frac{\partial}{\partial t} \nabla^2 u$$

and similarly for the other two equations (6-4)

$$\rho \frac{\partial^2 v}{\partial t^2} = \left(k + \frac{1}{3}\nu \frac{\partial}{\partial t}\right) \frac{\partial \Delta}{\partial y} + \nu \frac{\partial}{\partial t} \nabla^2 v$$

$$\rho \frac{\partial^2 w}{\partial t^2} = \left(k + \frac{1}{3}\nu \frac{\partial}{\partial t}\right) \frac{\partial \Delta}{\partial z} + \nu \frac{\partial}{\partial t} \nabla^2 w$$

To obtain the wave equations, we follow the same procedure as that given before for Eqs. (1-31) and (1-32) from (1-30). Differentiate Eqs. (6-4) by $\partial/\partial x$, $\partial/\partial y$, and $\partial/\partial z$, respectively, and add. This gives

$$\rho \frac{\partial^2 \Delta}{\partial t^2} = \left(k + \frac{4}{3}\nu \frac{\partial}{\partial t}\right) \nabla^2 \Delta \tag{6-5}$$

Differentiate the second of Eqs. (6-4) by $\partial/\partial z$ and the third by $\partial/\partial y$ and subtract. This gives

$$\rho \frac{\partial^2 \theta_x}{\partial t^2} = \nu \frac{\partial}{\partial t} \nabla^2 \theta_x$$

and similarly for the other two combinations

$$\begin{aligned} \rho \frac{\partial^2 \theta_y}{\partial t^2} &= \nu \frac{\partial}{\partial t} \nabla^2 \theta_y \\ \rho \frac{\partial^2 \theta_z}{\partial t^2} &= \nu \frac{\partial}{\partial t} \nabla^2 \theta_z \end{aligned} \tag{6-6}$$

We now have two possible waves. We have, much as in the case of a perfect fluid, a dilatational wave altered by the viscous effects. We have also introduced an equivoluminal viscous wave determined solely by the viscous effects.

Let us examine these two types of waves for simple harmonic plane wave propagation. For plane waves propagated in the x direction the displacements u, v, and w will be functions of x and t only. Equations (6-4) reduce to

$$\begin{aligned} \rho \frac{\partial^2 u}{\partial t^2} &= \left(k + \frac{4}{3}\nu \frac{\partial}{\partial t}\right) \frac{\partial^2 u}{\partial x^2} \\ \rho \frac{\partial^2 v}{\partial t^2} &= \nu \frac{\partial}{\partial t} \frac{\partial^2 v}{\partial x^2} \\ \rho \frac{\partial^2 w}{\partial t^2} &= \nu \frac{\partial}{\partial t} \frac{\partial^2 w}{\partial x^2} \end{aligned} \tag{6-7}$$

For a simple harmonic wave we look for solutions of the form

$$\begin{aligned} u &= e^{i(\omega t - m_1 x)} \\ v,\, w &= e^{i(\omega t - m_2 x)} \end{aligned} \tag{6-8}$$

Substituting for the u displacement, we obtain the following conditional equation for m_1.

$$\begin{aligned} \omega^2\rho &= m_1{}^2 k + i\frac{4}{3}\,\omega m_1{}^2\nu \\ m_1{}^2 &= \omega^2\left(c^2 + i\frac{4}{3}\frac{\omega\nu}{\rho}\right)^{-1} \\ m_1 &= \frac{\omega}{c}\left(1 + i\frac{4}{3}\frac{\omega v}{\rho c^2}\right)^{-\frac{1}{2}} \end{aligned} \tag{6-9}$$

where c is the velocity of propagation of dilatational waves for a perfect fluid defined by

$$c^2 \equiv \frac{k}{\rho} \tag{6-10}$$

Since the coefficient of viscosity is generally a small number, Eq. (6-9) can usually be approximated by

$$m_1 \approx \frac{\omega}{c} - i\frac{2}{3}\frac{\omega^2\nu}{\rho c^3} \tag{6-11}$$

The displacement u is then

$$u = e^{-(2\nu/3\rho c^3)\omega^2 x} e^{i\omega[t-(x/c)]} \tag{6-12}$$

The velocity of propagation is the same as that for a perfect fluid. In the general case (6-9) the velocity will be a function of frequency. This is an example of *material dispersion.* The dispersion problems discussed in Chaps. 3 and 4 are examples of *geometrical dispersion.* In addition there is an exponential attenuation factor proportional to the square of the frequency. If we should substitute typical values of ν, ρ, and c for the ocean into Eq. (6-12), we should find that the viscous attenuation factor is unimportant for most problems for frequencies below about ten thousand cycles per second. Substituting for the v, w displacements, we obtain the following conditional equation for m_2.

$$\begin{aligned} \omega^2\rho &= i\omega m_2{}^2\nu \\ m_2{}^2 &= -i\frac{\omega\rho}{\nu} \\ m_2 &= \left(\frac{\omega\rho}{2\nu}\right)^{\frac{1}{2}} - i\left(\frac{\omega\rho}{2\nu}\right)^{\frac{1}{2}} \\ &= \frac{\omega}{c_2} - i\left(\frac{\omega\rho}{2\nu}\right)^{\frac{1}{2}} \end{aligned} \tag{6-13}$$

where c_2 is defined by

$$c_2^2 \equiv \frac{2\nu\omega}{\rho} \tag{6-14}$$

The displacements v, w are then

$$v,\ w = e^{-(\omega\rho/2\nu)^{1/2}x} e^{i\omega[t-(x/c_2)]} \tag{6-15}$$

The quantity c_2 is the velocity of propagation of the equivoluminal viscous wave. The propagation is dispersive. Since ν is small, the velocity of propagation will be small; and the attenuation factor will be large. This wave is sensible only in the immediate vicinity of the source and only at very low frequencies.

Absorption, heat conduction

Another type of loss associated with the propagating medium itself is that of the transfer of vibrational sound energy to heat which is subsequently conducted away to the surrounding medium.

Heat flow by conduction is proportional to the negative of the temperature gradient in the direction of flow and to the cross section normal to the flow. We may then write for heat flow in the x direction

$$\frac{\delta q}{\delta t} = -\kappa \frac{\delta T}{\delta x}\,\delta A \tag{6-16}$$

where κ is a constant of proportionality and is referred to as the thermal conductivity. Consider a small rectangular parallelepiped of sides δx, δy, δz and a temperature T at its center. The temperature on the left-hand face may be expressed as $T - \frac{1}{2}(\partial T/\partial x)\delta x$ and the temperature on the right-hand face as $T + \frac{1}{2}(\partial T/\partial x)\delta x$. From (6-16) the heat flow through the left-hand face will then be

$$\frac{dq_i}{dt} = -\kappa \frac{\partial}{\partial x}\left(T - \frac{1}{2}\frac{\partial T}{\partial x}\,\delta x\right)\delta y\ \delta z$$

and the heat flow through the right-hand face

$$\frac{dq_o}{dt} = -\kappa \frac{\partial}{\partial x}\left(T + \frac{1}{2}\frac{\partial T}{\partial x}\,\delta x\right)\delta y\ \delta z$$

The net increase of heat per unit time to the volume $\delta x\ \delta y\ \delta z$ from heat flow into and out of the faces normal to the x direction will then be simply

$$\frac{dq_i}{dt} - \frac{dq_o}{dt} = \kappa \frac{\partial^2 T}{\partial x^2}\,\delta x\ \delta y\ \delta z \tag{6-17}$$

The total increase of heat per unit time to a unit mass from heat flow through all the faces will then be

$$\frac{dQ}{dt} = \frac{\kappa}{\rho} \nabla^2 T \tag{6-18}$$

This is the desired heat conduction equation. We shall carry through the derivation for a perfect gas rather than for a liquid. One reason for such a procedure is that we know the equation of state for a perfect gas [Eq. (1-43)]. This allows us to obtain the solution for the temperature variations associated with the heat conduction and the cross terms of the temperature variations propagated with the sound velocity and the displacement variations propagated with the heat conduction velocity. Under certain simplifying assumptions we would have been able to obtain the same attenuation coefficient [Eq. (6-36)] for a liquid as we shall obtain here for a perfect gas. From Eqs. (1-53) (1-43), and (1-55) we have for the heat differential for small volume changes of a unit mass of a perfect gas

$$\begin{aligned} dQ &= (s_P - s_V)\frac{T}{V}\,dV + c_V\,dT \\ &= (\gamma - 1)s_V \frac{P}{R}\,dV + s_V\,dT \\ &\approx (\gamma - 1)s_V \frac{P_0}{R}\,dV + s_V\,dT \\ &= (\gamma - 1)s_V \frac{T_0}{V_0}\,dV + s_V\,dT \end{aligned} \tag{6-19}$$

Equating Eqs. (6-18) and (6-19) we obtain

$$\frac{\partial T}{\partial t} + (\gamma - 1)\frac{T_0}{V_0}\frac{\partial V}{\partial t} = \frac{\kappa}{\rho s_V}\nabla^2 T \tag{6-20}$$

Since, as before, we are concerned solely with small volume changes, we may write

$$V = V_0(1 + \Delta) \tag{6-21}$$

$$T = T_0(1 + \eta) \tag{6-22}$$

From the general gas law (1-43) we shall have

$$\begin{aligned} \frac{P}{P_0} &= \frac{V_0 T}{V T_0} \\ p = P - P_0 &= P_0\left(\frac{V_0 T}{V T_0} - 1\right) \\ &= P_0\left(\frac{1 + \eta}{1 + \Delta} - 1\right) \\ &\approx P_0(\eta - \Delta) \end{aligned} \tag{6-23}$$

neglecting second-order terms. Substituting (6-21) and (6-22) into Eq. (6-20) and (6-23) into the equations of motion (1-28) we obtain

$$\frac{\partial \eta}{\partial t} + (\gamma - 1)\frac{\partial \Delta}{\partial t} = \frac{\kappa}{\rho_0 s_V}\nabla^2\eta$$

$$\frac{\partial \eta}{\partial t} + (\gamma - 1)\frac{\partial \Delta}{\partial t} = \alpha\nabla^2\eta \tag{6-24}$$

and

$$\rho_0\frac{\partial^2 \Delta}{\partial t^2} = -\nabla^2 p = P_0\nabla^2\Delta - P_0\nabla^2\eta$$

$$\frac{\partial^2 \Delta}{\partial t^2} = b^2\nabla^2\Delta - b^2\nabla^2\eta \tag{6-25}$$

where b is the isothermal sound velocity (1-47) defined by

$$b^2 \equiv \frac{P_0}{\rho_0} \tag{6-26}$$

and α is the thermal diffusivity defined by

$$\alpha \equiv \frac{\kappa}{\rho_0 s_V} \tag{6-27}$$

and where, consistent with the above, ρ is approximated by ρ_0. We now have two equations relating the dilatation Δ and the temperature variation η.

Let us examine the possible types of waves for simple harmonic plane wave propagation. For plane waves propagating in the x direction Δ and η will be functions of x and t only. Equations (6-25) and (6-24) reduce to

$$\frac{\partial^2 u}{\partial t^2} = b^2\frac{\partial^2 u}{\partial x^2} - b^2\frac{\partial \eta}{\partial x} \tag{6-28}$$

$$\frac{\partial \eta}{\partial t} + (\gamma - 1)\frac{\partial^2 u}{\partial x\,\partial t} = \alpha\frac{\partial^2 \eta}{\partial x^2} \tag{6-29}$$

For a simple harmonic wave we look for solutions of the form

$$u,\ \eta = e^{i(\omega t - mx)} \tag{6-30}$$

Substituting into (6-28) and (6-29) we obtain

$$\begin{aligned} (\omega^2 - b^2m^2)u + ib^2m\eta &= 0 \\ (\gamma - 1)\omega m u + (i\omega + \alpha m^2)\eta &= 0 \end{aligned} \tag{6-31}$$

If u and η are to have a value, the determinant of their coefficients must vanish. This gives the conditional equation for m,

$$\begin{aligned} (\omega^2 - b^2m^2)(i\omega + \alpha m^2) - ib^2m^2(\gamma - 1)\omega &= 0 \\ \alpha b^2m^4 + (i\omega c^2 - \alpha\omega^2)m^2 - i\omega^3 &= 0 \end{aligned} \tag{6-32}$$

where c is the adiabatic sound velocity (1-58) defined by

$$c^2 \equiv \gamma \frac{P_0}{\rho_0} = \gamma b^2 \tag{6-33}$$

If $\alpha = 0$, no heat flow, the solution of (6-32) is

$$m^2 = \frac{\omega^2}{c^2} \tag{6-34}$$

The velocity of sound propagation is the adiabatic velocity. If $\alpha = \infty$' isothermal conditions, the solution of (6-32) is

$$m^2 = \frac{\omega^2}{b^2} \tag{6-35}$$

The velocity of sound propagation is the isothermal velocity. Since the coefficient of thermal conductivity is generally small and of the order of the coefficient of viscosity, one of the roots of (6-32) will be of the order of (6-34). The other will be very large and of the order of $\alpha^{-\frac{1}{2}}$. The roots are therefore approximately

$$\begin{aligned}
m_1{}^2 &= \frac{\omega^2}{c^2} - i\frac{\alpha}{\omega c^2} m^2(\omega^2 - b^2m^2) \\
&\approx \frac{\omega^2}{c^2}\left[1 - i\frac{\alpha\omega}{c^2}\left(1 - \frac{b^2}{c^2}\right)\right] \\
&= \frac{\omega^2}{c^2}\left[1 - i\frac{\alpha\omega}{\gamma c^2}(\gamma - 1)\right] \\
m_1 &= \frac{\omega}{c}\left[1 - i\frac{\alpha\omega}{\gamma c^2}(\gamma - 1)\right]^{\frac{1}{2}} \\
&\approx \frac{\omega}{c} - i\frac{\alpha(\gamma - 1)\omega^2}{2\gamma c^3}
\end{aligned} \tag{6-36}$$

and

$$\begin{aligned}
\alpha b^2 m_2{}^4 + i\omega c^2 m_2{}^2 &\approx 0 \\
m_2{}^2 &= -i\frac{\omega c^2}{\alpha b^2} \\
m_2 &= \left(\frac{\omega\gamma}{2\alpha}\right)^{\frac{1}{2}} - i\left(\frac{\omega\gamma}{2\alpha}\right)^{\frac{1}{2}} \\
&= \frac{\omega}{c_2} - i\left(\frac{\omega\gamma}{2\alpha}\right)^{\frac{1}{2}}
\end{aligned} \tag{6-37}$$

where c_2 is defined by

$$c_2{}^2 \equiv \frac{2\alpha\omega}{\gamma} \tag{6-38}$$

The displacement u and the temperature variation η are then

$$\begin{aligned} u &= A_1 e^{-[\alpha(\gamma-1)/2\gamma c^3]\omega^2 x} e^{i\omega[t-(x/c)]} + A_2 e^{-(\omega\gamma/2\alpha)^{1/2} x} e^{i\omega[t-(x/c_2)]} \\ \eta &= B_1 e^{-[\alpha(\gamma-1)/2\gamma c^3]\omega^2 x} e^{i\omega[t-(x/c)]} + B_2 e^{-(\omega\gamma/2\alpha)^{1/2} x} e^{i\omega[t-(x/c_2)]} \end{aligned} \tag{6-39}$$

The coefficients B_1 and B_2 can be determined in terms of A_1 and A_2 from Eq. (6-31). The displacement and temperature consist of two waves. The velocity of propagation of the sound wave is the adiabatic velocity. It includes an attenuation factor proportional to the square of the frequency. If we should substitute typical values of α, γ, and c from the ocean into Eq. (6-39), we would find that the heat conduction attenuation factor is of the order of one-half the viscous attenuation factor and generally unimportant for frequencies below about ten thousand cycles per second. The quantity c_2 is the velocity of propagation of the thermal wave. The propagation is dispersive. Since α is small, the velocity of propagation is small; and the attenuation factor will be large. This wave is sensible only in the immediate vicinity of the source and only at low frequencies. The similarity of the form of the solutions in this and the previous section is apparent. The similarity arises through the fact that the dissipative force in both cases is proportional to the particle velocity. Nevertheless, the two sections are dealing with two different types of phenomena.

Propagation in a porous medium

In a number of problems the propagating medium cannot be considered to be composed of one component only but rather of two or more components. A sediment is an example of such a medium. The sediment is porous; and if it is under water, the pore spaces are filled with water. The propagating medium consists of two components, the particulate sedimentary matter and the pore-filling fluid. The physical characteristics of the medium pertinent to sound transmission are described by some combination of the individual properties of the two components and the coupling properties between them.

Let us determine the velocity of sound transmission in an ideal mixture of two components. We assume that the medium has no rigidity and no viscosity. This approximates the case of a suspension of one component in a second. We shall designate the density, bulk modulus, and sound velocity of the suspended material by ρ_2, k_2, and c_2 and those of the suspending fluid by ρ_1, k_1, and c_1. From (1-34) and (1-45) we have that the velocity of sound propagation in a perfect fluid is

$$c^2 = \frac{k}{\rho} = \frac{dP}{-\rho(dV/V_0)} = \frac{dP}{d\rho} \tag{6-40}$$

for small volume changes. Consider a given volume of the mixture and let V_1, m_1, and V_2, m_2 be the partial volumes and partial masses of the two components. Then for an ideal mixture the density will be

$$\begin{aligned}\rho &= \frac{m_1 + m_2}{V_1 + V_2} \\ &= \frac{\eta + 1}{(\eta/\rho_1) + (1/\rho_2)}\end{aligned} \tag{6-41}$$

where the parameter η is defined by

$$\eta = \frac{m_1}{m_2} \tag{6-42}$$

and ρ_1 and ρ_2 are given by

$$\begin{aligned}\rho_1 &= \frac{m_1}{V_1} \\ \rho_2 &= \frac{m_2}{V_2}\end{aligned} \tag{6-43}$$

During the passage of a sound wave ρ_1 and ρ_2 will vary but η necessarily will remain a constant. We may then write

$$\begin{aligned}\frac{d\rho}{dP} &= (\eta + 1)\left(\frac{\eta}{\rho_1} + \frac{1}{\rho_2}\right)^{-2}\left(\frac{\eta}{\rho_1^2}\frac{d\rho_1}{dP} + \frac{1}{\rho_2^2}\frac{d\rho_2}{dP}\right) \\ c^2 = \frac{dP}{d\rho} &= \frac{(\eta\rho_2 + \rho_1)^2}{(\eta + 1)[\eta\rho_2^2(dP/d\rho_1)^{-1} + \rho_1^2(dP/d\rho_2)^{-1}]} \\ &= \frac{c_1^2c_2^2(\eta\rho_2 + \rho_1)^2}{(\eta + 1)(\eta\rho_2^2c_2^2 + \rho_1^2c_1^2)}\end{aligned} \tag{6-44}$$

since by definition from (6-40) the sound velocities of each component are

$$\begin{aligned}c_1^2 &= \frac{dP}{d\rho_1} \\ c_2^2 &= \frac{dP}{d\rho_2}\end{aligned} \tag{6-45}$$

It is generally more convenient to define the mixture in terms of the porosity g. The porosity is defined as the ratio of the volume fraction of fluid to total volume. From (6-42) and (6-43) the porosity will be in terms of η

$$g = \frac{V_1}{V_1 + V_2} = \frac{\eta(\rho_2/\rho_1)}{\eta(\rho_2/\rho_1) + 1} \tag{6-46}$$

$$\eta = \frac{\rho_1}{\rho_2}\frac{g}{1 - g} \tag{6-47}$$

Substituting (6-47) into (6-44) we obtain

$$c^2 = \frac{\rho_1{}^2\rho_2{}^2c_1{}^2c_2{}^2}{[g\rho_1 + (1-g)\rho_2][g\rho_1\rho_2{}^2c_2{}^2 + (1-g)\rho_2\rho_1{}^2c_1{}^2]}$$

$$= \frac{1}{[g\rho_1 + (1-g)\rho_2][g(\rho_1c_1{}^2)^{-1} + (1-g)(\rho_2c_2{}^2)^{-1}]}$$

$$= \frac{1}{[g\rho_1 + (1-g)\rho_2][gk_1{}^{-1} + (1-g)k_2{}^{-1}]} \tag{6-48}$$

For $g = 1$, all fluid, Eq. (6-48) reduces to $c_1{}^2$; and for $g = 0$, all suspended matter, Eq. (6-48) reduces to $c_2{}^2$ as expected. In the usual case both the density and velocity of the suspended matter are greater than the density and velocity of the fluid. Thus, there will be a greater contrast between

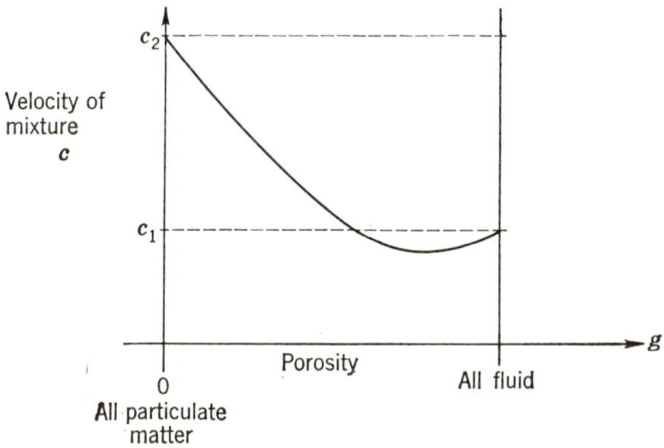

FIG. 6-1.

$\rho_1c_1{}^2$ and $\rho_2c_2{}^2$ (k_1 and k_2) than that between ρ_1 and ρ_2. As the porosity of the mixture is decreased from unity, the contribution to the first bracket will be more important than that to the second; and the velocity of the mixture will decrease. With continuing decrease in porosity the $(1 - g)$ terms become dominant; the velocity increases and approaches c_2. We have the interesting result that the velocity of a mixture can be lower than the individual velocities of either of its components. In a physical sense the effect of adding particulate matter is first to load the mixture without an appreciable change in the bulk modulus, thus decreasing the sound velocity. Figure 6-1 is a graph of the sound velocity of the mixture versus porosity.

If we were to consider an actual sediment, two alterations would have to be included in the above derivation. First, a sediment generally exhibits some rigidity. There is usually some bonding or frame connection between the particulate matter. We should not be able to use the equation for a perfect fluid but would have to return to the equation for an elastic medium. Second, viscosity must be included. This will not be the viscosity coefficient for an unbounded fluid given in the first section of this chapter. The viscous effect here is that of the motion of the fluid within the pore spaces and will be dependent on the size of the pore spaces and the interconnections between the pores. The viscosity coefficient here will be considerably larger than that of the example cited above. Dispersion will be pronounced, and the attenuation will be larger. Both will increase with increasing frequency. The example just discussed may be considered valid in the limit of zero frequency, considering viscous effects alone. An interesting result from such an examination of the viscous effects is that a high-porosity sediment which has a velocity

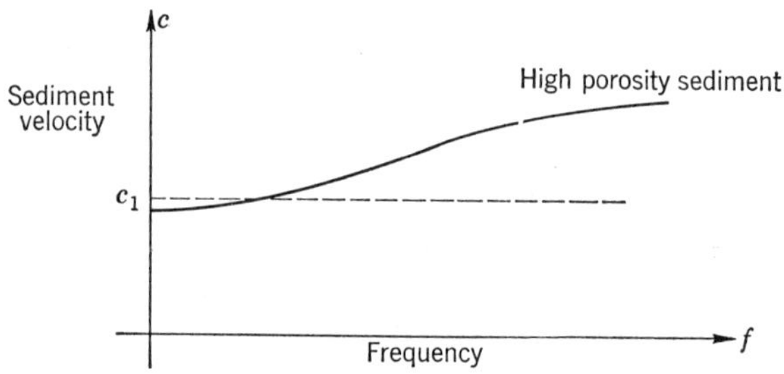

FIG. 6-2.

lower than that of the contained fluid at low frequencies will in general have a velocity higher than that of the contained fluid at high frequencies, as illustrated in Fig. 6-2.

General theory for scattering and diffraction

Problems involving scattering or diffraction are of a somewhat different kind than the transmission problems that we have been investigating so far. In scattering we are interested in the secondary waves resulting from incidence of a primary wave on small regions of different acoustical properties than the transmitting medium. In diffraction we are interested in the wave motion beyond an obstacle which has cut off portions of the advancing wave front. Both may be considered as problems involving unusual source distributions. In scattering problems the anomalous region acts as a source of secondary waves under the influence of an incident primary wave. In diffraction problems the dissected wave front is the source (initial) condition for the description of propagation in the medium ahead of it.

In the transmission problems of the previous chapters the source condition has generally been relatively simple. The complications have arisen through variations in the transmitting medium. Such problems have been the ones of major emphasis in this book. For the solution of scattering and diffraction problems the properties of the medium itself are generally taken to be constant. The complications arise through the introduction of complex initial conditions.

The general theory that will be derived here appropriately belongs in Chap. 2. However, since we shall make use of it only for the description of scattering and diffraction phenomena, it is given in this chapter. We shall need the following two mathematical relations in the derivation and, subsequently, Gauss's theorem

$$\iiint \nabla^2\varphi \, dV = \iint \frac{\partial \varphi}{\partial n} \, d\sigma \tag{6-49}$$

or

$$\iiint \left[\frac{\partial}{\partial x}\left(\psi \frac{\partial \varphi}{\partial x}\right) + \frac{\partial}{\partial y}\left(\psi \frac{\partial \varphi}{\partial y}\right) + \frac{\partial}{\partial z}\left(\psi \frac{\partial \varphi}{\partial z}\right) \right] dV = \iint \psi \frac{\partial \varphi}{\partial n} \, d\sigma \tag{6-50}$$

and Green's theorem, which follows directly from (6-50) by interchange of φ and ψ and subtraction,

$$\iiint [\psi\nabla^2\varphi - \varphi\nabla^2\psi] \, dV = \iint \left[\psi \frac{\partial \varphi}{\partial n} - \varphi \frac{\partial \psi}{\partial n} \right] d\sigma \tag{6-51}$$

The left-hand integrals are taken over an enclosed volume V bounded by a surface σ. The right-hand integral is taken over the surface σ. The derivatives $\partial/\partial n$ are taken normal to the surface σ in an outward sense.

Consider the simple wave equation for a liquid in terms of the displacement or velocity potential φ,

$$\rho \frac{\partial^2 \varphi}{\partial t^2} = k\nabla^2\varphi \tag{6-52}$$

It is in the form of an equation of motion. The left-hand side is the product of the density and acceleration. The right-hand side represents a force per unit mass. If in the region of interest we have a source distribution, we must add its stress effects to the equation of motion (6-52). If $\Phi(x, y, z, t)$ represents the potential function of the source per unit volume appropriate to the right-hand side of (6-52), we must add a term $\rho\Phi$ per unit mass. Equation (6-52) becomes

$$\rho \frac{\partial^2 \varphi}{\partial t^2} = k\nabla^2\varphi + \rho\Phi \tag{6-53}$$

or

$$\frac{\partial^2 \varphi}{\partial t^2} = c^2\nabla^2\varphi + \Phi \tag{6-54}$$

where $c = (k/\rho)^{1/2}$ is the sound velocity. Introducing the *Lorentzian operator* defined by

$$\Box^2 = \nabla^2 - \frac{1}{c^2}\frac{\partial^2}{\partial t^2} \tag{6-55}$$

Eqs. (6-52) and (6-54) may be written as

$$\Box^2\varphi = 0 \tag{6-56}$$

$$\Box^2\varphi = -\frac{1}{c^2}\Phi \tag{6-57}$$

where Eq. (6-56) applies to a region of no sources and Eq. (6-57) to a region with a source distribution Φ. Equations (6-56) and (6-57) resemble the determining equations of potential theory, Laplace's and Poisson's equations, respectively, where here the time-dependent Lorentzian operator is appropriate rather than the Laplacian operator. Their form suggests possible solutions similar to those of potential theory.

From Chap. 1 we know that the propagation out from a simple point source with spherical symmetry is given by

$$\varphi = \frac{1}{r} f\left(t - \frac{r}{c}\right) \tag{6-58}$$

We should like to relate f to Φ. The quantity Φ is the source function per unit volume. We may consider a point source to be represented by ΦT where T is a small volume about the source. In the limit as T approaches zero, Φ approaches infinity in such a manner that their product remains finite. With this in mind let us take a volume integral within a small sphere about the point source of Eq. (6-54),

$$-\iiint \frac{\partial^2\varphi}{\partial t^2}\,dV + c^2\iiint \nabla^2\varphi\,dV = -\iiint \Phi\,dV \tag{6-59}$$

In the limit as r approaches zero, the first integral disappears. Applying Gauss's theorem (6-49) to the remainder of the left-hand side, we are left with

$$\iint \frac{\partial\varphi}{\partial n}\,d\sigma = \iint \frac{\partial\varphi}{\partial r} r^2\,d\Omega = -\frac{1}{c^2}\iiint \Phi\,dV$$

$$\iint r^2\left(-\frac{1}{r^2}f + \frac{1}{r}f'\right)d\Omega = -\frac{1}{c^2}\iiint \Phi\,dV \tag{6-60}$$

where Ω is the solid angle measured from the center of the sphere. In the limit as r approaches zero, the second integral on the left-hand side of (6-60) disappears, leaving finally

$$f = \frac{T}{4\pi c^2}\Phi \tag{6-61}$$

Equation (6-58) can then be written in terms of Φ as

$$\varphi = \frac{T}{4\pi c^2 r}\,\Phi\left(t - \frac{r}{c}\right) \tag{6-62}$$

Equation (6-62) is the solution for a source distribution Φ over a small volume T. If we now consider a source distribution spread throughout space with observing point at the origin of the coordinate system, the solution will be

$$\varphi = \frac{1}{4\pi c^2}\iiint \Phi\left(t - \frac{r}{c}\right)\frac{dV}{r} \tag{6-63}$$

In a number of problems we are interested in restricting the volume of integration V. We must then include the contributions coming into the volume from the region beyond. These contributions will be represented by a surface integral over the bounding surface of the volume of interest. From Eq. (6-63) we see, as expected, that the contribution to our observing point at a time t from a source at a distance r is determined by the source action at an earlier time τ defined by

$$\tau = t - \frac{r}{c} \tag{6-64}$$

where r, the distance from the source to the observing point at the center of the coordinate system, is as usual.

$$r = (x^2 + y^2 + z^2)^{1/2} \tag{6-65}$$

We shall be interested in time functions of τ as well as of t. For convenience we shall designate a function of τ by a bracket symbol [ ]. Thus, if

$$\varphi = \varphi(x, y, z, t)$$

the quantity $[\varphi]$ will mean

$$[\varphi] = \varphi(x, y, z, \tau)$$

We shall also be interested in derivatives with respect to the dependent variable r. For convenience we shall indicate such a derivative by $\delta/\delta r$. With this notation in mind we obtain from (6-64) and (6-65) the following relations,

$$\varphi(0, 0, 0, \tau) = \varphi(0, 0, 0, t) \tag{6-66}$$

$$\frac{\partial[\varphi]}{\partial\tau} = \left[\frac{\partial\varphi}{\partial t}\right] \tag{6-67}$$

$$\frac{\partial[\varphi]}{\partial x} = \left[\frac{\partial\varphi}{\partial x}\right] - \frac{x}{rc}\left[\frac{\partial\varphi}{\partial t}\right] \tag{6-68}$$

$$\frac{\delta[\varphi]}{\delta r} = \frac{x}{r}\frac{\partial[\varphi]}{\partial x} + \frac{y}{r}\frac{\partial[\varphi]}{\partial y} + \frac{z}{r}\frac{\partial[\varphi]}{\partial z} \tag{6-69}$$

and substituting (6-68) into (6-69).

$$\frac{\delta[\varphi]}{\delta r} + \frac{1}{c}\left[\frac{\partial \varphi}{\partial t}\right] = \frac{x}{r}\left[\frac{\partial \varphi}{\partial x}\right] + \frac{y}{r}\left[\frac{\partial \varphi}{\partial y}\right] + \frac{z}{r}\left[\frac{\partial \varphi}{\partial z}\right] \tag{6-70}$$

To derive the required solution, we need the expression

$$\begin{aligned}\frac{\partial}{\partial x}\left\{\frac{1}{r}\left[\frac{\partial \varphi}{\partial x}\right]\right\} &= \frac{1}{r}\frac{\partial}{\partial x}\left[\frac{\partial \varphi}{\partial x}\right] - \frac{x}{r^3}\left[\frac{\partial \varphi}{\partial x}\right] \\ &= \frac{1}{r}\left[\frac{\partial^2 \varphi}{\partial x^2}\right] - \frac{x}{r^2 c}\left[\frac{\partial^2 \varphi}{\partial x\, \partial t}\right] - \frac{x}{r^3}\left[\frac{\partial \varphi}{\partial x}\right]\end{aligned} \tag{6-71}$$

using (6-68). Adding the two similar expressions with respect to y and z to (6-71), we obtain

$$\begin{aligned}&\frac{\partial}{\partial x}\left\{\frac{1}{r}\left[\frac{\partial \varphi}{\partial x}\right]\right\} + \frac{\partial}{\partial y}\left\{\frac{1}{r}\left[\frac{\partial \varphi}{\partial y}\right]\right\} + \frac{\partial}{\partial z}\left\{\frac{1}{r}\left[\frac{\partial \varphi}{\partial z}\right]\right\} \\ &\quad = \frac{1}{r}\left\{\left[\frac{\partial^2 \varphi}{\partial x^2}\right] + \left[\frac{\partial^2 \varphi}{\partial y^2}\right] + \left[\frac{\partial^2 \varphi}{\partial z^2}\right]\right\} - \frac{1}{rc}\left\{\frac{x}{r}\left[\frac{\partial^2 \varphi}{\partial x\, \partial t}\right] + \frac{y}{r}\left[\frac{\partial^2 \varphi}{\partial y\, \partial t}\right]\right. \\ &\quad \left. + \frac{z}{r}\left[\frac{\partial^2 \varphi}{\partial z\, \partial t}\right]\right\} - \frac{1}{r^2}\left\{\frac{x}{r}\left[\frac{\partial \varphi}{\partial x}\right] + \frac{y}{r}\left[\frac{\partial \varphi}{\partial y}\right] + \frac{z}{r}\left[\frac{\partial \varphi}{\partial z}\right]\right\} \\ &\quad = -\frac{[\Phi]}{rc^2} + \frac{1}{rc^2}\left[\frac{\partial^2 \varphi}{\partial t^2}\right] - \frac{1}{rc}\left\{\frac{\delta}{\delta r}\left[\frac{\partial \varphi}{\partial t}\right] + \frac{1}{c}\left[\frac{\partial^2 \varphi}{\partial t^2}\right]\right\} - \frac{1}{r^2}\left\{\frac{\delta[\varphi]}{\delta r}\right. \\ &\quad \left. + \frac{1}{c}\left[\frac{\partial \varphi}{\partial t}\right]\right\} = -\frac{[\Phi]}{rc^2} - \frac{1}{r^2}\frac{\delta}{\delta r}\left\{\frac{r}{c}\left[\frac{\partial \varphi}{\partial t}\right] + [\varphi]\right\}\end{aligned} \tag{6-72}$$

using (6-54) and (6-70). Let us integrate Eq. (6-72) over a finite volume V, defined on the outside by a surface σ and on the inside by a small sphere s about the origin as center. The small inner surface will be collapsed to zero in the limit. The volume element of integration will be taken as $dV = r^2\, d\Omega\, dr$ where Ω is the solid angle measured from the origin. Applying Gauss's theorem (6-50) to the left-hand side of (6-72) we obtain

$$\begin{aligned}\iint_\sigma \frac{1}{r}\left[\frac{\partial \varphi}{\partial n}\right] d\sigma - \iint_s \left[\frac{\partial \varphi}{\partial n}\right] r\, d\Omega = -\iiint_\tau \frac{[\Phi]}{rc^2}\, dV - \iiint_\tau \frac{\delta}{\delta r}\left\{\frac{r}{c}\left[\frac{\partial \varphi}{\partial t}\right]\right. \\ \left. + [\varphi]\right\} dr\, d\Omega\end{aligned} \tag{6-73}$$

The second integral on the left-hand side disappears in the limit as s approaches zero. The integration with respect to r in the last integral may be made immediately, leaving

$$-\iint_\sigma \left\{\frac{r}{c}\left[\frac{\partial \varphi}{\partial t}\right] + [\varphi]\right\} d\Omega + \iint_s \left\{\frac{r}{c}\left[\frac{\partial \varphi}{\partial t}\right] + [\varphi]\right\} d\Omega \tag{6-74}$$

In the limit as s approaches zero, the second integral becomes, using (6-66),

$$4\pi\varphi(0, 0, 0, t) \tag{6-75}$$

The differential of the solid angle $d\Omega$ may be replaced by

$$d\Omega = \frac{d\sigma}{r^2}\cos\theta \tag{6-76}$$

in terms of the differential of the surface area $d\sigma$, where θ is the angle between the normal n to σ and r and is given by

$$\cos\theta = \frac{\partial r}{\partial n} \tag{6-77}$$

Substituting (6-74), (6-75), (6-76), and (6-77) into (6-73), we obtain the final solution for the disturbance observed at the origin at a time t.

$$\varphi = \varphi(0, 0, 0, t) = \frac{1}{4\pi c^2}\iiint \frac{[\Phi]}{r}\,dV + \frac{1}{4\pi}\iint \left\{\frac{1}{r}\left[\frac{\partial\varphi}{\partial n}\right] + \frac{1}{r^2}[\varphi]\frac{\partial r}{\partial n} + \frac{1}{rc}\left[\frac{\partial\varphi}{\partial t}\right]\frac{\partial r}{\partial n}\right\} d\sigma \tag{6-78}$$

This solution is referred to as *Kirchhoff's solution*. It gives the resulting disturbance at the origin at a time t in terms of the source functions throughout the volume V and the disturbances entering through the surface σ at times τ.

For scattering problems we are interested in the secondary waves emitted from a small region of anomalous acoustical properties under the influence of an incident primary wave. The surface integral disappears, and we are left with only the volume integral. For diffraction problems the wave front dissected by the obstacles is the initial condition. The surface of integration is made to coincide with the wave front. The volume integral disappears, and we are left with only the surface integral.

Scattering from a small object

Let us examine the secondary waves scattered from a small anomalous region of differing acoustical properties under the influence of an incident primary wave. We shall take the incident wave to be a simple harmonic plane wave propagating in the negative x direction. With reference to Fig. 6-3 the bulk modulus, density, and sound velocity of the anomalous region are designated by m_2, ρ_2, and c_2 and those of the normal transmitting medium by m_1, ρ_1, and c_1. From Chap. 1 we have for the equations of motion in a fluid,

$$\rho \frac{\partial^2 u}{\partial t^2} = -\frac{\partial p}{\partial x} = m \frac{\partial \Delta}{\partial x}$$
$$\rho \frac{\partial^2 v}{\partial t^2} = -\frac{\partial p}{\partial y} = m \frac{\partial \Delta}{\partial y} \tag{6-79}$$
$$\rho \frac{\partial^2 w}{\partial t^2} = -\frac{\partial p}{\partial z} = m \frac{\partial \Delta}{\partial z}$$

For simple harmonic motion the solutions are of the form

$$u, v, w = e^{i\omega t}\bar{u}, \bar{v}, \bar{w} \tag{6-80}$$

Equation (6-79) becomes

$$m \frac{\partial \bar{\Delta}}{\partial x} + \omega^2 \rho \bar{u} = 0 \tag{6-81}$$

and similarly for the other two equations. For a primary plane wave propagating in the negative x direction at some distance from the anomal-

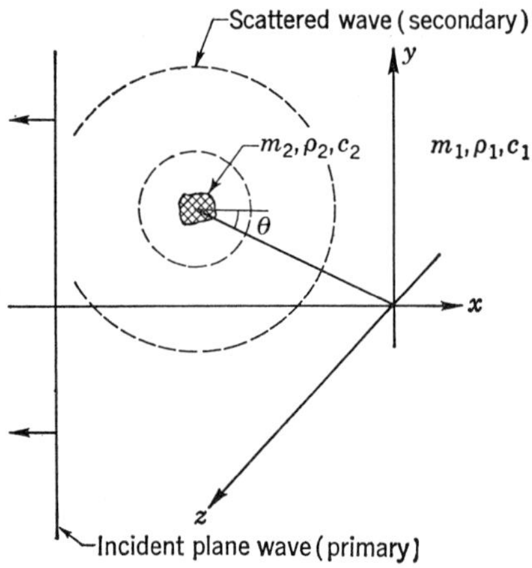

FIG. 6-3.

ous region, v_0 and w_0 are zero. The equations of motion reduce to the single equation

$$m_1 \frac{\partial \bar{\Delta}_0}{\partial x} + \omega^2 \rho_1 \bar{u}_0 = 0 \tag{6-82}$$

and the solution to

$$u_0 = e^{i(\omega t - k_1 x)} \tag{6-83}$$

In the anomalous region both u_0 and Δ_0 are functions of y and z as well as x and t. In this region there are an extra dilatational stress given by $(m_2 - m_1)\Delta_0$ and an extra acceleration in the x direction given by $(\rho_2 - \rho_1)u_0$ over that necessary to define the primary plane wave motion. This stress and this acceleration are the source functions for the secondary waves. From (6-81) the equations of motion for the secondary waves will then be

$$m_1 \frac{\partial \bar{\Delta}}{\partial x} + (m_2 - m_1) \frac{\partial \bar{\Delta}_0}{\partial x} + \omega^2 \rho_1 \bar{u} + \omega^2(\rho_2 - \rho_1) u_0 = 0$$
$$m_1 \frac{\partial \bar{\Delta}}{\partial y} + (m_2 - m_1) \frac{\partial \bar{\Delta}_0}{\partial y} + w^2 \rho_1 \bar{v} = 0 \qquad (6\text{-}84)$$
$$m_1 \frac{\partial \bar{\Delta}}{\partial z} + (m_2 - m_1) \frac{\partial \bar{\Delta}_0}{\partial z} + w^2 \rho_1 \bar{w} = 0$$

where the added terms are, of course, defined only for the anomalous region. Differentiating Eqs. (6-84) by $\partial/\partial x$, $\partial/\partial y$, $\partial/\partial z$, respectively, and adding, we obtain the wave equation

$$\nabla^2 \bar{\Delta} + \frac{\omega^2}{c_1{}^2} \bar{\Delta} = -\frac{\omega^2(\rho_2 - \rho_1)}{m_1} \frac{\partial \bar{u}_0}{\partial x} - \frac{(m_2 - m_1)}{m_1} \nabla^2 \bar{\Delta}_0 \qquad (6\text{-}85)$$

of the form (6-54). Its solution is given by (6-63) or (6-78). The integration shall be carried over a small volume completely enclosing the anomalous region. The surface integral of (6-78) disappears. For simple harmonic motion we have the relations

$$[\Phi] = \Phi(x, y, z, \tau) = \Phi\left(x, y, z, t - \frac{r}{c}\right) = e^{-ikr} e^{i\omega t} \Phi(x, y, z)$$
$$= e^{-ikr} \Phi(x, y, z, t) = e^{-ikr} \Phi \qquad (6\text{-}86)$$

Substituting (6-86) and (6-85) into (6-78) we obtain

$$\bar{\varphi} = \frac{1}{4\pi c^2} \iiint \frac{e^{-ikr}}{r} \bar{\Phi}\, dV \qquad (6\text{-}87)$$

$$\bar{\Delta} = \frac{1}{4\pi} \iiint \frac{e^{-ikr}}{r} \left\{ \frac{(m_2 - m_1)}{m_1} \nabla^2 \bar{\Delta}_0 + \frac{\omega^2(\rho_2 - \rho_1)}{m_1} \frac{\partial \bar{u}_0}{\partial x} \right\} dV \qquad (6\text{-}88)$$

Applying Green's theorem (6-51) to the first integral of (6-88), we have

$$P = \frac{(m_2 - m_1)}{4\pi m_1} \iiint \frac{e^{-ik_1 r}}{r} \nabla^2 \bar{\Delta}_0\, dV$$
$$= \frac{(m_2 - m_1)}{4\pi m_1} \iiint \bar{\Delta}_0 \nabla^2 \left(\frac{e^{-ik_1 r}}{r} \right) dV + \frac{m_2 - m_1}{4\pi m_1} \iint \left\{ \frac{e^{-ik_1 r}}{r} \frac{\partial \Delta_0}{\partial n} \right.$$
$$\left. - \Delta_0 \frac{\partial}{\partial n} \left(\frac{e^{-ik_1 r}}{r} \right) \right\} d\sigma \qquad (6\text{-}89)$$

Over the surface σ, $m_2 - m_1$ is zero so that the surface integral disappears. We have for the Laplacian of a spherically symmetrical function

$$\nabla^2 \left(\frac{e^{-ik_1 r}}{r} \right) = \left(\frac{\partial^2}{\partial r^2} + \frac{2}{r} \frac{\partial}{\partial r} \right) \left(\frac{e^{-ik_1 r}}{r} \right) = -k_1{}^2 \frac{e^{-ik_1 r}}{r} \qquad (6\text{-}90)$$

Substituting into (6-89) we obtain

$$P = -k_1^2 \frac{(m_2 - m_1)}{4\pi m_1} \iiint \bar{\Delta}_0 \frac{e^{-ik_1 r}}{r}\, dV$$
$$\approx -\frac{k_1^2(m_2 - m_1)\bar{\Delta}_0 T}{4\pi m_1 r} e^{-ik_1 r} \tag{6-91}$$

where T is the total volume of the anomalous region. The evaluation of Eq. (6-91) is a good approximation if the dimensions of the anomalous region are small compared with the wavelength, $\lambda_1 = 2\pi/k_1$. For the second integral of (6-88) let us integrate with respect to x first and integrate by parts. The integrated term evaluated on the surface σ will vanish since $\rho_2 - \rho_1$ is zero there. We are left with

$$Q = \frac{\omega^2(\rho_2 - \rho_1)}{4\pi m_1} \iiint \frac{e^{-ik_1 r}}{r} \frac{\partial \bar{u}_0}{\partial x}\, dV$$
$$= \frac{\omega^2(\rho_2 - \rho_1)}{4\pi m_1} \iiint \bar{u}_0 \left(\frac{-ik_1 e^{-ik_1 r}}{r} - \frac{e^{-ik_1 r}}{r^2} \right) \frac{\partial r}{\partial x}\, dV \tag{6-92}$$

For $\partial r/\partial x$ we have

$$\frac{\partial r}{\partial x} = \frac{\partial}{\partial x}(x^2 + y^2 + z^2)^{1/2} = \frac{x}{r} = \cos\theta \tag{6-93}$$

where θ is the angle between x and r measured from the anomalous region. Substituting into (6-92) we obtain

$$Q = -\frac{\omega^2(\rho_2 - \rho_1)}{4\pi m_1} \iiint \bar{u}_0 \cos\theta \left(\frac{ik_1 e^{-ik_1 r}}{r} + \frac{e^{-ik_1 r}}{r^2} \right) dV$$
$$\approx -\frac{ik_1\omega^2(\rho_2 - \rho_1)\bar{u}_0 \cos\theta\, T}{4\pi m_1 r} e^{-ik_1 r} \tag{6-94}$$

The evaluation of Eq. (6-94) is a good approximation if the dimensions of the anomalous region are small compared with λ_1 and if λ_1 is small compared with the distance to the observing point r. To relate $\bar{u}_0$ to $\bar{\Delta}_0$ we have from (6-83) and the definition of the dilatation for an undisturbed primary wave propagating in the negative x direction,

$$\Delta_0 = \frac{\partial u_0}{\partial x} = ik_1 u_0 \tag{6-95}$$

Substituting (6-91), (6-94), and (6-95) back into (6-88) we obtain finally for the dilatation of the secondary wave in terms of the primary wave,

$$\bar{\Delta} = -\frac{k_1^2 T}{4\pi r} \left(\frac{m_2 - m_1}{m_1} + \frac{\rho_2 - \rho_1}{\rho_1} \cos\theta \right) \bar{\Delta}_0 e^{-ik_1 r}$$
$$\Delta = -\frac{\omega^2 T}{4\pi c_1^2 r} \left(\frac{m_2 - m_1}{m_1} + \frac{\rho_2 - \rho_1}{\rho_1} \cos\theta \right) \bar{\Delta}_0 e^{i(\omega t - k_1 r)} \tag{6-96}$$

The amplitude of the secondary wave is proportional to the amplitude of the incident wave, the difference in bulk moduli, the difference in densities, the volume of the anomalous region, and the square of the frequency of the incident wave. The intensity is then, of course, proportional to ω^4. Such a frequency dependence for the scattered wave is often referred to as *Rayleigh scattering*. The secondary wave is spherically divergent away from the anomalous region. Its amplitude falls off with distance as r^{-1}, and its phase is given by $(\omega t - k_1 r)$.

There is a dependence of the amplitude of the scattered wave on the polar angle θ. The scattering phenomenon may be considered to consist of two parts. There is a dilatational effect dependent on the difference in the bulk moduli of the two regions. The dilatational effect gives a uniform scattering in all directions. Such a secondary source is referred to as a *simple* (*or single*) *source*. There is also an accelerational (or inertial) effect in the direction of propagation of the primary wave dependent on the difference in the densities of the two regions. The accelerational effect gives a directional scattering which varies as the cosine of the angle between the primary and secondary waves. Such a secondary source is referred to as a *double source*. An example of a double source is a simple source and its image (negative) in a free surface taken in the limit as the distance d between them approaches zero where the amplitude $\pm A$ of each is considered to approach infinity in such a manner that the product Ad remains finite. Following the algebra through one obtains a directional dependence of $\cos \theta$. This latter example is the one usually taken for the definition of a double source.

For problems in which the approximation that the dimensions of the anomalous region are small compared with the wavelength cannot be made, the integrals of (6-89) and (6-92) must be evaluated. The shape of the anomalous region becomes important, and exact evaluations except in the case of simple objects such as spheres and infinite cylinders become exceedingly difficult. For these two objects the formulation is straightforward. The evaluations in terms of cylindrical and spherical Bessel functions and Legendre polynomials involve heavy algebra.

Diffraction

In diffraction problems all the volume sources are taken exterior to the volume of integration. The surface of integration is taken to coincide with the dissected wave front of the incident wave at the diffracting obstacle and is considered to close around the observing point at infinity. Equation (6-78) reduces to

$$\varphi = \frac{1}{4\pi} \iint \left\{ \frac{1}{r} \left[\frac{\partial \varphi}{\partial n} \right] + \frac{1}{r^2} [\varphi] \frac{\partial r}{\partial n} + \frac{1}{rc} \left[\frac{\partial \varphi}{\partial t} \right] \frac{\partial r}{\partial n} \right\} d\sigma \qquad \text{(6-97)}$$

This is the fundamental equation for diffraction. Let us derive the expression for the diffraction equation when the source exterior to σ is a simple harmonic point source. Then the disturbance φ from this source on σ will be in real variable form

$$\varphi = \frac{A}{r_1} \cos (\omega t - k_1 r) = \frac{A}{r_1} \cos \left[2\pi \left(ft - \frac{r_1}{\lambda} \right) \right] \tag{6-98}$$

where A is the ampitude of the source and r_1 the distance from the source to the surface σ. From (6-97) we shall have the following expressions

$$\frac{\partial \varphi}{\partial n} = \frac{\partial \varphi}{\partial r_1} \frac{\partial r_1}{\partial n} = \left\{ - \frac{A}{r_1^2} \cos \left[2\pi \left(ft - \frac{r_1}{\lambda} \right) \right] + \frac{2\pi A}{\lambda r_1} \sin \left[2\pi \left(ft - \frac{r_1}{\lambda} \right) \right] \right\} \frac{\partial r_1}{\partial n} \tag{6-99}$$

$$\frac{\partial \varphi}{\partial t} = - \frac{2\pi f A}{r_1} \sin \left[2\pi \left(ft - \frac{r_1}{\lambda} \right) \right] = - \frac{2\pi A c}{\lambda r_1} \sin \left[2\pi \left(ft - \frac{r_1}{\lambda} \right) \right] \tag{6-100}$$

We shall assume that r and r_1 are much larger than λ. Then the second term of (6-97) and the first term of (6-99) may be neglected with respect to the other terms. The derivatives of r and r_1 with respect to n are simply, as before, the cosines of r and r_1 with the outward drawn normal to σ.

$$\begin{aligned} \frac{\partial r_1}{\partial n} &= \cos \widehat{r_1 n} \equiv \cos \theta_1 \\ \frac{\partial r}{\partial n} &= \cos \widehat{rn} \equiv \cos \theta \end{aligned} \tag{6-101}$$

Substituting (6-99), (6-100), and (6-101) into (6-97) with the approximation $\lambda \ll r$, r_1 and remembering that a bracketed quantity refers to replacing t by $t - (r/c)$, we obtain

$$\varphi = \frac{A}{2\lambda} \iint \frac{\cos \theta_1 - \cos \theta}{r_1 r} \sin \left[2\pi \left(ft - \frac{r_1 + r}{\lambda} \right) \right] d\sigma \tag{6-102}$$

Equation (6-102) can be considered a more exact form of Huygens' principle than that which was given in Chap. 2. In particular it includes an amplitude dependence for the Huygens wavelet as a function of angle. If $d\sigma$ corresponded to a portion of a wave front emitted from the point source, $\cos \theta_1$ would be unity. An elemental portion of the wave front would emit a wavelet of the form

$$d\varphi = \frac{A}{2\lambda r_1 r} (1 - \cos \theta) \sin \left[2\pi \left(ft - \frac{r_1 + r}{\lambda} \right) \right] d\sigma \tag{6-103}$$

The amplitude in the forward direction of propagation, $\theta = 180$, is maximum. The amplitude backward is zero.

The evaluation of the integrals (6-97) or (6-102) is in general quite difficult. We shall consider only a relatively simple approximate evaluation which is useful in a number of problems. For an incident plane wave the r_1 amplitude dependence does not appear. For a plane wave incident to a plane obstacle the r_1 phase dependence is proportional to distance along σ normal to the line of intersection of the two planes. For a plane wave incident normal to a plane obstacle there is no dependence on r_1. Equation (6-102) becomes, then, for this latter case

$$\varphi = \frac{A}{2\lambda} \iint \frac{1 - \cos\theta}{r} \sin\left[2\pi\left(ft - \frac{r}{\lambda}\right)\right] d\sigma \qquad (6\text{-}104)$$

For diffraction from an edge (Fig. 6-4) we should expect that the portions of the wave front near the diffracting edge would give the major contribution within the shadow zone defined by ray theory. We shall follow this

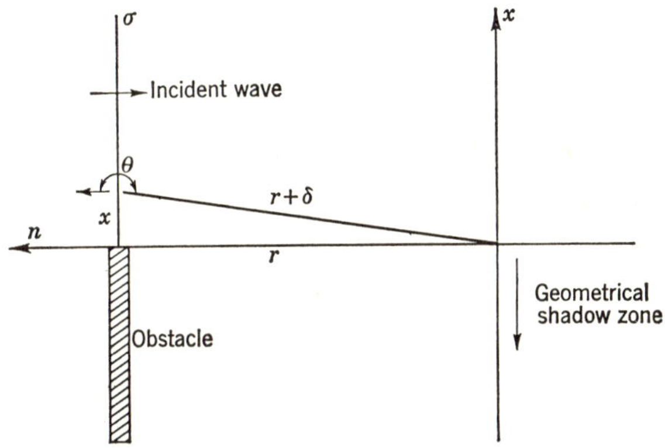

FIG. 6-4.

assumption in the derivation that follows. With reference to Fig. 6-4 we have then

$$\cos\theta \approx -1 \qquad \delta \approx \frac{x^2}{2r} \qquad (6\text{-}105)$$

The integration along σ normal to x is constant. We may then consider diverging cylindrical wavelets of amplitude dependence $(\lambda r)^{-1/2}$ rather than $(\lambda r)^{-1}$. Equation (6-104) becomes

$$\begin{aligned}\varphi &\approx \frac{A}{(\lambda r)^{1/2}} \int \sin\left[2\pi\left(ft - \frac{r+\delta}{\lambda}\right)\right] dx \\ &= \frac{A}{(\lambda r)^{1/2}} \left\{\sin\left[2\pi\left(ft - \frac{r}{\lambda}\right)\right] \int \cos\left(\frac{\pi x^2}{\lambda r}\right) dx \right. \\ &\qquad \left. - \cos\left[2\pi\left(ft - \frac{r}{\lambda}\right)\right] \int \sin\left(\frac{\pi x^2}{\lambda r}\right) dx\right\} \end{aligned} \qquad (6\text{-}106)$$

Changing the variable of integration to y where

$$x^2 = \frac{\lambda r}{2} y^2 \tag{6-107}$$

we have

$$\begin{aligned} \varphi &= \frac{A}{\sqrt{2}} (a^2 + b^2)^{1/2} \sin \left[2\pi \left(ft - \frac{r}{\lambda} \right) - \theta \right] \\ &= \frac{A}{\sqrt{2}} (a^2 + b^2)^{1/2} \cos \left[2\pi \left(ft - \frac{r}{\lambda} \right) + \eta \right] \end{aligned} \tag{6-108}$$

where

$$a \equiv \int \cos \left(\frac{\pi y^2}{2} \right) dy \tag{6-109}$$

$$b \equiv \int \sin \left(\frac{\pi y^2}{2} \right) dy$$

and

$$\tan \theta \equiv \frac{b}{a} = \cot \eta \tag{6-110}$$

The quantities a and b are referred to as the *Fresnel integrals*. The quantity $(a^2 + b^2)^{1/2}/\sqrt{2}$ is the ratio of the amplitudes of the diffracted to incident waves. The quantity η is the phase difference between the diffracted and incident waves over the geometrical ray path phase. The Fresnel integrals provide a convenient means for estimating diffraction effects from a variety of plane obstacles.

A convenient graphical means for carrying out such an evaluation is through the use of the *Cornu spiral*. Let us plot a versus b, both evaluated from 0 to y. Such a plot (Fig. 6-5) is referred to as the Cornu spiral. The spirals converge on the points (0.5, 0.5) and (−0.5, −0.5). The magnitude of the line between any two points on the spiral is equal to $(a^2 + b^2)^{1/2}$ for the limits of integration indicated by the end points of the line. The corresponding value of η is the angle this line makes with the b axis.

For no obstacles the integration extends from $y = -\infty$ to $y = \infty$. The quantity $(a^2 + b^2)^{1/2} = \sqrt{2}$ and $\eta = 45$ deg. We see that the approximate evaluation by Fresnel integrals predicts the correct amplitude, unity, for this extremum case. The phase change, however, is incorrect by 45 deg. For a diffracting straight edge (Fig. 6-4) the amplitude is one-half at the geometrical shadow boundary. The limits of integration here extend from $y = 0$ to $y = \infty$. Progressing into the shadow zone the lower limit of integration climbs up the right-hand spiral. The amplitude decreases continuously. Progressing into the ray insonified region the lower limit of integration moves down the left-hand spiral. The amplitude oscillates with decreasing amplitude of oscillation and

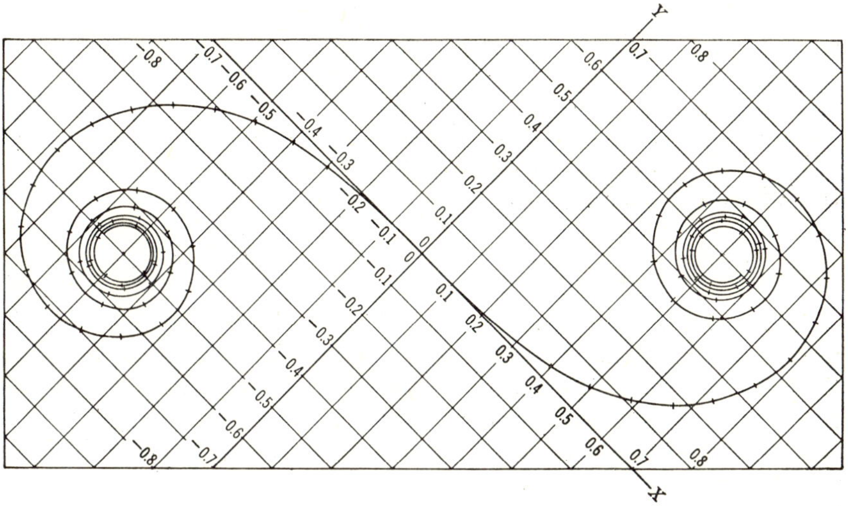

FIG. 6-5. (*From G. S. Monk, "Light,"* McGraw-Hill Book Company, Inc., New York, 1937.)

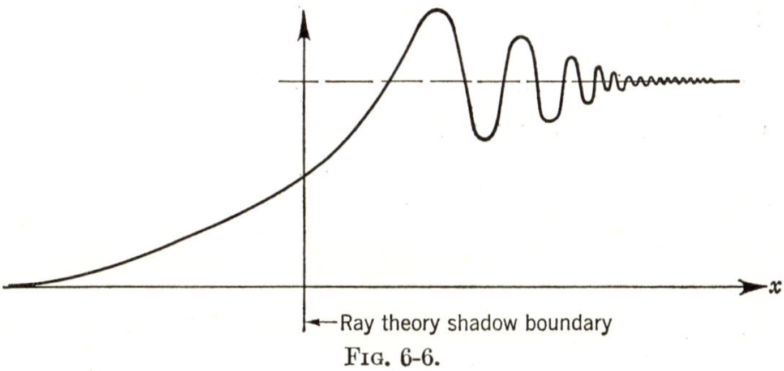

FIG. 6-6.

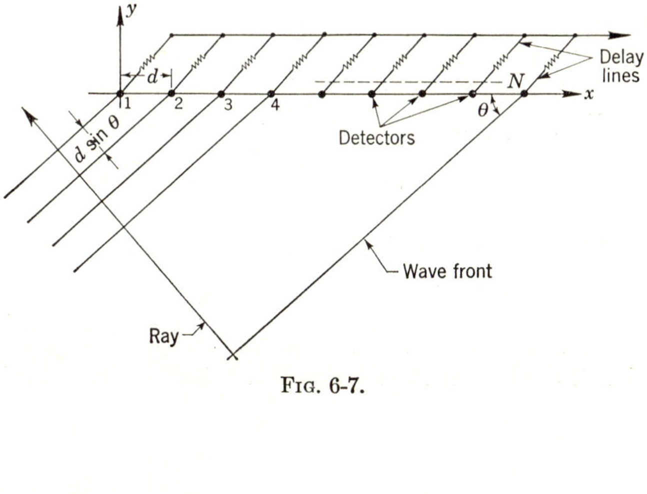

FIG. 6-7.

increasing frequency of oscillation about the value unity. Figure 6-6 is a sketch of the amplitude of the diffracted wave as a function of distance. The diffracted wave can extend for a considerable distance into the geometrical shadow boundary of ray theory, particularly at low frequencies. The diffraction pattern from a slit or from a series of slits can be found in a similar manner from Fig. 6-5 with proper regard to the limits of integration of the Fresnel integrals.

Reception to a linear array

As a final problem we shall consider the reception of a simple harmonic plane wave to a linear array of detectors. This is the inverse problem to the Fraunhofer diffraction grating. The formulation and resultant equations are exactly the same in the two cases.

We wish to examine the composite signal obtained by adding the individual signals from each detector. We shall be particularly interested in the directivity pattern of the array. For a simple harmonic plane wave propagating in the negative x and positive y directions at an angle θ to the x axis as shown in Fig. 6-7, the composite signal to a set of N detectors spaced d units apart is

$$\begin{aligned}\varphi &= e^{i\omega t} + e^{i(\omega t + kd\sin\theta)} + e^{i(\omega t + 2kd\sin\theta)} + \cdots + e^{i[\omega t + (N-1)kd\sin\theta]} \\ &= e^{i\omega t}[1 + e^{i\alpha} + e^{i2\alpha} + \cdots + e^{i(N-1)\alpha}] \qquad (6\text{-}111)\end{aligned}$$

where, for convenience, α is defined by

$$\alpha \equiv kd \sin\theta \qquad (6\text{-}112)$$

The bracketed quantity of (6-111) is a finite geometric series with constant multiplier $e^{i\alpha}$. We may express it by its sum as a series so that (6-111) becomes

$$\begin{aligned}\varphi &= e^{i\omega t}\left(\frac{e^{iN\alpha} - 1}{e^{i\alpha} - 1}\right) = e^{i\omega t}\,\frac{e^{iN(\alpha/2)} - e^{-iN(\alpha/2)}}{e^{i(\alpha/2)} - e^{-i(\alpha/2)}}\,e^{i(N-1)(\alpha/2)} \\ &= \frac{\sin[N(\alpha/2)]}{\sin(\alpha/2)}\,e^{i(N-1)(\alpha/2)}e^{i\omega t} \qquad (6\text{-}113)\end{aligned}$$

The $\sin[N(\alpha/2)]/\sin(\alpha/2)$ term is the ratio of the amplitude of the composite signal to the amplitude of the signal at a single detector. The $e^{i(N-1)(\alpha/2)}$ term is the phase change of the composite signal with respect to the signal at the first detector.

If at the same time we had imposed electronically or otherwise a delay time τ between successive detectors before compositing such that the signal at detector 1 is added to the signal at detector 2 at an earlier time τ

and so on, the resultant would be

$$\begin{aligned}\varphi &= e^{i(\omega t} + e^{i[\omega(t-\tau)+kd\sin\theta]} + e^{i[\omega(t-2\tau)+2kd\sin\theta]} \\ &\qquad + \cdots + e^{i\{\omega[t-(N-1)\tau]+(N-1)kd\sin\theta\}} \\ &= e^{i\omega t}[1 + e^{i(\alpha-\beta)} + e^{i2(\alpha-\beta)} + \cdots + e^{i(N-1)(\alpha-\beta)}] \\ &= \frac{\sin[N(\alpha-\beta)/2]}{\sin[(\alpha-\beta)/2]} e^{i(N-1)[(\alpha-\beta)/2]} e^{i\omega t} \end{aligned} \tag{6-114}$$

where, for convenience, β is defined by

$$\beta \equiv \omega\tau \tag{6-115}$$

The amplitude in both cases is given by an expression of the form

$$\frac{\sin(\pi N v)}{\sin(\pi v)} \tag{6-116}$$

where N is an integer. The intensity is proportional to the square of (6-116). Its numerator will have zeros at

$$Nv = 0, 1, 2, \ldots, N, N+1, N+2, \ldots, 2N, 2N+1, 2N+2, \ldots. \tag{6-117}$$

These will also be zeros for (6-116) except at those values

$$Nv = 0, N, 2N, \ldots. \tag{6-118}$$

where the denominator is also zero. At these values of Nv the value of the expression (6-116) will be, using L'Hospital's rule,

$$\lim_{Nv=0,N,2N,\ldots} \left[\frac{(d/dv)\sin(\pi N v)}{(d/dv)\sin(\pi v)}\right] = \lim_{Nv=0,N,2N\ldots} \left[\frac{\pi N\cos(\pi N v)}{\pi\cos(\pi v)}\right] = \pm N \tag{6-119}$$

This is a maximum. It is the maximum amplitude that the composite signal can have and corresponds to constructive reception at all detectors. These maxima points (6-118) are referred to as the *principal maxima*. In addition there will be of necessity $N - 2$ alternating subsidiary maxima and minima between each pair of principal maxima (or minima) since from (6-117) we see there are $N - 1$ zeros. By our usual definition of amplitude we are interested in the principal value of (6-116), i.e., its value without regard to sign. Both maxima and minima of (6-116) are maxima of the amplitude. The maxima and minima of (6-116) are located at solutions of

$$\begin{gathered}\frac{d}{dv}\left[\frac{\sin(\pi N v)}{\sin(\pi v)}\right] = 0 \\ N\sin(\pi v)\cos(\pi N v) - \sin(\pi N v)\cos(\pi v) = 0 \\ N\tan(\pi v) = \tan(\pi N v)\end{gathered} \tag{6-120}$$

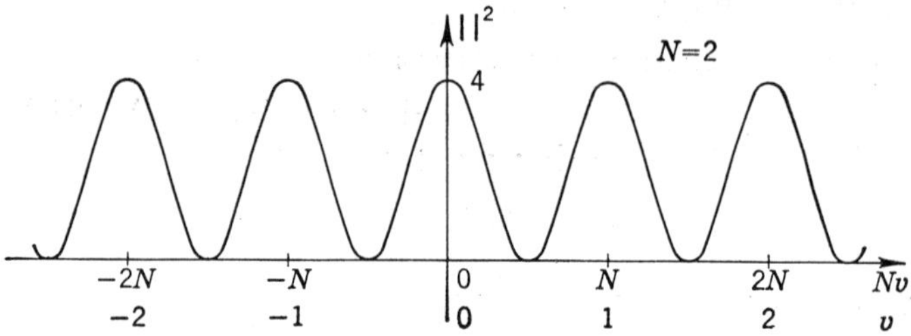

FIG. 6-8.

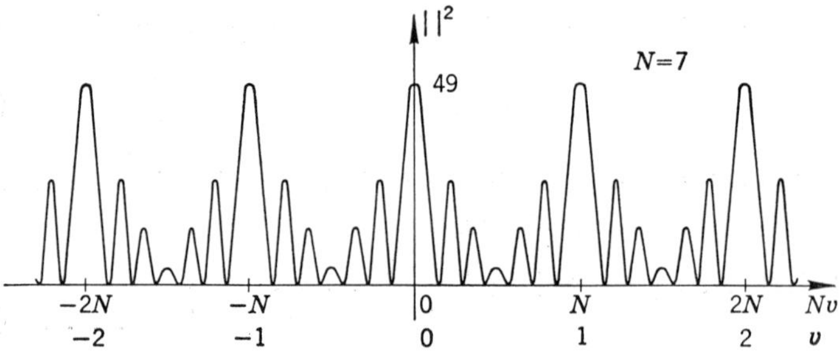

FIG. 6-9.

Figures 6-8 and 6-9 are plots of the square of (6-116) versus Nv for $N = 2$ and $N = 7$. The sequence is repetitive in N.

From (6-113) and (6-114) we shall have respectively

$$\text{Intensity } \alpha \left| \frac{\sin [N(\alpha/2)]}{\sin (\alpha/2)} \right|^2 = \left| \frac{\sin [N(kd \sin \theta/2)]}{\sin (kd \sin \theta/2)} \right|^2 = \left| \frac{\sin [N\pi(d \sin \theta/\lambda)]}{\sin [\pi(d \sin \theta/\lambda)]} \right|^2 \tag{6-121}$$

and

$$\text{Intensity } \alpha \left| \frac{\sin [N\pi(d \sin \theta - c\tau)/\lambda]}{\sin [\pi(d \sin \theta - c\tau)/\lambda]} \right|^2 \tag{6-122}$$

From (6-116) we see that v is given by

$$v = \frac{d \sin \theta}{\lambda} \tag{6-123}$$

or

$$v = \frac{d \sin \theta - c\tau}{\lambda} \tag{6-124}$$

respectively. From (6-123) we have for $\theta = 0$, $v = 0$ and for $\theta = \pm 90$ deg, $v = \pm d/\lambda$. An interesting choice for d is

$$d = \frac{\lambda}{2} \tag{6-125}$$

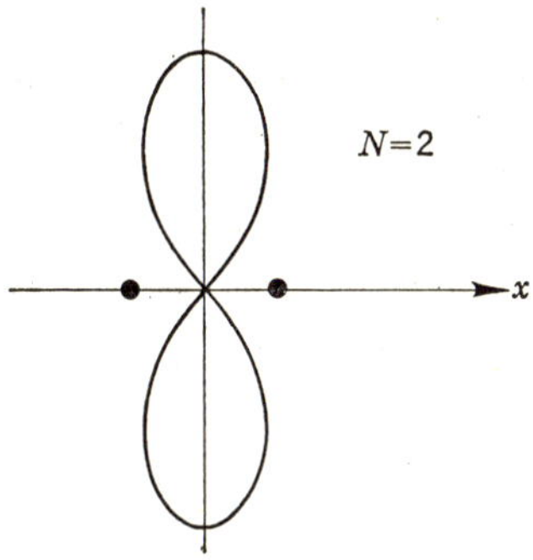

FIG. 6-10.

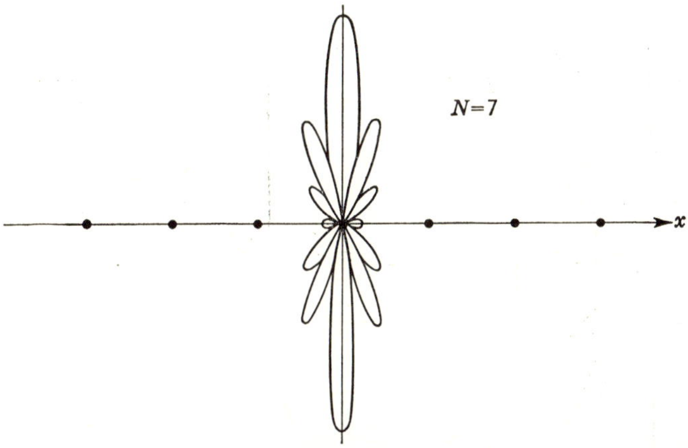

FIG. 6-11.

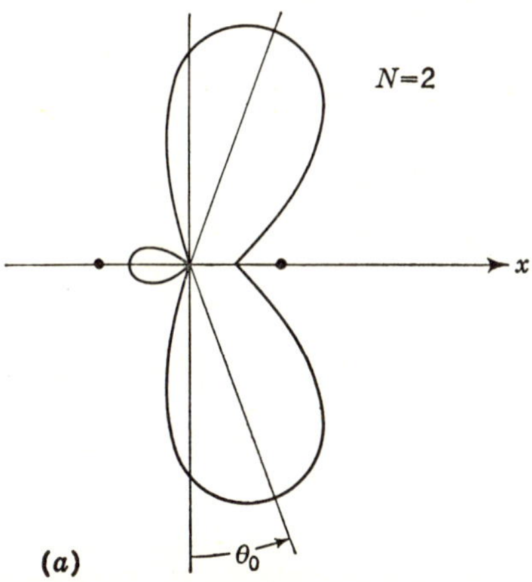

FIG. 6-12*a*.

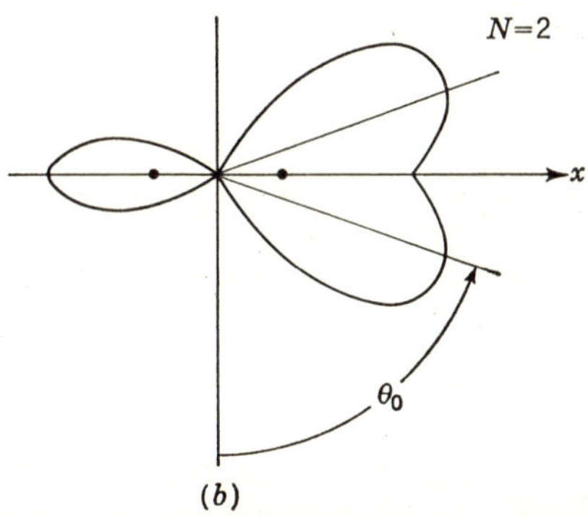

FIG. 6-12*b*.

The directivity pattern will then cover one diffraction pattern. For d greater than $\lambda/2$ there will be more than one principal maximum as a function of angle of incidence. Figures 6-10 and 6-11 are polar plot sketches of the directivity pattern versus θ for $N = 2$ and $N = 7$ when (6-125) holds. The directivity pattern in three dimensions forms a figure of revolution about the x axis. If we also included a particular delay time, $\tau = \tau_0$, the principal maxima would be turned to an angle θ_0 given by

$$d \sin \theta_0 = c\tau_0$$

$$\theta_0 = \sin^{-1}\left(\frac{c\tau_0}{d}\right)$$

from (6-124). This corresponds to shifting the zero horizontally in Figs.

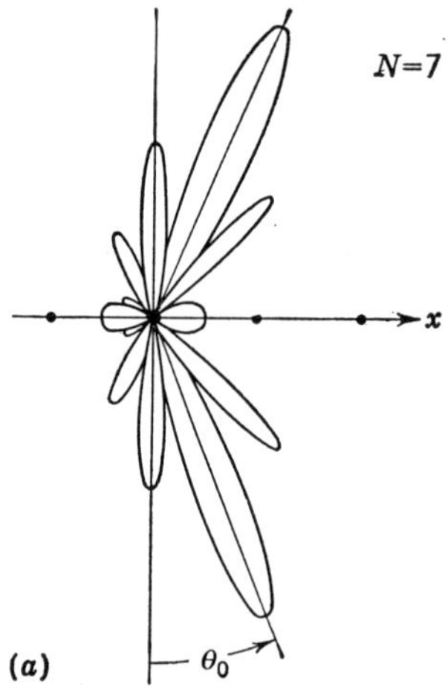

FIG. 6-13*a*.

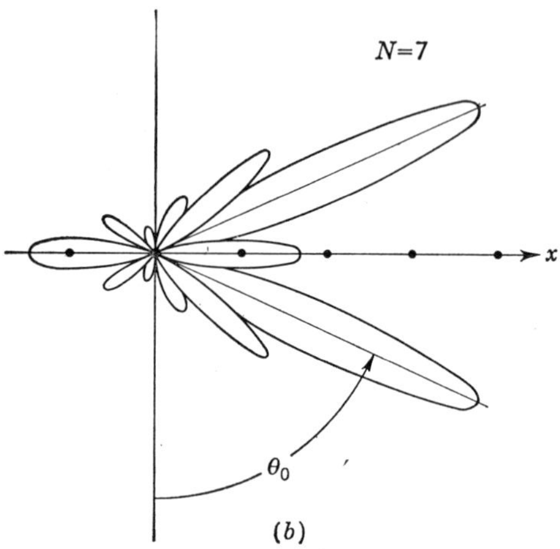

FIG. 6-13*b*.

6-8 and 6-9. Figures 6-12 and 6-13 are sketches of the directivity patterns of Figs. 6-10 and 6-11 shifted through an angle θ_0.

BIBLIOGRAPHY

Ament, W. S.: Sound Propagation in a Gross Mixture, *J. Acoust. Soc. Amer.*, vol. 25, pp. 638–641, 1953.

Anderson, V. C.: Sound Scattering from a Fluid Sphere, *J. Acoust. Soc. Amer.*, vol. 22, pp. 426–431, 1950.

Biot, M. A.: Theory of Propagation of Elastic Waves in a Fluid-saturated Porous Solid, *J. Acoust. Soc. Amer.*, vol. 28, pp. 168–191, 1956.

Chambré, P. L.: Speed of a Plane Wave in a Gross Mixture, *J. Acoust. Soc. Amer.*, vol. 26, pp. 329–331, 1954.

Lamb, H.: "Hydrodynamics," Cambridge University Press, New York, 1932, and Dover Publications, New York, 1945.

Machlup, S.: A Theoretical Model for Sound Scattering by Marine Crustaceans, *J. Acoust. Soc. Amer.*, vol. 24, pp. 290–293, 1952.
Markham, J. J., R. T. Beyer, and R. B. Lindsay: Absorption of Sound in Fluids, *Revs. Mod. Phys.*, vol. 23, pp. 353–411, 1951.
Morse, P. M.: "Vibration and Sound," 2d ed., McGraw-Hill Book Company, Inc., New York, 1948.
Morse, R. W.: Acoustic Propagation in Granular Media, *J. Acoust. Soc. Amer.*, vol. 24, pp. 696–700, 1952.
Page, L.: "Introduction to Theoretical Physics," D. Van Nostrand Company, Inc., Princeton, N.J., 1935.
Sommerfeld, A.: "Optics," Academic Press, Inc., New York, 1954.
Strutt, J. W. (Lord Rayleigh): "The Theory of Sound," The Macmillan Company, London, 1894, and Dover Publications, New York, 1945.
Vigoreux, P.: "Ultrasonics," John Wiley & Sons, Inc., New York, 1951.
Wilson, W.: "Theoretical Physics," Methuen & Co., Ltd., London, 1931.
Wood, A. B.: "A Textbook of Sound," The Macmillan Company, New York, 1955.
Zwikker, C., and C. W. Kosten: "Sound Absorbing Materials," Elsevier Publishing Company, Amsterdam, 1949.

GENERAL REFERENCES

It might be of some interest to the reader to have a list of general references on sound. A short list is given below. The books by Wood and by Kinsler and Frey are general references and are of a more elementary, descriptive nature. The books by Morse and McLachlan have particular emphasis on vibrations. The books by Bullen and Kolsky are on elastic wave propagation. The book by Ewing, Jardetzky, and Press is a more advanced text on elastic wave propagation. The books by Rayleigh and Lamb are the standard references to much of the basic theory of sound. The book by Beranek is on experimental sound. The book by Horton emphazises sonar.

Beranek, W.: "Acoustic Measurements," John Wiley & Sons, Inc., New York, 1949.
Bullen, K. E.: "Introduction to the Theory of Seismology," Cambridge University Press, New York, 1947.
Ewing, W. M., F. Press, and W. S. Jardetzky: "Elastic Waves in Layered Media," McGraw-Hill Book Company, Inc., New York, 1957.
Horton, W.: "Fundamentals of Sonar," U.S. Naval Institute, Annapolis, 1957.
Kinsler, L. E., and A. R. Frey: "Fundamentals of Acoustics," John Wiley & Sons, Inc., New York, 1950.
Kolsky, H.: "Stress Waves in Solids," Oxford University Press, New York, 1953.
Lamb, H.: "Hydrodynamics," Cambridge University Press, New York, 1932, and Dover Publications, New York, 1945.
McLachlan, N. W.: "Theory of Vibrations," Dover Publications, New York, 1951.
Morse, P. M.: "Vibration and Sound," 2d ed., McGraw-Hill Book Company, Inc., New York, 1948.
Strutt, J. W. (Lord Rayleigh): "The Theory of Sound," The Macmillan Company, London, 1894, and Dover Publications, New York, 1945.
Wood, A. B.: "A Textbook of Sound," The Macmillan Company, New York, 1955.

Index